Continuous Functions

To Claire and Patricia,
By your gaiety, "joie de vivre", and femininity,
you have embellished my life,
and you have allowed me to conserve the tenacity
needed for this endeavor

Analysis for PDEs Set

coordinated by
Jacques Blum

Volume 2

Continuous Functions

Jacques Simon

iSTE

WILEY

First published 2020 in Great Britain and the United States by ISTE Ltd and John Wiley & Sons, Inc.

ISTE Ltd
27-37 St George's Road
London SW19 4EU
UK

www.iste.co.uk

John Wiley & Sons, Inc.
111 River Street
Hoboken, NJ 07030
USA

www.wiley.com

FSC www.fsc.org MIX Paper from responsible sources FSC® C013604

Library of Congress Control Number: 2020933955

British Library Cataloguing-in-Publication Data
A CIP record for this book is available from the British Library
ISBN 978-1-78630-010-2

Contents

Introduction

Objective. This book is the second of six volumes in a series dedicated to the mathematical tools for solving partial differential equations derived from physics:

Volume 1: *Banach, Fréchet, Hilbert and Neumann Spaces*;

Volume 2: *Continuous Functions*;

Volume 3: *Distributions*;

Volume 4: *Lebesgue and Sobolev Spaces*;

Volume 5: *Traces*;

Volume 6: *Partial Differential Equations.*

This second volume is devoted to the partial differentiation of functions and the construction of primitives, which is its inverse mapping, and to their properties, which will be useful for constructing distributions and studying partial differential equations later.

Target audience. We intended to find simple methods that require a minimal level of knowledge to make these tools accessible to the largest audience possible – PhD candidates, advanced students[1] and engineers – without losing generality and even generalizing some standard results, which may be of interest to some researchers.

1 **Students?** What might I have answered if one of my MAS students in 1988 had asked for more details about the *de Rham duality theorem* that I used to obtain the pressure in the Navier–Stokes equations? Perhaps I could say that "Jacques-Louis LIONS, my supervisor, wrote that it follows from the de Rham cohomology theorem, of which I understand neither the statement, nor the proof, nor why it implies the result that we are using." What a despicably unscientific appeal to authority!
This question was the starting point of this work: writing proofs that I can explain to my students for every result that I use. It took me 5 years to find the "elementary" proof of the *orthogonality theorem*

Originality. The construction of primitives, the Cauchy integral and the weighting with which they are obtained are performed for a function taking values in a *Neumann space*, that is, a space in which every Cauchy sequence converges.

Neumann spaces. The sequential completeness characterizing these spaces is the most general property of E that guarantees that the integral of a continuous function taking values in E will belong to it, see *Case where E is not a Neumann space* (§ 4.3, p. 92). This property is more general than the more commonly considered property of completeness, that is the convergence of all Cauchy filters; for example, if E is an infinite-dimensional Hilbert space, then E-weak is a Neumann space but is not complete [Vol. 1, Property (4.11), p. 82].

Moreover, sequential completeness is more straightforward than completeness.

Semi-norms. We use families of semi-norms, instead of the equivalent notion of locally convex topologies, to be able to define differentiability (p. 73) by comparing the semi-norms of a variation of the variable to the semi-norms of the variation of the value. A section on *Familiarization with Semi-normed Spaces* can be found on p. xiii. Semi-norms can be manipulated in a similar fashion to normed spaces, except that we are working with several semi-norms instead of a single norm.

Primitives. We show that any continuous field $q = (q_1, \ldots, q_d)$ on an open set Ω of $\mathbb{R}^d$ has a primitive f, namely that $\nabla f = q$, if and only if it is orthogonal to the divergence-free test fields, that is, if $\int_\Omega q \cdot \psi = 0_E$ for every $\psi = (\psi_1, \ldots, \psi_d)$ such that $\nabla \cdot \psi = 0$. This is the *orthogonality theorem* (Theorem 9.2).

When Ω is simply connected, for a primitive f to exist, it is necessary and sufficient for q to have local primitives. This is the *local primitive gluing theorem* (Theorem 9.4). On any such open set, it is also necessary and sufficient that it verifies Poincaré's condition $\partial_i q_j = \partial_j q_i$ for every i and j to be satisfied if the field is $\mathcal{C}^1$ (Theorem 9.10), or a weak version of this condition, $\int_\Omega q_j \partial_i \varphi = \int_\Omega q_i \partial_j \varphi$ for every test function φ, if the field is continuous (Theorem 9.11).

We explicitly determine all primitives (Theorem 9.17) and construct one that depends continuously on q (Theorem 9.18).

Integration. We extend the Cauchy integral to uniformly continuous functions taking values in a Neumann space, because this will be an essential tool for constructing primitives.

(Theorem 9.2, p. 194) on the existence of the primitives of a field q. I needed a way to obtain $\int_\Gamma q \cdot d\ell = 0$ for every closed path Γ from the condition $\int_\Omega q \cdot \psi = 0$ for every divergence-free ψ. It gave me the greatest mathematical satisfaction I have ever experienced to explicitly construct an incompressible tubular flow (see p. 184). Twenty-five years later, I am finally ready to answer any other questions from my (very persistent) students.

The properties established here for continuous functions will also be used to extend them to integrable distributions in Volume 4, by continuity or transposition. Indeed, one of the objectives of the *Analysis for PDEs* series is to extend integration and Sobolev spaces to take values in Neumann spaces. However, it seemed more straightforward to first construct distributions (in Volume 3) using just continuous functions before introducing integrable distributions (in Volume 4), which play the role usually fulfilled by *classes of almost everywhere equal integrable functions.*

Weighting. The weighted function $f \diamond \mu$ of a function f defined on an open set Ω by the weight μ, a real function with compact support D, is a function defined on the open set $\Omega_D = \{x \in \mathbb{R}^d : x + D \subset \Omega\}$ by $(f \diamond \mu)(x) = \int_{\mathring{D}} f(x+y)\, \mu(y)\, \mathrm{d}y$. This concept will be repeatedly useful. It plays an analogous role to convolution, which is equivalent to it up to a symmetry of μ when $\Omega = \mathbb{R}^d$.

Novelties. Many results are natural extensions of previous results, but the following seemed most noteworthy:

— The construction of the topology of the space $\mathcal{K}(\Omega; E)$ of continuous functions with compact support using the semi-norms $\|f\|_{\mathcal{K}(\Omega;E);q} = \sup_{x\in\Omega} q(x)\|f(x)\|_{E;\nu}$ indexed by $q \in \mathcal{C}^+(\Omega)$ and $\nu \in \mathcal{N}_E$ (Definition 1.17). This is equivalent to and much simpler than the inductive limit topology of the $\mathcal{C}_K(\Omega; E)$.

— The fact that if a function $f \in \mathcal{C}(\Omega)$ satisfies $\sup_{x\in\Omega} q(x)|f(x)| < \infty$ for every $q \in \mathcal{C}^+(\Omega)$, then its support is compact (Theorem 1.22). This is the basis for defining the semi-norms of $\mathcal{D}(\Omega)$ in Volume 3.

— The *concentration theorem* for the integral and the construction of an incompressible *tubular flow* (Theorems 8.18 and 8.17), which are key steps in our construction of the primitives of a field taking values in a Neumann space, as it is explained in the comment *Utility of the concentration theorem*, p. 186.

Prerequisites. The proofs in the main body of the text only use definitions and results established in Volume 1, whose statements are recalled either in the text or in the Appendix. Detailed proofs are given, including arguments that may seem trivial to experienced readers, and the theorem numbers are systematically referenced.

Comments. Comments with a smaller font size than the main body of the text appeal to external results or results that have not yet been established. The Appendix on *Reminders* is also written with a smaller font size, since its contents are assumed to be familiar.

Historical notes. Wherever possible, the origin of the concepts and results is given as a footnote[2].

---

2 **Appeal to the reader.** Many important results lack historical notes because I am not familiar with their origins. I hope that my readers will forgive me for these omissions and any injustices they may discover. And I encourage the scholars among you to notify me of any improvements for future editions!

Navigation through the book:

— The **Table of Contents** at the start of the book lists the topics discussed.

— The **Table of Notations**, p. xv, specifies the meaning of the notation in case there is any doubt.

— The **Index**, p. 243, provides an alternative access to specific topics.

— All hypotheses are stated directly within the theorems themselves.

— The numbering scheme is shared across every type of statement to make results easier to find by number (for instance, Theorem 2.9 is found between the statements 2.8 and 2.10, which are a definition and a theorem, respectively).

Acknowledgments. Enrique FERNÁNDEZ-CARA suggested to me a large number of improvements to various versions of this work. Jérôme LEMOINE was kind enough to proofread the countless versions of the book and correct just as many mistakes and oversights.

Olivier BESSON, Fulbert MIGNOT, Nicolas DEPAUW, and Didier BRESCH also provided many improvements, in form and in substance.

Pierre DREYFUSS gave me insight into the necessity of simply connected domains for the existence of primitives with Poincaré's condition, as explained on p. 209 in the comment *Is simple connectedness necessary for gluing together local primitives?*

Joshua PEPPER spent much time discussing about the best way to adapt this work in English.

Thank you, my friends.

Jacques SIMON
Chapdes-Beaufort
April 2020

Familiarization with Semi-normed Spaces

A **semi-normed space** E is a vector space endowed with a family $\{\| \, \|_{E;\nu} : \nu \in \mathcal{N}_E\}$ of semi-norms.
— The set $\mathcal{N}_E$ indexing the semi-norms is, *a priori*, arbitrary.
— A **normed** space is the special case where this family simply consists of a single norm.
— Every locally convex topological vector space can be endowed with a family of semi-norms that generates its topology (Neumann's theorem).
— We only consider **separated** spaces, namely in which $\|u\|_{E;\nu} = 0$ for every $\nu \in \mathcal{N}_E$, then $u = 0_E$.

Working with semi-normed spaces:
— $u_n \to u$ in E means that $\|u_n - u\|_{E;\nu} \to 0$ for every $\nu \in \mathcal{N}_E$.
— U is **bounded** in E means that $\sup_{u \in U} \|u\|_{E;\nu} < \infty$ for every $\nu \in \mathcal{N}_E$.
— T is **continuous** from F into E at the point u means that, for every $\nu \in \mathcal{N}_E$ and $\epsilon > 0$, there exists a finite set M of $\mathcal{N}_F$ and $\eta > 0$ such that $\sup_{\mu \in M} \|v - u\|_{F;\mu} \leq \eta$ implies $\|T(v) - T(u)\|_{E;\nu} \leq \epsilon$.

Examples — real-valued function spaces:
— The space $\mathcal{C}_b(\Omega)$ of continuous and bounded functions is endowed with the norm $\|f\|_{\mathcal{C}_b(\Omega)} = \sup_{x \in \Omega} |f(x)|$.
— $\mathcal{C}(\Omega)$ is endowed with the semi-norms $\|f\|_{\mathcal{C}(\Omega);K} = \sup_{x \in K} |f(x)|$ indexed by the compact sets $K \subset \Omega$.
— $L^p(\Omega)$ is endowed with the norm $\|f\|_{L^p(\Omega)} = (\int_\Omega |f|^p)^{1/p}$.
— $L^p_{\text{loc}}(\Omega)$ is endowed with the semi-norms $\|f\|_{L^p_{\text{loc}}(\Omega);\omega} = (\int_\omega |f|^p)^{1/p}$ indexed by the bounded open sets ω such that $\overline{\omega} \subset \Omega$.

Examples — abstract-valued function spaces:
— $\mathcal{C}_b(\Omega; E)$ is endowed with the semi-norms $\|f\|_{\mathcal{C}_b(\Omega;E);\nu} = \sup_{x \in \Omega} \|f(x)\|_{E;\nu}$ indexed by $\nu \in \mathcal{N}_E$.
— $\mathcal{C}(\Omega; E)$ is endowed with the semi-norms $\|f\|_{\mathcal{C}(\Omega;E);K,\nu} = \sup_{x \in K} \|f(x)\|_{E;\nu}$ indexed by the compact sets $K \subset \Omega$ and $\nu \in \mathcal{N}_E$.
— $L^p(\Omega; E)$ is endowed with the semi-norms $\|f\|_{L^p(\Omega;E);\nu} = (\int_\Omega \|f\|^p_{E;\nu})^{1/p}$ indexed by $\nu \in \mathcal{N}_E$.

Examples — weak space, dual space:
— E-weak is endowed with the semi-norms $\|e\|_{E\text{-weak};e'} = |\langle e', e\rangle|$ indexed by $e' \in E'$.
— E' is endowed with the semi-norms $\|e'\|_{E';B} = \sup_{e \in B} |\langle e', e\rangle|$ indexed by the bounded sets B of E.
— E'-weak is endowed with the semi-norms $\|e'\|_{E'\text{-weak};e''} = |\langle e'', e'\rangle|$ indexed by $e'' \in E''$.
— E'-*weak is endowed with the semi-norms $\|e'\|_{E'\text{-*weak};e} = |\langle e', e\rangle|$ indexed by $e \in E$.

Neumann spaces and others:
— A **sequentially complete** space is a space in which every Cauchy sequence converges.
— A **Neumann** space is a sequentially complete separated semi-normed space.
— A **Fréchet** space is a sequentially complete metrizable semi-normed space.
— A **Banach** space is a sequentially complete normed space.

Advantages of using semi-norms rather than topology:
— Semi-norms allow the definition of $L^p(\Omega; E)$ (by raising the semi-norms of E to the power p).
— They allow the definition of the differentiability of a mapping from a semi-normed space into another (by comparing the semi-norms of an increase in the variable to the semi-norms of the increase in the value).
— They are easy to manipulate: working with them is just like working with normed spaces, the main difference being that there are several semi-norms or norms instead of a single norm.
— Some definitions are simpler, for example that of a bounded set U: "$\sup_{v \in U} \|v\|_{E;\nu} < \infty$ for any semi-norm $\| \ \|_{E;\nu}$ of E" would be expressed, in terms of topology, in the more abstract form "for any open set V containing 0_E, there is $t > 0$ such that $tU \subset V$".

Notations

SPACES OF FUNCTIONS

OPERATIONS ON A FUNCTION f

$\operatorname{supp} f$	support	4
Lf or $L \circ f$	composition with the linear mapping L	55

DERIVATIVES OF A FUNCTION f

f' or $\mathrm{d}f/\mathrm{d}x$	derivative of a function of a single real variable	32
$\partial_i f$	partial derivative: $\partial_i f = \partial f / \partial x_i$	37
$\partial^\beta f$	derivative of order β: $\partial^\beta f = \partial_1^{\beta_1} \dots \partial_d^{\beta_d} f$	41
β	positive multi-integer: $\beta = (\beta_1, \dots, \beta_d)$, $\beta_i \geq 0$	37
$\lvert\beta\rvert$	differentiability order: $\lvert\beta\rvert = \lvert\beta_1\rvert + \dots \lvert\beta_d\rvert$	37
$\partial^0 f$	derivative of order 0: $\partial^0 f = f$	37
∇f	gradient: $\nabla f = (\partial_1 f, \dots, \partial_d f)$	32
$\mathrm{d}f$	differential	74
q	field: $q = (q_1, \dots, q_d)$	176
$\nabla \cdot q$	divergence: $\nabla \cdot q = \partial_1 q_1 + \dots \partial_d q_d$	183
$\nabla^{-1} q$	primitive that depends continuously on q	211
q^*	explicit primitive: $q^*(x) = \int_{\Gamma(a,x)} q \cdot \mathrm{d}\ell$	192

INTEGRALS AND PATHS

$\int_\omega f$	Cauchy integral	89
$\mathbb{S}_\omega^n f$	approximate integral	89
$\int_{S_r} f \, \mathrm{d}s$	surface integral over a sphere	213
$\int_\Gamma q \cdot \mathrm{d}\ell$	line integral of a vector field along a path	177
Γ	path	173
$[\Gamma]$	image of a path: $[\Gamma] = \{\Gamma(t) : t_\mathrm{i} \leq t \leq t_\mathrm{e}\}$	173
$\overleftarrow{\Gamma}$	reverse path	174
$\Gamma_{\{a\}}$	path consisting of a single point	179
$\Gamma_{\overrightarrow{a,x}}$	rectilinear path	179
$\overrightarrow{\cup}$	path concatenation	174
$\mathcal{T}$	tube around a path: $\mathcal{T} = [\Gamma] + B$	183
H	homotopy	186
$[H]$	image of a homotopy	186

SEPARATED SEMI-NORMED SPACES

E	separated semi-normed space	1
$\Vert\ \Vert_{E;\nu}$	semi-norm of E of index ν	1
$\mathcal{N}_E$	set indexing the semi-norms of E	1
$\underset{\leftrightarrows}{\equiv}$	equality of families of semi-norms	7
$\underset{\leftrightarrows}{=}$	topological equality	7
$\underset{\leftrightarrows}{\approx}$	topological equality up to an isomorphism	23
$\underset{\rightarrow}{\subseteq}$	topological inclusion	7
E-weak	space E endowed with pointwise convergence in E'	236
E'	dual of E	236
E^d	Euclidean product $E \times \dots \times E$	31
$E_1 \times \cdots \times E_\ell$	product of spaces	59
$\widehat{E}$	sequential completion of E	93
$\mathring{U}$	interior of the set U	229
$\overline{U}$	closure of U	229

∂U	boundary of U	229
$[u, v]$	closed segment: $[u, v] = \{tu + (1-t)v : 0 \leq t \leq 1\}$	34
$\mathcal{L}(E; F)$	space of continuous linear mappings	56
$\mathcal{L}^\ell(E_1 \times \ldots \times E_\ell; F)$	space of continuous multilinear mappings	59

POINTS AND SETS IN $\mathbb{R}^d$

$\mathbb{R}^d$	Euclidean space: $\mathbb{R}^d = \{x = (x_1, \ldots, x_d) : \forall i,\ x_i \in \mathbb{R}\}$	232
$\|x\|$	Euclidean norm: $\|x\| = (x_1^2 + \cdots + x_d^2)^{1/2}$	232
$x \cdot y$	Euclidean scalar product: $x \cdot y = x_1 y_1 + \ldots + x_d y_d$	232
$\mathbf{e}_i$	ith basis vector of $\mathbb{R}^d$	32
Ω	domain on which a function f is defined	3
Ω_D	domain of $f \diamond \mu$: $\Omega_D = \{x : x + D \subset \Omega\}$, and its figure	148, 149
$\Omega_{1/n}$	Ω with a neighborhood of the boundary of size $1/n$ removed	13, 158
$\Omega_{1/n}^n$	$\Omega_{1/n}$ truncated by $\|x\| < n : \Omega_{1/n}^n = \{x : \|x\| < n,\ B(x, 1/n) \subset \Omega\}$	13
$\Omega_{1/n}^{*a}$	part of $\Omega_{1/n}$ which is star-shaped with respect to a, and its figure	197, 200
Ω_r^n	potato-shaped set: $\Omega_r^n = \{x : \|x\| < n,\ B(x, r) \subset \Omega\}$	13
κ_n	crown-shaped set: $\kappa_n = \Omega_{1/(n+2)}^{n+2} \setminus \overline{\Omega_{1/n}^n}$	166, 167
ω	subset of $\mathbb{R}^d$	3
$\|\omega\|$	Lebesgue measure of the open set ω	84
σ	negligible subset of $\mathbb{R}^d$	107
$B(x, r)$	closed ball $B(x, r) = \{y \in \mathbb{R}^d : \|y - x\| \leq r\}$	13
$\mathring{B}(x, r)$	open ball $\mathring{B}(x, r) = \{y \in \mathbb{R}^d : \|y - x\| < r\}$	13
v_d	measure of the unit ball: $v_d = \|\mathring{B}(0, 1)\|$	107
$C(x, \rho, r)$	open crown $C(x, \rho, r) = \{y \in \mathbb{R}^d : \rho < \|y - x\| < r\}$	107
$S(x, r)$	sphere: $S(x, r) = \{y \in \mathbb{R}^d : \|y - x\| = r\}$	109
$\Delta_{s,n}$	closed cube of edge length 2^{-n} centered at 2^{-ns}	84
$P(v^1, \ldots, v^d)$	open parallelepiped with edges $v^1, \ldots, v^d$	115

OTHER SETS

$\mathbb{N}^*$	set of natural numbers: $\mathbb{N}^* = \{0, 1, 2, \ldots\}$	227
$\mathbb{N}$	set of non-zero natural numbers: $\mathbb{N} = \{1, 2, \ldots\}$	227
$\mathbb{Z}$	set of integers: $\mathbb{Z} = \{\ldots, -2, -1, 0, 1, 2, \ldots\}$	227
$\mathbb{Q}$	set of rational numbers	227
$\mathbb{R}$	space of real numbers	228
$[\![m, n]\!]$	integer interval: $[\![m, n]\!] = \{i \in \mathbb{N}^* : m \leq i \leq n\}$	227
$[\![m, \infty]\!]$	extended integer interval: $[\![m, \infty]\!] = \{i \in \mathbb{N}^* : i \geq m\} \cup \{\infty\}$	227
(a, b)	open interval: $(a, b) = \{x \in \mathbb{R} : a < x < b\}$	228
$[a, b]$	closed interval: $[a, b] = \{x \in \mathbb{R} : a \leq x \leq b\}$	228
$\Subset$	compact inclusion in $\mathbb{R}^d$	80
$\subset$	algebraic inclusion	
$\setminus$	set difference: $U \setminus V = \{u \in U : u \notin V\}$	
$\times$	product: $U \times V = \{(u, v) : u \in U,\ v \in V\}$	
$\emptyset$	empty set	

SPECIAL FUNCTIONS

det	determinant	112
e	exponential number	238

TYPOGRAPHY

▮	end of statement
$\square$	end of proof or remark

Chapter 1

Spaces of Continuous Functions

This chapter is dedicated to the properties of spaces of continuous functions taking values in a semi-normed space E that we will need later. Definitions of the space $\mathcal{C}(\Omega; E)$ of continuous functions, the space $\mathbf{C}(\Omega; E)$ of uniformly continuous functions and variants of these spaces are given in § 1.2. We then compare these spaces (§ 1.3) and study their completions (§ 1.4) and metrizability properties (§ 1.5). The space $\mathcal{K}(\Omega; E)$ of functions with compact support is investigated in § 1.6. We also study continuous extensions (§ 1.8), separation of variables (§ 1.9) and sequential compactness (§ 1.10) in these spaces.

These topics are more necessary than original, with the exception of our construction of the topology of $\mathcal{K}(\Omega; E)$ using the semi-norms $\|f\|_{\mathcal{K}(\Omega;E);q} = \sup_{x\in\Omega} q(x)\|f(x)\|_{E;\nu}$ indexed by $q \in \mathcal{C}^+(\Omega)$ and $\nu \in \mathcal{N}_E$. These semi-norms yield properties that are usually obtained using the *inductive limit topology of the* $\mathcal{C}_K(\Omega; E)$. A useful tool for this, which is also new, is Theorem 1.23: if a family $\mathcal{F}$ of functions in $\mathcal{C}(\Omega)$ satisfies $\sup_{f\in\mathcal{F}} \sup_{x\in\Omega} q(x)|f(x)| < \infty$ for every $q \in \mathcal{C}^+(\Omega)$, their supports are all included in the same compact set.

1.1. Notions of continuity

We reserve the term **function** for mappings defined on a subset of $\mathbb{R}^d$, which is written as Ω in general.

Let us begin by defining separated semi-normed spaces[1] (the definitions of vector spaces and semi-norms are recalled in the Appendix, § A.2). We will then consider functions taking values in these spaces.

DEFINITION 1.1.– *A* ***semi-normed space*** *is a vector space E endowed with a family of semi-norms $\{\|\ \|_{E;\nu} : \nu \in \mathcal{N}_E\}$.*

1 **History of the notion of semi-normed space.** John VON NEUMANN introduced semi-normed spaces in 1935 [59] (with a superfluous countability condition). He also showed [59, Theorem 26, p. 19] that they coincide with the locally convex topological vector spaces that Andrey KOLMOGOROV had previously introduced in 1934 [49, p. 29].

Any such space is said to be **separated** *if* $u = 0_E$ *is the only element such that* $\|u\|_{E;\nu} = 0$ *for every* $\nu \in \mathcal{N}_E$.

A **normed space** *is a vector space* E *endowed with a norm* $\| \ \|_E$. ▮

Caution. Definition 1.1 is general but not universal. For Laurent SCHWARTZ [67, p. 240], a semi-normed space is a space endowed with a *filtering* family of semi-norms (Definition 2.21). This definition is equivalent, since every family is equivalent to a filtering family [Vol. 1, Theorem 3.15]. For Nicolas BOURBAKI [12, editions published after 1981, Chapter III, p. III.1] and Robert EDWARDS [32, p. 80], a semi-normed space is a space endowed with a *single* semi-norm, which drastically changes the meaning. □

Let us define various notions relating to the continuity[2] of a function taking values in a semi-normed space. These are special cases of Definition 1.24 of a continuous mapping from a semi-normed space into another.

DEFINITION 1.2.– *Let* f *be a function from a subset* Ω *of* $\mathbb{R}^d$ *into a separated semi-normed space* E *with a family of semi-norms* $\{\| \ \|_{E;\nu} : \nu \in \mathcal{N}_E\}$.

(a) *We say that* f *is* **continuous at the point** x *of* Ω *if, for every* $\nu \in \mathcal{N}_E$ *and* $\epsilon > 0$, *there exists* $\eta > 0$ *such that, if* $y \in \Omega$ *and* $|y - x| \leq \eta$, *then*

$$\|f(y) - f(x)\|_{E;\nu} \leq \epsilon.$$

We say that f *is* **continuous** *if it is continuous at every point of* Ω.

(b) *We say that* f *is* **uniformly continuous** *if, for every* $\nu \in \mathcal{N}_E$ *and* $\epsilon > 0$, *there exists* $\eta > 0$ *such that, if* x *and* y *belong to* Ω *and* $|y - x| \leq \eta$, *then*

$$\|f(y) - f(x)\|_{E;\nu} \leq \epsilon.$$

(c) *We say that* f *is* **sequentially continuous at the point** x *of* Ω *if, for every sequence* $(x_n)_{n\in\mathbb{N}}$ *in* Ω *such that* $x_n \to x$ *in* $\mathbb{R}^d$, *we have* $f(x_n) \to f(x)$ *in* E.

We say that f *is* **sequentially continuous** *if it is sequentially continuous at every point of* Ω.

---

2 **History of the notion of continuous mapping.** Augustin CAUCHY defined sequential continuity for a real function on an interval in 1821 in [20]. Bernard Placidus Johann Nepomuk BOLZANO also contributed to the emergence of this notion.

History of the notion of uniformly continuous function. Eduard HEINE defined the notion of uniform continuity of functions on (a subset of) $\mathbb{R}^d$ in 1870 in [46]. This notion had previously been implicitly used by Augustin CAUCHY in 1823 to define the integral of a real function [21, pp. 122–126] and was later explicitly used by Peter DIRICHLET.

(d) *We say that f is* **bounded** *if its image $f(\Omega) = \{f(x) : x \in \Omega\}$ is a bounded set (of E), or in other words, if, for every $\nu \in \mathcal{N}_E$,*

$$\sup_{x\in\Omega} \|f(x)\|_{E;\nu} < \infty. \blacksquare$$

We say that a sequence $(u_n)_{n\in\mathbb{N}}$ in a separated semi-normed space E **converges** to a limit $u \in E$, and we denote $u_n \to u$ as $n \to \infty$, if, for every semi-norm $\| \; \|_{E;\nu}$ of E,

$$\|u_n - u\|_{E;\nu} \to 0 \quad \text{as } n \to \infty.$$

Recall that a function is continuous if and only if it is sequentially continuous (Theorem A.29[3], since $\mathbb{R}^d$ is a normed space).

1.2. Spaces $\mathcal{C}(\Omega; E)$, $\mathcal{C}_b(\Omega; E)$, $\mathcal{C}_K(\Omega; E)$, $\mathbf{C}(\Omega; E)$ and $\mathbf{C}_b(\Omega; E)$

Addition of functions taking values in a vector space and multiplication by a scalar $t \in \mathbb{R}$ are defined by

$$(f+g)(x) \stackrel{\text{def}}{=} f(x) + g(x), \quad (tf)(x) \stackrel{\text{def}}{=} tf(x). \tag{1.1}$$

Let us first define spaces of continuous functions[4].

DEFINITION 1.3.– *Let $\Omega \subset \mathbb{R}^d$ and E be a separated semi-normed space with a family of semi-norms $\{\| \; \|_{E;\nu} : \nu \in \mathcal{N}_E\}$.*

(a) *We denote by $\mathcal{C}(\Omega; E)$ the vector space of continuous functions from Ω into E endowed with the semi-norms and indexed by the compact sets $K \subset \Omega$ and $\nu \in \mathcal{N}_E$,*

$$\|f\|_{\mathcal{C}(\Omega;E);K,\nu} \stackrel{\text{def}}{=} \sup_{x\in K} \|f(x)\|_{E;\nu}.$$

(b) *We denote by $\mathcal{C}_b(\Omega; E)$ the vector space of continuous and bounded functions from Ω into E endowed with the semi-norms and indexed by $\nu \in \mathcal{N}_E$,*

$$\|f\|_{\mathcal{C}_b(\Omega;E);\nu} \stackrel{\text{def}}{=} \sup_{x\in\Omega} \|f(x)\|_{E;\nu}. \blacksquare$$

3 **Theorem A.29.** Theorems with numbers in the form A.n can be found in the Appendix.

4 **History of the notion of function space.** Bernhard RIEMANN introduced the concept of (infinite-dimensional) function space in 1892 in his inaugural lecture *On the Hypotheses Which Lie at the Bases of Geometry* [64, p. 276] (see the extract cited in [14, p. 176]).

Justification. Addition and scalar multiplication of functions make $\mathcal{C}(\Omega; E)$ and $\mathcal{C}_b(\Omega; E)$ vector spaces.

In (a), the mapping $f \mapsto \sup_{x \in K} \|f(x)\|_{E;\nu}$ is a semi-norm on $\mathcal{C}(\Omega; E)$, since the upper envelope of a family of semi-norms is a semi-norm whenever it is everywhere finite (Theorem A.6). This is indeed the case here, since, for every x, the mapping $f \mapsto \|f(x)\|_{E;\nu}$ is a semi-norm on $\mathcal{C}(\Omega; E)$ and, for each f, $\sup_{x \in K} \|f(x)\|_{E;\nu} < \infty$ because continuous functions are bounded on compact sets (Theorem A.34).

In (b), $\sup_{x \in \Omega} \|f(x)\|_{E;\nu}$ similarly defines a semi-norm on $\mathcal{C}_b(\Omega; E)$. □

These spaces are written as $\mathcal{C}(\Omega)$ and $\mathcal{C}_b(\Omega)$, respectively, in the case where $E = \mathbb{R}$.

The topology with which we have endowed $\mathcal{C}(\Omega; E)$ is said to be the **topology of uniform convergence on compact sets**, and the topology of $\mathcal{C}_b(\Omega; E)$ is said to be the **topology of uniform convergence**.

Let us now define spaces of uniformly continuous functions.

DEFINITION 1.4.– *Let $\Omega \subset \mathbb{R}^d$ and E be a separated semi-normed space.*

(a) *We denote by $\mathbf{C}(\Omega; E)$ the vector space of uniformly continuous functions from Ω into E.*

(b) *We denote by $\mathbf{C}_b(\Omega; E)$ the vector space of uniformly continuous and bounded functions from Ω into E endowed with the semi-norms of $\mathcal{C}_b(\Omega; E)$.* ▮

We denote these spaces by $\mathbf{C}(\Omega)$ and $\mathbf{C}_b(\Omega)$, respectively, in the case where $E = \mathbb{R}$.

Absence of a topology on $\mathbf{C}(\Omega; E)$. We shall not endow $\mathbf{C}(\Omega; E)$ with semi-norms, since we do not require them. □

Next, let us define the support of a function (recall that $^{-}$ denotes the closure).

DEFINITION 1.5.– *The **support** of a function f from a subset Ω of $\mathbb{R}^d$ into a separated semi-normed space E is the set*

$$\operatorname{supp} f \stackrel{\text{def}}{=} \overline{\{x \in \Omega : f(x) \neq 0_E\}} \cap \Omega. ▮$$

Let us state some properties of the support of a function defined on an open set[5].

THEOREM 1.6.– *Let f be a function from a subset Ω of $\mathbb{R}^d$ into a separated semi-normed space E. If Ω is an open set:*

(a) *Then* $\operatorname{supp} f = \Omega \setminus \mathcal{O}$, *where $\mathcal{O}$ is the largest open subset of $\mathbb{R}^d$ on which $f = 0_E$; in other words, $\mathcal{O}$ is the interior of $\{x \in \Omega : f(x) = 0_E\}$.*

(b) *If f is also continuous, then $\{x \in \Omega : f(x) \neq 0_E\}$ is open and included in the interior* $\overset{\circ}{\operatorname{supp}} f$ *of the support of f, and $f = 0_E$ on $\Omega \setminus \overset{\circ}{\operatorname{supp}} f$.* ▮

Proof of Theorem 1.6. (a) Let

$$\mathcal{O} = \Omega \setminus \operatorname{supp} f = \Omega \cap (\mathbb{R}^d \setminus \overline{\{x \in \Omega : f(x) \neq 0_E\}}).$$

This is an open set, as a finite intersection of open sets (Theorem A.10), and f is zero on it.

If U is another open set included in Ω on which f is zero, then the set $\{x \in \Omega : f(x) \neq 0_E\}$ is included in $\mathbb{R}^d \setminus U$. Since the latter is closed,

$$\operatorname{supp} f \subset \overline{\{x \in \Omega : f(x) \neq 0_E\}} \subset \mathbb{R}^d \setminus U.$$

Hence, $U \subset \Omega \cap (\mathbb{R}^d \setminus \operatorname{supp} f) = \mathcal{O}$. Therefore, $\mathcal{O}$ is indeed the largest such open set U or, in other words, the interior (Definition A.8) of $\{x \in \Omega : f(x) = 0_E\}$.

(b) If $f(x) \neq 0_E$, there exists a semi-norm on the space E of values of f such that $\|f(x)\|_{E;\nu} = a > 0$. If f is continuous, there exists $\eta > 0$ such that $y \in \Omega$ and $|y - x| \leq \eta$ imply $\|f(y) - f(x)\|_{E;\nu} \leq a/2$, so $\|f(y)\|_{E;\nu} \geq a/2$, and hence $f(y) \neq 0_E$. Since Ω is open, it contains a ball $B(x, \eta')$. Thus, the ball $B(x, r)$, where $r = \inf\{\eta, \eta'\}$ is included in the set $\{x \in \Omega : f(x) \neq 0_E\}$, which shows that this set is indeed open.

The set $\{x \in \Omega : f(x) \neq 0_E\}$ is an open set included in the support of f and so is included in the interior $\overset{\circ}{\operatorname{supp}} f$ of the support. Therefore, $f = 0_E$ outside of this set. □

Let us finally define spaces of continuous functions with support in an arbitrary compact set. Recall that, in $\mathbb{R}^d$, a compact set is a closed and bounded set by the Borel–Lebesgue theorem [Theorem A.23 (b)].

DEFINITION 1.7.– *Let $\Omega \subset \mathbb{R}^d$ and E be a separated semi-normed space.*

5 **Numbering of statements.** The same numbering scheme is used for all statements – Definitions 1.1–1.5, Theorem 1.6, Definition 1.7, etc. –, to make it easier to find a given result by number. For example, the reader will struggle to find Theorems 1.1–1.5, because these numbers were assigned to definitions.

(a) *Given a compact subset K of $\mathbb{R}^d$ included in Ω, we denote*

$$\mathcal{C}_K(\Omega; E) \stackrel{\text{def}}{=} \{f \in \mathcal{C}(\Omega; E) : \operatorname{supp} f \subset K\},$$

endowed with the semi-norms of $\mathcal{C}_b(\Omega; E)$.

(b) *Given a compact subset D of $\mathbb{R}^d$ (not necessarily included in Ω), we denote*

$$\mathbf{C}_D(\Omega; E) \stackrel{\text{def}}{=} \{f \in \mathbf{C}(\Omega; E) : \operatorname{supp} f \subset D\},$$

endowed with the semi-norms of $\mathcal{C}_b(\Omega; E)$. ▮

Justification. (a) The vector space $\mathcal{C}_K(\Omega; E)$ can be endowed with the semi-norms of $\mathcal{C}_b(\Omega; E)$ because every function f in this space is bounded, since f is zero outside of K and bounded on K, noting that continuous functions are bounded on compact sets (Theorem A.34).

(b) The vector space $\mathbf{C}_D(\Omega; E)$ can be endowed with the semi-norms of $\mathcal{C}_b(\Omega; E)$ because every function f in this space is bounded, since f is zero outside of D and bounded on $\Omega \cap D$. Indeed, since D is precompact (Theorem A.19 (a)), the subset $\Omega \cap D$ is also precompact (Theorem A.20), and so is its image $f(\Omega \cap D)$ (Theorem A.33), which is therefore bounded (Theorem A.19 (a)). □

We denote these spaces by $\mathcal{C}_K(\Omega)$ and $\mathbf{C}_D(\Omega)$, respectively, in the case where $E = \mathbb{R}$.

Caution about behavior at the boundary. When Ω is an open set, functions in $\mathcal{C}_K(\Omega; E)$ must be zero on some neighborhood of the boundary $\partial\Omega$, since K is included in Ω.

By contrast, functions in $\mathbf{C}_D(\Omega; E)$ do not necessarily vanish on a neighborhood of $\partial\Omega$ (regardless of whether Ω is open, unless it is the whole of $\mathbb{R}^d$) when D is not included in Ω. To highlight this difference, we have chosen notation that differentiates between arbitrary compact sets D and compact sets K that are included in Ω. □

Utility of the spaces $\mathcal{C}_K(\Omega; E)$ and $\mathbf{C}_D(\Omega; E)$. The space $\mathcal{C}_K(\Omega; E)$, or, more precisely, its generalization $\mathcal{C}_K^m(\Omega; E)$, will be useful for our study of distributions in Volume 3, since the space $\mathcal{D}(\Omega)$ of test functions is the union of the $\mathcal{C}_K^\infty(\Omega)$.

The space $\mathbf{C}_D(\Omega; E)$ will be useful for our study of the Cauchy integral of continuous functions (for example in Theorem 4.22), since $\mathcal{B}(\Omega; E)$ (Definition 4.7) is the union of the $\mathbf{C}_D(\Omega; E)$. □

1.3. Comparison of spaces of continuous functions

Let us first define the topological equalities and inclusions of separated semi-normed spaces. We will then discuss inclusions between spaces of continuous functions.

DEFINITION 1.8.– *Let* $\{\| \|_{1;\nu} : \nu \in \mathcal{N}_1\}$ *and* $\{\| \|_{2;\mu} : \mu \in \mathcal{N}_2\}$ *be two families of semi-norms on the same vector space* E. *We say that the first family* ***dominates*** *the second if, for every* $\mu \in \mathcal{N}_2$, *there exists a finite set* N_1 *in* $\mathcal{N}_1$ *and* $c_1 \in \mathbb{R}$ *such that, for every* $u \in E$,

$$\|u\|_{2;\mu} \leq c_1 \sup_{\nu \in N_1} \|u\|_{1;\nu}.$$

We say that the two families are ***equivalent*** *if each family dominates the other. We also say that they* ***generate the same topology***. ∎

Terminology. The *topology* of E is the set of its open sets. We can say that two families of semi-norms *generate the same topology* instead of saying that they *are equivalent* because equivalence of two families of semi-norms implies equality of their open sets [Vol. 1, Theorem 3.4] and the converse also holds [Vol. 1, Theorem 7.14 (a) and 8.2 (a), with $L = T =$ Identity]. □

DEFINITION 1.9.– *Let* E *and* F *be two semi-normed spaces.*

(a) *We write* $E \underset{\Leftrightarrow}{\equiv} F$ *if* $E = F$ *and if their additions, multiplications and families of semi-norms coincide, or, in other words, if they have the same vector space structures and the same semi-norms.*

(b) *We say that* E *is* ***topologically equal*** *to* F, *written* $E \underset{\Leftrightarrow}{=} F$, *if* $E = F$, *their additions and multiplications coincide, and their families of semi-norms are equivalent.*

(c) *We say that* E *is* ***topologically included*** *in* F, *written* $E \underset{\rightarrow}{\subseteq} F$, *if* E *is a vector subspace of* F *and if the family of semi-norms of* E *dominates the family of restrictions to* E *of the semi-norms of* F. *In other words, if, for every* $\mu \in \mathcal{N}_F$, *there exists a finite set* N *in* $\mathcal{N}_E$ *and* $c \in \mathbb{R}$ *such that, for every* $u \in E$,

$$\|u\|_{F;\mu} \leq c \sup_{\nu \in N} \|u\|_{E;\nu}.$$

(d) *We say that* E *is a* ***topological subspace*** *of* F *if it is a* ***vector subspace*** *of* F, *i.e. a vector space under the addition and multiplication of* E, *and it is endowed with the restrictions of the semi-norms of* F, *or, more generally, with a family equivalent to this family of restrictions.* ∎

Let us now show that spaces of continuous functions are separated and compare them.

THEOREM 1.10.– *Let* $\Omega \subset \mathbb{R}^d$, E *be a separated semi-normed space,* K *a compact set included in* Ω *and* D *a compact subset of* $\mathbb{R}^d$. *Then:*

(a) *$\mathcal{C}(\Omega;E)$, $\mathcal{C}_{\mathrm{b}}(\Omega;E)$, $\mathcal{C}_K(\Omega;E)$, $\mathbf{C}_{\mathrm{b}}(\Omega;E)$ and $\mathbf{C}_D(\Omega;E)$ are separated semi-normed spaces.*

(b) *$\mathcal{C}_K(\Omega;E)$, $\mathbf{C}_{\mathrm{b}}(\Omega;E)$ and $\mathbf{C}_D(\Omega;E)$ are closed and hence sequentially closed topological subspaces of $\mathcal{C}_{\mathrm{b}}(\Omega;E)$.*

(c)
$$\mathcal{C}_K(\Omega;E) \underset{\leftrightarrows}{\equiv} \mathbf{C}_K(\Omega;E) \underset{\to}{\subsetneq} \mathbf{C}_{\mathrm{b}}(\Omega;E) \underset{\to}{\subsetneq} \mathcal{C}_{\mathrm{b}}(\Omega;E) \underset{\to}{\subsetneq} \mathcal{C}(\Omega;E).$$

(d) *If Ω is bounded,*
$$\mathbf{C}_{\mathrm{b}}(\Omega;E) = \mathbf{C}(\Omega;E).$$

(e) *If Ω is compact,*
$$\mathbf{C}_{\mathrm{b}}(\Omega;E) \underset{\leftrightarrows}{\equiv} \mathcal{C}_{\mathrm{b}}(\Omega;E) \underset{\leftrightarrows}{\equiv} \mathcal{C}(\Omega;E). \blacksquare$$

Proof. (a) These spaces are semi-normed spaces (Definition 1.1) by construction.

The space $\mathcal{C}(\Omega;E)$ is separated (Definition 1.1) because, if every semi-norm of one of its elements f is zero, then (Definition 1.3 (a)), for every $x \in \Omega$ and $\nu \in \mathcal{N}_E$, we have $\|f(x)\|_{E;\nu} = 0$, so $f(x) = 0_E$ because E is separated and therefore f is zero. The same proof works for $\mathcal{C}_{\mathrm{b}}(\Omega;E)$ (Definition 1.3 (b)) and hence for the spaces $\mathcal{C}_K(\Omega;E)$, $\mathbf{C}_{\mathrm{b}}(\Omega;E)$ and $\mathbf{C}_D(\Omega;E)$, since they are endowed with the semi-norms of $\mathcal{C}_{\mathrm{b}}(\Omega;E)$.

(b) The vector spaces $\mathcal{C}_K(\Omega;E)$ and $\mathbf{C}_D(\Omega;E)$ are included in $\mathcal{C}_{\mathrm{b}}(\Omega;E)$, as we saw earlier in the justification of Definition 1.7, and so is $\mathbf{C}_{\mathrm{b}}(\Omega;E)$. These spaces are topological subspaces (Definition 1.9 (d)) of $\mathcal{C}_{\mathrm{b}}(\Omega;E)$ because they are endowed with its semi-norms. Let us check that they are closed.

Closedness of $\mathcal{C}_K(\Omega;E)$ in $\mathcal{C}_{\mathrm{b}}(\Omega;E)$. Let us show that the complement is open. Thus, let $f \in \mathcal{C}_{\mathrm{b}}(\Omega;E) \setminus \mathcal{C}_K(\Omega;E)$. There exists $x \in \Omega \setminus K$ such that $f(x) \neq 0_E$ and hence $\nu \in \mathcal{N}_E$ such that $\|f(x)\|_{E;\nu} = a > 0$. Therefore, $\|g - f\|_{\mathcal{C}_{\mathrm{b}}(\Omega;E);\nu} \leq a/2$ implies $g(x) > 0$, and so $g \in \mathcal{C}_{\mathrm{b}}(\Omega;E) \setminus \mathcal{C}_K(\Omega;E)$, showing that the complement is open.

Closedness of $\mathbf{C}_{\mathrm{b}}(\Omega;E)$ in $\mathcal{C}_{\mathrm{b}}(\Omega;E)$. Let us show that the complement is open. Thus, let $f \in \mathcal{C}_{\mathrm{b}}(\Omega;E) \setminus \mathbf{C}_{\mathrm{b}}(\Omega;E)$. There exists $\nu \in \mathcal{N}_E$, $a > 0$ and, for all $n \in \mathbb{N}$, x_n and y_n in Ω such that $|x_n - y_n| \leq 1/n$ and $\|f(x_n) - f(y_n)\|_{E;\nu} \geq a$. Therefore, $\|g - f\|_{\mathcal{C}_{\mathrm{b}}(\Omega;E);\nu} \leq a/3$ implies $\|g(x_n) - g(y_n)\|_{E;\nu} \geq a/3$, and so $g \in \mathcal{C}_{\mathrm{b}}(\Omega;E) \setminus \mathbf{C}_{\mathrm{b}}(\Omega;E)$, showing that the complement is open.

Closedness of $\mathbf{C}_D(\Omega;E)$ in $\mathcal{C}_{\mathrm{b}}(\Omega;E)$. The set $\mathbf{C}_D(\Omega;E)$ is the intersection of $\mathbf{C}_{\mathrm{b}}(\Omega;E)$ and $\mathcal{C}_D(\Omega;E)$, which are closed (for the latter, consider the above proof with D instead of K), and so it is itself closed (Theorem A.10).

(c) *Identity* $\mathcal{C}_K(\Omega; E) \underset{\leftrightarrow}{\equiv} \mathbf{C}_K(\Omega; E)$. Algebraic equality is satisfied because every function in $\mathcal{C}_K(\Omega; E)$ is uniformly continuous on the compact set K by Heine's theorem (Theorem A.34) and therefore on the whole of Ω (since it vanishes on ∂K). Topological identity follows because these two spaces are both endowed with the semi-norms of $\mathcal{C}_b(\Omega; E)$.

Inclusion $\mathbf{C}_K(\Omega; E) \underset{\to}{\subsetneq} \mathbf{C}_b(\Omega; E)$. Algebraic inclusion follows from the fact that every function in $\mathbf{C}_K(\Omega; E)$ is bounded on the compact set K (Heine's theorem again) and is therefore bounded on the whole of Ω. Topological inclusion follows because both spaces are endowed with the semi-norms of $\mathcal{C}_b(\Omega; E)$.

Inclusion $\mathbf{C}_b(\Omega; E) \underset{\to}{\subsetneq} \mathcal{C}_b(\Omega; E)$. This topological inclusion follows from the fact that every uniformly continuous function is continuous, noting that $\mathbf{C}_b(\Omega; E)$ is endowed with the semi-norms of $\mathcal{C}_b(\Omega; E)$.

Inclusion $\mathcal{C}_b(\Omega; E) \underset{\to}{\subsetneq} \mathcal{C}(\Omega; E)$. This topological inclusion follows from the fact that, by Definition 1.3 of the semi-norms of $\mathcal{C}(\Omega; E)$ and $\mathcal{C}_b(\Omega; E)$, for every $f \in \mathcal{C}_b(\Omega; E)$ and $\nu \in \mathcal{N}_E$ and every compact set $K \subset \Omega$,

$$\|f\|_{\mathcal{C}(\Omega;E);K,\nu} = \sup_{x \in K} \|f(x)\|_{E;\nu} \leq \sup_{x \in \Omega} \|f(x)\|_{E;\nu} = \|f\|_{\mathcal{C}_b(\Omega;E);\nu}.$$

(d) *Case where Ω is bounded.* In this case, $\mathbf{C}_b(\Omega; E) = \mathbf{C}(\Omega; E)$ because, given that Ω is precompact (Theorem A.23 (b)), uniformly continuous functions on Ω are bounded. Indeed, their images are precompact (Theorem A.33) and hence bounded (Theorem A.19 (a)).

(e) *Case where Ω is compact. Equality* $\mathbf{C}_b(\Omega; E) = \mathcal{C}_b(\Omega; E) = \mathcal{C}(\Omega; E)$. This algebraic equality follows from the fact that every continuous function on a compact set is uniformly continuous and bounded on this set (Heine's theorem once again, Theorem A.34).

Identity $\mathbf{C}_b(\Omega; E) \underset{\leftrightarrow}{\equiv} \mathcal{C}_b(\Omega; E)$. This identity follows from the algebraic equality because both spaces are endowed with the semi-norms of $\mathcal{C}_b(\Omega; E)$.

Equality $\mathcal{C}_b(\Omega; E) \underset{\leftrightarrows}{=} \mathcal{C}(\Omega; E)$. Here, we can take $K = \Omega$ in Definition 1.3 (a), which gives $\|f\|_{\mathcal{C}_b(\Omega;E);\nu} = \|f\|_{\mathcal{C}(\Omega;E);\Omega,\nu}$ and hence $\mathcal{C}(\Omega; E) \underset{\to}{\subsetneq} \mathbf{C}_b(\Omega; E)$ (because $\mathbf{C}_b(\Omega; E)$ is endowed with the semi-norms of $\mathcal{C}_b(\Omega; E)$). The topological equality follows because the inverse inclusion holds by (c). □

Non-closedness of $\mathbf{C}(\Omega)$ in $\mathcal{C}(\Omega)$. If Ω is an open set (or more generally a set that is not compact),

$$\mathbf{C}(\Omega) \text{ is not (sequentially) closed in } \mathcal{C}(\Omega). \tag{1.2}$$

For example, the functions f_n defined on $(0, 1)$ by $f_n(x) = \inf\{n, 1/x\}$ are uniformly continuous and, as $n \to \infty$, they converge in $\mathcal{C}((0, 1))$ to the function $f(x) = 1/x$, which is not uniformly continuous. □

1.4. Sequential completeness of spaces of continuous functions

Let us define sequential completeness.

DEFINITION 1.11.– *A* ***Neumann space*** *is a separated semi-normed space that is sequentially complete, namely one in which every Cauchy sequence converges.*

A sequence $(u_n)_{n\in\mathbb{N}}$ in a separated semi-normed space E is said to be ***Cauchy*** *if, for every semi-norm $\|\ \|_{E;\nu}$ of E,*

$$\sup_{m\geq n} \|u_m - u_n\|_{E;\nu} \to 0 \quad \text{as } n \to \infty. \ \blacksquare$$

Terminology. We named these spaces *Neumann spaces* in Volume 1 in honor of John VON NEUMANN, who introduced them in 1935 [59]. Thus, readers will need to recall this definition before using it elsewhere. □

Let us show that spaces of continuous functions are sequentially complete whenever the space E of values is itself sequentially complete[6].

THEOREM 1.12.– *Let $\Omega \subset \mathbb{R}^d$, K be a compact set included in Ω, and D a compact subset of $\mathbb{R}^d$.*

If E is a Neumann space, then $\mathcal{C}(\Omega; E)$, $\mathcal{C}_b(\Omega; E)$, $\mathcal{C}_K(\Omega; E)$, $\mathrm{C}_b(\Omega; E)$ and $\mathrm{C}_D(\Omega; E)$ are also Neumann spaces. ▮

Proof. Let $\{\|\ \|_{E;\nu} : \nu \in \mathcal{N}_E\}$ be the family of semi-norms of E.

Sequential completeness of $\mathcal{C}(\Omega; E)$. Let $(f_n)_{n\in\mathbb{N}}$ be a Cauchy sequence in $\mathcal{C}(\Omega; E)$. By Definition 1.3 (a) of the semi-norms of $\mathcal{C}(\Omega; E)$, for every compact set $K' \subset \Omega$, every $\nu \in \mathcal{N}_E$ and $\epsilon > 0$, there exists $\ell \in \mathbb{N}$ such that

$$\sup_{n\geq\ell} \sup_{x\in K'} \|f_n(x) - f_\ell(x)\|_{E;\nu} = \sup_{n\geq\ell} \|f_n - f_\ell\|_{\mathcal{C}(\Omega;E);K'\nu} \leq \epsilon.$$

Then, for every $x \in \Omega$, $(f_n(x))_{n\in\mathbb{N}}$ is a Cauchy sequence in E; since E is a Neumann space, this sequence has a limit, which we shall denote as $f(x)$. Taking the limit as $n \to \infty$, the previous inequality yields

$$\sup_{x\in K'} \|f(x) - f_\ell(x)\|_{E;\nu} \leq \epsilon. \tag{1.3}$$

6 **History of Theorem 1.12.** In a series of unpublished but nonetheless hugely influential lectures, Karl WEIERSTRASS showed that the limit of a sequence of sequentially continuous real functions that converge uniformly is itself sequentially continuous. In other words, he showed that an inequality of type (1.3) implies the sequential continuity of f. This is the key point in the proof of the sequential completeness of $\mathcal{C}(\mathbb{R})$.

Let us prove by contradiction that f is sequentially continuous. Suppose not. Then there exists $\epsilon > 0$, $\nu \in \mathcal{N}_E$, $x \in \Omega$, and a sequence $(x_m)_{m\in\mathbb{N}}$ in Ω such that $x_m \to x$ and $\|f(x_m) - f(x)\|_{E;\nu} \geq 3\epsilon$. Applying the inequality (1.3) to the compact set $K' = \{x_m : m \in \mathbb{N}\} \cup \{x\}$ then gives, for every $m \in \mathbb{N}$,

$$\begin{aligned}
&\|f_\ell(x_m) - f_\ell(x)\|_{E;\nu} = \\
&\quad = \|f_\ell(x_m) - f(x_m) + f(x_m) - f(x) + f(x) - f_\ell(x)\|_{E;\nu} \geq \\
&\quad \geq \|f_\ell(x_m) - f(x_m)\|_{E;\nu} - \|f(x_m) - f(x)\|_{E;\nu} - \|f(x) - f_\ell(x))\|_{E;\nu} \geq \\
&\quad \geq \epsilon.
\end{aligned}$$

Therefore, f_ℓ is not sequentially continuous. This proves that f is sequentially continuous.

Hence, f is continuous (Theorem A.29, since $\mathbb{R}^d$ is a normed space), and thus $f \in \mathcal{C}(\Omega; E)$. Equation (1.3) then shows that $f_n \to f$ in $\mathcal{C}(\Omega; E)$, which proves that $\mathcal{C}(\Omega; E)$ is sequentially complete.

Sequential completeness of $\mathcal{C}_\mathrm{b}(\Omega; E)$. The space $\mathcal{C}_\mathrm{b}(\Omega; E)$ may be shown to be sequentially complete by repeating the above proof with Ω instead of K' (the boundedness of the limit f then follows from the inequality (1.3)).

Sequential completeness of $\mathcal{C}_K(\Omega; E)$, $\mathbf{C}_\mathrm{b}(\Omega; E)$ *and* $\mathbf{C}_D(\Omega; E)$. Since every closed topological subspace of a sequentially complete space is sequentially complete (Theorem A.27), the subspaces $\mathcal{C}_K(\Omega; E)$, $\mathbf{C}_\mathrm{b}(\Omega; E)$ and $\mathbf{C}_D(\Omega; E)$ of $\mathcal{C}_\mathrm{b}(\Omega; E)$ are sequentially complete (they are closed by Theorem 1.10 (b)). □

Another proof of the sequential completeness of $\mathbf{C}_\mathrm{b}(\Omega; E)$**.** The proof above can be revisited with Ω instead of K' to show that the limit f thus obtained is uniformly continuous by using two sequences $(x_m)_{m\in\mathbb{N}}$ and $(y_m)_{m\in\mathbb{N}}$ in Ω such that $|x_m - y_m| \leq 1/m$ instead of $x_m \to x$. □

1.5. Metrizability of spaces of continuous functions

Let us define metrizability[7].

Definition 1.13.– *We say that a semi-normed space is* ***metrizable*** *if it is separated and its family of semi-norms is countable or equivalent to a countable family of semi-norms.*

A ***Fréchet space*** *is a metrizable semi-normed space that is sequentially complete. A* ***Banach space*** *is a sequentially complete normed space.* ▮

7 **History of the notion of metrizable space.** Maurice Fréchet gave the general definition of a metric space in 1906 in [38].

Justification of the term "metrizable". We speak of a *metrizable* space because any countable family, or equivalently any sequence $(\| \ \|_k)_{k\in\mathbb{N}}$ of semi-norms, may be associated with a *metric* d, that generates the same topology, given by

$$d(u,v) \stackrel{\text{def}}{=} \sum_{k\in\mathbb{N}} 2^{-k} \frac{\|u-v\|_k}{1+\|u-v\|_k}.$$

Strictly speaking, Definition 1.13 defines a **separated countably semi-normable** space. We prefer to speak instead of a *metrizable* space by abuse of language because this equivalent notion is more familiar. A **metrizable** space is, more precisely, a space that is "topologically equal to a metric space." □

Superiority of a sequence of semi-norms over a metric. The semi-norms of a metrizable space E characterize its bounded sets by $\sup_{u\in U}\|u\|_{E;k} < \infty$ for every $k\in\mathbb{N}$ (Definition A.7).

By contrast, *if E is not normable* and d is a metric that generates its topology, its bounded sets are not characterized by $\sup_{u\in U} d(u, 0_E) < \infty$. Even worse, none of the "balls" $\{u\in E : d(u,z)\le r\}$ are bounded (if $r>0$). If they were, $\{u : d(u,z) < r\}$ would be a non-empty bounded open set. But the existence of such an open set is equivalent to normability by **Kolmogorov's theorem**[8]. □

Let us state some metrizability and normability properties of spaces of continuous functions.

THEOREM 1.14.– *Let $\Omega\subset\mathbb{R}^d$, E be a separated semi-normed space, K a compact set included in Ω and D a compact subset of $\mathbb{R}^d$. Then:*

(a) *If E is metrizable, then $\mathcal{C}_b(\Omega;E)$, $\mathcal{C}_K(\Omega;E)$, $\mathbf{C}_b(\Omega;E)$ and $\mathbf{C}_D(\Omega;E)$ are metrizable.*

(b) *If E is metrizable and Ω is open, then $\mathcal{C}(\Omega;E)$ is metrizable.*

(c) *If E is a normed space, then $\mathcal{C}_b(\Omega;E)$, $\mathcal{C}_K(\Omega;E)$, $\mathbf{C}_b(\Omega;E)$ and $\mathbf{C}_D(\Omega;E)$ are normed spaces.* ▮

Proof. (a) If E is metrizable, then, by Definition 1.13, we may endow it with a family of semi-norms indexed by a countable set $\mathcal{N}_E$. Therefore, $\mathcal{C}_b(\Omega;E)$ is metrizable because its semi-norms are indexed by $\mathcal{N}_E$ (Definition 1.3 (b)). Furthermore, $\mathbf{C}_b(\Omega;E)$, $\mathcal{C}_K(\Omega;E)$ and $\mathbf{C}_D(\Omega;E)$ are also metrizable, since they are endowed (Definitions 1.4 (b) and 1.7) with the semi-norms of $\mathcal{C}_b(\Omega;E)$.

(c) If E is a normed space, the family of semi-norms of $\mathcal{C}_b(\Omega;E)$ reduces to the norm $\|f\|_{\mathcal{C}_b(\Omega;E)} = \sup_{x\in\Omega}\|f(x)\|_E$. Thus, by definition, the spaces $\mathcal{C}_K(\Omega;E)$, $\mathbf{C}_b(\Omega;E)$ and $\mathbf{C}_D(\Omega;E)$ are endowed with this norm as well.

8 **History of Komogorov's theorem.** Andrey KOLMOGOROV showed in 1934 [49, p. 33] that a topological vector space is normable if and only if there exists a bounded convex neighborhood of the origin, which here is equivalent to the existence of a bounded open set.

(b) Suppose that Ω is open. For every $n \in \mathbb{N}$, let

$$\Omega^n_{1/n} \stackrel{\text{def}}{=} \{x \in \mathbb{R}^d : |x| < n,\ B(x, 1/n) \subset \Omega\},$$

where $B(x, 1/n)$ denotes the closed ball $\{y \in \mathbb{R}^d : |y - x| \leq 1/n\}$. As we shall check in Lemma 1.15, every compact set $K \subset \Omega$ is included in one of the compact sets $K_n = \overline{\Omega^n_{1/n}}$, so the family of semi-norms of $\mathcal{C}(\Omega; E)$ (Definition 1.3 (a)) is equivalent to the sub-family associated with the K_n, which is indexed by $\mathbb{N} \times \mathcal{N}_E$. If, moreover, E is metrizable, by Definition 1.13, we can choose $\mathcal{N}_E$ to be countable, in which case $\mathbb{N} \times \mathcal{N}_E$ is also countable (Theorem A.2), and so $\mathcal{C}(\Omega; E)$ is metrizable. □

We still need to show the following lemma.

LEMMA 1.15.– *Let Ω be an open subset of $\mathbb{R}^d$, $r > 0$, $n \in \mathbb{N}$, and*

$$\Omega_r \stackrel{\text{def}}{=} \{x \in \mathbb{R}^d : B(x, r) \subset \Omega\}, \quad \Omega^n_r \stackrel{\text{def}}{=} \{x \in \mathbb{R}^d : |x| < n,\ B(x, r) \subset \Omega\},$$

where $B(x, r) \stackrel{\text{def}}{=} \{y \in \mathbb{R}^d : |y - x| \leq r\}$.

Then, Ω_r and Ω^n_r are open, the closure $\overline{\Omega^n_r}$ is a compact set included in Ω, the $(\Omega^n_{1/n})_{n \in \mathbb{N}}$ forms an increasing open covering of Ω and every compact set K included in Ω is included in one of the $\overline{\Omega^n_{1/n}}$. ▮

Proof. The set $\Omega_r = \{x \in \mathbb{R}^d : B(x, r) \subset \Omega\}$ is open because if $x \in \Omega_r$, then the closed ball $B(x, r)$ is a compact set included in Ω, so the strong inclusion theorem (Theorem A.22) provides $s > 0$ such that $B(x, r) + B(0, s) \subset \Omega$, and hence $B(x, s) \subset \Omega_r$.

The intersection Ω^n_r with the open ball $\{x \in \mathbb{R}^d : |x| < n\}$ is therefore also open (Theorem A.10).

The closure $\overline{\Omega^n_r}$, which is closed by definition and bounded, is compact in $\mathbb{R}^d$ by the Borel–Lebesgue theorem (Theorem A.23 (b)). It is included in $\Omega^{n+1}_{r/2}$ and hence in Ω.

Finally, the sets $\Omega^n_{1/n}$ increase with n and form a covering of Ω. Every compact set $K \subset \Omega$ is therefore, by Definition A.17 (a), included in one of these sets, and, finally, in one of their closures $\overline{\Omega^n_{1/n}}$. □

Let us state a few consequences of the previous two theorems.

THEOREM 1.16.– *Let $\Omega \subset \mathbb{R}^d$, E be a separated semi-normed space, K a compact set included in Ω and D a compact subset of $\mathbb{R}^d$. Then:*

(a) *If E is a Fréchet space, then* $\mathcal{C}_b(\Omega; E)$, $\mathcal{C}_K(\Omega; E)$, $\mathbf{C}_b(\Omega; E)$ *and* $\mathbf{C}_D(\Omega; E)$ *are Fréchet spaces.*

(b) *If E is a Fréchet space and Ω is open, then* $\mathcal{C}(\Omega; E)$ *is a Fréchet space.*

(c) *If E is a Banach space, then* $\mathcal{C}_b(\Omega; E)$, $\mathcal{C}_K(\Omega; E)$, $\mathbf{C}_b(\Omega; E)$ *and* $\mathbf{C}_D(\Omega; E)$ *are Banach spaces.*

In particular:

(d) *If Ω is open, then* $\mathcal{C}(\Omega)$ *is a Fréchet space.*

(e) $\mathcal{C}_b(\Omega)$, $\mathcal{C}_K(\Omega)$, $\mathbf{C}_b(\Omega)$ *and* $\mathbf{C}_D(\Omega)$ *are Banach spaces.* ▮

Proof. (a), (b) *and* (c). This follows from the sequential completeness properties in Theorem 1.12 and the metrizability and normability properties in Theorem 1.14.

(e) *and* (d). This follows from (a), (b) and (c), since $\mathbb{R}$ is a Banach space (Theorem A.23 (a)) and hence a Fréchet space. □

Metrizability of $\mathcal{C}(\Omega)$. The properties of Theorems 1.14 (b) and 1.16 (c) and (d) still hold when:

$$\begin{cases} \textit{Ω is the union of a sequence $(K_n)_{n\in\mathbb{N}}$ of compact subsets of $\mathbb{R}^d$ such that,} \\ \textit{for any compact set $K \subset \Omega$, there exists n such that $K \subset K_n$.} \end{cases}$$

This property is, for example, satisfied whenever Ω is open or closed, but not for arbitrary subsets of $\mathbb{R}^d$. □

1.6. The space $\mathcal{K}(\Omega; E)$

Let us define the space of continuous functions with compact support[9]. The support of a function was defined on page 4. We denote

$$\mathcal{C}^+(\Omega) \stackrel{\text{def}}{=} \{q \in \mathcal{C}(\Omega) : \forall x \in \Omega,\ q(x) \geq 0\}.$$

DEFINITION 1.17.– *Let $\Omega \subset \mathbb{R}^d$, E be a separated semi-normed space, and $\{\| \ \|_{E;\nu} : \nu \in \mathcal{N}_E\}$ its family of semi-norms.*

We denote by $\mathcal{K}(\Omega; E)$ the vector space of continuous functions from Ω into E with compact support endowed with the semi-norms, indexed by $q \in \mathcal{C}^+(\Omega)$ and $\nu \in \mathcal{N}_E$,

$$\|f\|_{\mathcal{K}(\Omega;E);q,\nu} \stackrel{\text{def}}{=} \sup_{x\in\Omega} q(x)\|f(x)\|_{E;\nu}. \ ▮$$

---

9 **History of the semi-norms of $\mathcal{K}(\Omega; E)$.** The semi-norms in Definition 1.17, which are new to the best of the author's knowledge, provide a very simple way [SIMON, 75, forthcoming] of generating the inductive limit topology of the $\mathcal{C}_K(\Omega; E)$ with which BOURBAKI endowed $\mathcal{K}(\Omega; E)$ [11, Chapter III, § 1, p. 41].

Justification. The continuous functions with compact support form a vector subspace of $\mathcal{C}(\Omega; E)$ because the support of the sum of two functions is included in the union of their supports, which is compact if the supports of both functions are compact.

The mapping $f \mapsto \sup_{x\in\Omega} q(x)\|f(x)\|_{E;\nu}$ is a semi-norm on $\mathcal{K}(\Omega; E)$, since the upper envelope of a set of semi-norms is itself a semi-norm whenever it is everywhere finite (Theorem A.6). This is indeed the case here, since, on the one hand, for every x, the mapping $f \mapsto q(x)\|f(x)\|_{E;\nu}$ is a semi-norm on $\mathcal{K}(\Omega; E)$ and, on the other hand, for each f, $\sup_{x\in\Omega} q(x)\|f(x)\|_{E;\nu}$ is finite, since q and $\|f\|_{E;\nu}$ are bounded on the support of f (because continuous functions are bounded on compact sets by Theorem A.34). □

Notation $\mathcal{K}(\Omega; E)$**.** This notation was used by BOURBAKI [11, Chapter III, § 1, p. 40]. □

Benefit of the semi-norms of $\mathcal{K}(\Omega; E)$**.** The semi-norms newly introduced in Definition 1.17 yield the properties usually obtained from the *inductive limit topology of the* $\mathcal{C}_K(\Omega; E)$. In particular, the functions in a bounded subset of $\mathcal{K}(\Omega; E)$ all have supports included in the same compact set K included in Ω (Theorem 1.23, which is also new). These semi-norms generate the inductive limit topology [SIMON, 75, forthcoming] and are simpler. □

Motivation of the study of $\mathcal{K}(\Omega; E)$**.** We study the space $\mathcal{K}(\Omega; E)$ because we will frequently need continuous approximations with compact support of functions or distributions. More importantly, the properties of $\mathcal{K}(\Omega)$ stated in Theorems 1.22 and 1.23 are the basis of those of the space $\mathcal{D}(\Omega)$ that will be used to construct the space $\mathcal{D}'(\Omega; E)$ of distributions in Volume 3. □

Let us state a condition to guarantee that the support of a function is compact.

THEOREM 1.18.– *The support of a function from a subset Ω of $\mathbb{R}^d$ into a separated semi-normed space is compact if and only if it is included in a compact subset of Ω.* ▮

Proof. By Definition 1.5, the support of f is $\overline{\{x \in \Omega : f(x) \neq 0_E\}} \cap \Omega$, where E is the semi-normed space in question, so it is not necessarily closed.

If $\operatorname{supp} f \subset K \subset \Omega$, where K is compact, then

$$\overline{\{x \in \Omega : f(x) \neq 0_E\}} \subset \overline{K} = K \subset \Omega,$$

so $\operatorname{supp} f = \overline{\{x \in \Omega : f(x) \neq 0_E\}}$. This set is closed (by definition) and bounded (like K) in $\mathbb{R}^d$, and thus compact by the Borel–Lebesgue theorem (Theorem A.23 (b)).

The converse is clear (taking $K = \operatorname{supp} f$). □

Let us state a few elementary properties.

THEOREM 1.19.– *Let $\Omega \subset \mathbb{R}^d$, E be a separated semi-normed space, and K a compact set included in Ω.*

Then $\mathcal{K}(\Omega; E)$ is a separated semi-normed space and

$$\mathcal{C}_K(\Omega; E) \subsetneq \mathcal{K}(\Omega; E) \subsetneq \mathbf{C}_b(\Omega; E). \blacksquare$$

Proof. Let $\{\| \ \|_{E;\nu} : \nu \in \mathcal{N}_E\}$ be the family of semi-norms of E.

Semi-normed space. The space $\mathcal{K}(\Omega; E)$ is a semi-normed space by Definition 1.17. It is separated (Definition 1.1) because if every semi-norm of one of its elements f is zero, then

$$\|f\|_{\mathcal{K}(\Omega;E);1_\Omega,\nu} = \sup_{x\in\Omega} \|f(x)\|_{E;\nu} = 0,$$

where 1_Ω is the constant function of value 1 on Ω. This holds for every $\nu \in \mathcal{N}_E$, so $f(x) = 0_E$ for every $x \in \Omega$ because E is separated, and therefore f is zero.

Inclusion $\mathcal{C}_K(\Omega; E) \subsetneq \mathcal{K}(\Omega; E)$. Let $f \in \mathcal{C}_K(\Omega; E)$. Its support is included in the compact set K and hence is compact (Theorem 1.18), so $f \in \mathcal{K}(\Omega; E)$. Furthermore, for every $q \in \mathcal{C}^+(\Omega)$ and $\nu \in \mathcal{N}_E$, we have, by Definitions 1.17 and 1.3,

$$\|f\|_{\mathcal{K}(\Omega;E);q,\nu} = \sup_{x\in\Omega} q(x)\|f(x)\|_{E;\nu} \leq \sup_{x\in K} q(x) \sup_{x\in\Omega} \|f(x)\|_{E;\nu} = c_q\|f\|_{\mathcal{C}_b(\Omega;E);\nu},$$

where $c_q = \sup_{x\in K} q(x)$, which is finite by Heine's theorem (Theorem A.34). This gives us the topological inclusion (Definition 1.9 (c)), since, by Definition 1.7 (a), $\mathcal{C}_K(\Omega; E)$ is endowed with the semi-norms of $\mathcal{C}_b(\Omega; E)$.

Inclusion $\mathcal{K}(\Omega; E) \subsetneq \mathbf{C}_b(\Omega; E)$. Let $f \in \mathcal{K}(\Omega; E)$. The function f is continuous with compact support, so it is uniformly continuous and bounded on its support, again by Heine's theorem, and hence bounded everywhere. Thus, $f \in \mathbf{C}_b(\Omega; E)$. The topological inclusion follows from the fact that, by Definition 1.4 (b), $\mathbf{C}_b(\Omega; E)$ is endowed with the semi-norms of $\mathcal{C}_b(\Omega; E)$ and, for every $\nu \in \mathcal{N}_E$,

$$\|f\|_{\mathcal{C}_b(\Omega;E);\nu} = \sup_{x\in\Omega} \|f(x)\|_{E;\nu} = \|f\|_{\mathcal{K}(\Omega;E);1_\Omega,\nu}. \tag{1.4}$$

□

Let us observe that the topologies of $\mathcal{C}(\Omega; E)$, $\mathcal{C}_b(\Omega; E)$ and $\mathcal{K}(\Omega; E)$ coincide on $\mathcal{C}_K(\Omega; E)$.

THEOREM 1.20.– *Let $\Omega \subset \mathbb{R}^d$, E be a separated semi-normed space, and K a compact set included in Ω.*

Then the topologies of $\mathcal{C}(\Omega; E)$, $\mathcal{C}_b(\Omega; E)$, $\mathcal{C}_K(\Omega; E)$ and $\mathcal{K}(\Omega; E)$ coincide on $\mathcal{C}_K(\Omega; E)$. ▮

Proof. According to the inclusion $\mathcal{C}_K(\Omega; E) \subsetneq \mathcal{K}(\Omega; E) \subsetneq \mathcal{C}_b(\Omega; E) \subsetneq \mathcal{C}(\Omega; E)$ (Theorems 1.19 and 1.10 (c)), it suffices to show that the family of semi-norms of $\mathcal{C}(\Omega; E)$ dominates that of $\mathcal{C}_K(\Omega; E)$ on the latter space. Given that the space $\mathcal{C}_K(\Omega; E)$ (Definition 1.7 (a)) is endowed with the semi-norms of $\mathcal{C}_b(\Omega; E)$, the desired result follows from the fact that for every $f \in \mathcal{C}_K(\Omega; E)$ and $\nu \in \mathcal{N}_E$,

$$\|f\|_{\mathcal{C}_b(\Omega;E);\nu} = \sup_{x\in\Omega} \|f(x)\|_{E;\nu} = \sup_{x\in K} \|f(x)\|_{E;\nu} = \|f\|_{\mathcal{C}(\Omega;E);K,\nu}. \ \square$$

Let us show that $\mathcal{K}(\Omega; E)$ is sequentially complete whenever E is a sequentially complete normed space.

THEOREM 1.21.– *For every $\Omega \subset \mathbb{R}^d$:*

If E is a Banach space, then $\mathcal{K}(\Omega; E)$ is a Neumann space.

In particular, $\mathcal{K}(\Omega)$ is a Neumann space. ▮

Proof. Let E be a Banach space. Observe that for all $f \in \mathcal{K}(\Omega; E)$ and $q \in \mathcal{C}^+(\Omega)$, the function qf belongs to $\mathcal{K}(\Omega; E)$ and hence to $\mathcal{C}_b(\Omega; E)$ by Theorem 1.19. Furthermore, by the definition of their semi-norms (Definitions 1.17 and 1.3 (b)),

$$\|f\|_{\mathcal{K}(\Omega;E);q} = \sup_{x\in\Omega} q(x)\|f(x)\|_E = \|qf\|_{\mathcal{C}_b(\Omega;E)} \tag{1.5}$$

(the index ν is absent because E only has one semi-norm, namely its norm).

Now, let $(f_n)_{n\in\mathbb{N}}$ be a Cauchy sequence in $\mathcal{K}(\Omega; E)$. By (1.4), this sequence is Cauchy in $\mathcal{C}_b(\Omega; E)$, which is sequentially complete (Theorem 1.12), so it has a limit f in this space. By (1.5), for every $q \in \mathcal{C}^+(\Omega)$, the sequence $(qf_n)_{n\in\mathbb{N}}$ is Cauchy in $\mathcal{C}_b(\Omega; E)$ and therefore has a limit in this space, which must necessarily be qf. Hence, $qf \in \mathcal{C}_b(\Omega; E)$ and so, again by (1.5),

$$\sup_{x\in\Omega} q(x)\|f(x)\|_E < \infty. \tag{1.6}$$

This implies that the support of the function $\|f\|_E$ is compact by Theorem 1.22, which we will prove below. Indeed, this function $\|f\|_E$ is continuous, since

$\big|\|f(y)\|_E - \|f(x)\|_E\big| \le \|f(y) - f(x)\|_E$ (because norms are contractions, see Theorem A.5). Since the supports of f and $\|f\|_E$ coincide, it follows that $f \in \mathcal{K}(\Omega; E)$. Therefore, once again by (1.5),

$$\|f_n - f\|_{\mathcal{K}(\Omega;E);q} = \|q(f_n - f)\|_{\mathcal{C}_b(\Omega;E)} \to 0.$$

Hence, $f_n \to f$ in $\mathcal{K}(\Omega; E)$, which is therefore sequentially complete, i.e. a Neumann space.

This is, in particular, true for $\mathcal{K}(\Omega)$, since $\mathbb{R}$ is a Banach space (Theorem A.23 (a)). □

It remains to be checked that the property (1.6) implies the compactness of the support of $\|f\|_E$. More precisely, we need to establish the following property[10].

THEOREM 1.22.– *Let $f \in \mathcal{C}(\Omega)$, where $\Omega \subset \mathbb{R}^d$, such that, for every $q \in \mathcal{C}^+(\Omega)$,*

$$\sup_{x\in\Omega} q(x)\,|f(x)| < \infty.$$

Then the support of f is compact. ▮

Proof. Let

$$K \stackrel{\text{def}}{=} \overline{\{x \in \Omega : f(x) \neq 0\}}.$$

Let us first show by contradiction that

$$K \subset \Omega. \tag{1.7}$$

Suppose that K is not included in Ω. Then, since in $\mathbb{R}^d$ the closure coincides with the set of limits of convergent sequences (Theorem A.25 (b)), there exists $z \in K \setminus \Omega$ and a sequence $(z_m)_{m\in\mathbb{N}}$ of Ω converging to z such that, for every $m \in \mathbb{N}$,

$$f(z_m) \neq 0.$$

Since $z \notin \Omega$, taking a subsequence if necessary, we can choose $r_m > 0$ for each $m \in \mathbb{N}$ such that the ball $B_m = \{x : |x - z_m| \le r_m\}$ does not contain z and is disjoint from the B_i for $i \le m-1$. Then we define a function q on the whole of $\mathbb{R}^d$ as follows:

$$q(x) = \begin{cases} \dfrac{m}{|f(z_m)|}\left(1 - \dfrac{|x - z_m|}{r_m}\right) & \text{on } B_m, \\ 0 & \text{outside of the } B_m. \end{cases} \tag{1.8}$$

10 **History of Theorem 1.22.** This property, newly introduced here, is equivalent to a special case of a result by Nicolas BOURBAKI relating to the inductive limit topology of the $\mathcal{C}_K(\Omega; E)$ which is specified in footnote 11.

This function is everywhere continuous except at z, therefore its restriction to Ω is continuous. For every $m \in \mathbb{N}$, it satisfies

$$q(z_m)\,|f(z_m)| = m.$$

This contradicts the hypothesis that qf is bounded. Therefore, (1.7) is satisfied, that is, K is included in Ω.

Hence, by Definition 1.5 of the support,

$$\operatorname{supp} f = K \cap \Omega = K.$$

Next, let us show that

$$K \text{ is bounded.}$$

If not, we can choose the z_m and the r_m as above, but with $|z_m| \to \infty$ instead of $z_m \to z$. Then q is continuous at every point and, once again, $q(z_m)\,|f(z_m)| = m$, which contradicts the hypothesis that qf is bounded. Therefore, K is bounded.

Since K is also closed in $\mathbb{R}^d$ by definition, it is compact by the Borel–Lebesgue theorem (Theorem A.23 (b)). The support of f being equal to K, it is indeed compact. □

Let us generalize this result to a family of functions that will be useful later[11].

THEOREM 1.23.– *Let $\mathcal{F} \subset C(\Omega)$, where $\Omega \subset \mathbb{R}^d$, such that, for every $q \in C^+(\Omega)$,*

$$\sup_{f \in \mathcal{F}} \sup_{x \in \Omega} q(x)\,|f(x)| < \infty.$$

Then all the functions of $\mathcal{F}$ have their support included in the same compact subset of Ω. ▌

Proof. Let

$$K \overset{\text{def}}{=} \overline{\cup_{f \in \mathcal{F}} \operatorname{supp} f}.$$

We will first show by contradiction that

$$K \subset \Omega.$$

11 **History of Theorem 1.23.** This new property is equivalent to a result by BOURBAKI [11, Chapter III, § 1, Proposition 2, (ii), p. 42]: the support of every function in a bounded subset of $\mathcal{K}(\Omega; E)$ endowed with the inductive limit topology of the $C_K(\Omega; E)$ is included in some compact set $K \subset \Omega$. The proof given by BOURBAKI, which uses topological arguments (infinite direct topological sum and strict inductive limits), is completely different from the proof given here.

Suppose that K is not included in Ω. Then, since in $\mathbb{R}^d$ the closure coincides with the set of limits of convergent sequences (Theorem A.25 (b)), there exists $z \in K \setminus \Omega$ and a sequence $(y_m)_{m\in\mathbb{N}}$ converging to z such that, for every $m \in \mathbb{N}$, there exists $f_m \in \mathcal{F}$ such that y_m belongs to the support of f_m. By Definition 1.5 of the latter,

$$y_m \in \overline{\{x \in \Omega : f_m(x) \neq 0\}}.$$

Again by Theorem A.25 (b), there exists $z_m \in \Omega$ such that $|z_m - y_m| \leq 1/m$ and

$$f_m(z_m) \neq 0.$$

Then $z_m \to z$ and $z \notin \Omega$ thus, taking a subsequence if necessary, we can choose $r_m > 0$ for each $m \in \mathbb{N}$ such that the ball $B_m = \{x : |x - z_m| \leq r_m\}$ does not contain z and is disjoint from the B_i for $i \leq m-1$. Defining q as in (1.8) but with $f_m(z_m)$ instead of $f(z_m)$, we obtain a function which is everywhere continuous except at z. Its restriction to Ω is continuous and satisfies, for all $m \in \mathbb{N}$,

$$q(z_m)\,|f_m(z_m)| = m.$$

This contradicts the hypothesis that the qf are bounded as a whole. Thus, K is indeed included in Ω.

Now, let us show that

$$K \text{ is bounded.}$$

If not, we can choose the y_m, z_m and r_m, and therefore q, as above but with $|y_m| \to \infty$ instead of $y_m \to z$. Then $|z_m| \to \infty$ instead of $z_m \to z$, thus q is now everywhere continuous and satisfies again $q(z_m)\,|f_m(z_m)| = m$. This contradicts the hypothesis that the qf are bounded as a whole. Therefore, K is indeed bounded.

By construction, K is also closed in $\mathbb{R}^d$, so it is compact by the Borel–Lebesgue theorem (Theorem A.23 (b)), and it contains the support of all the f in $\mathcal{F}$. □

Non-metrizability of $\mathcal{K}(\Omega)$. If Ω is non-empty, then $\mathcal{K}(\Omega)$ is not metrizable [SIMON, 75], despite being the union of the normed spaces $\mathcal{C}_{K_n}(\Omega)$, where the K_n are compact sets whose union is Ω. This illustrates that the countable union of normed spaces is not necessarily metrizable. Similarly, $\mathcal{K}(\Omega; E)$ is not metrizable if $E \neq \{0\}$. □

Characterization of $\mathcal{K}(\Omega; E)$ when E is a normed space. Theorem 1.22 implies the following property (by applying it to $\|f\|_E$ and using equality (1.5)). *If E is a normed space:*

$$f \in \mathcal{K}(\Omega; E) \quad \Leftrightarrow \quad f \in \mathcal{C}(\Omega; E) \textit{ and } qf \textit{ is bounded for every } q \in \mathcal{C}^+(\Omega).\ \square$$

1.7. Continuous mappings

Let us define the notion of continuity of a mapping from a separated semi-normed space into another, which generalizes the definition of a continuous function (Definition 1.2).

DEFINITION 1.24.– *Let E and F be two separated semi-normed spaces with families of semi-norms $\{\| \ \|_{E;\nu} : \nu \in \mathcal{N}_E\}$ and $\{\| \ \|_{F;\mu} : \mu \in \mathcal{N}_F\}$.*

A mapping T from a subset X of E into F is said to be ***continuous at the point*** *u if, for every $\mu \in \mathcal{N}_F$ and $\epsilon > 0$, there exists a finite subset N of $\mathcal{N}_E$ and $\eta > 0$ such that:*

$$v \in X, \ \sup_{\nu \in N} \|v - u\|_{E;\nu} \leq \eta \quad \Rightarrow \quad \|T(v) - T(u)\|_{F;\mu} \leq \epsilon.$$

We say that T is ***continuous*** *if it is continuous at every point of X.* ▮

Let us characterize the continuous linear mappings (the definition of a linear mapping is recalled in the Appendix, Definition A.39).

THEOREM 1.25.– *Let E and F be two separated semi-normed spaces with families of semi-norms $\{\| \ \|_{E;\nu} : \nu \in \mathcal{N}_E\}$ and $\{\| \ \|_{F;\mu} : \mu \in \mathcal{N}_F\}$.*

A linear mapping L from E into F is continuous if and only if, for every $\mu \in \mathcal{N}_F$, there exists a finite subset N of $\mathcal{N}_E$ and $c \geq 0$ such that: for every $u \in E$,

$$\|Lu\|_{F;\mu} \leq c \sup_{\nu \in N} \|u\|_{E;\nu}. \ ▮$$

Proof. If L is linear, then $L(0_E) = 0_F$. If L is also continuous, then, by Definition 1.24, its continuity at 0_E implies that, for every $\nu \in \mathcal{N}_E$, there exists a finite subset N of $\mathcal{N}_E$ and $\eta > 0$ such that, if $\sup_{\nu \in N} \|v\|_{E;\nu} \leq \eta$, then $\|Lv\|_{F;\mu} \leq 1$.

If $\sup_{\nu \in N} \|u\|_{E;\nu} \neq 0$, by choosing $v = \eta u / \sup_{\nu \in N} \|u\|_{E;\nu}$, we obtain

$$\|L(v)\|_{F;\mu} = \frac{\eta}{\sup_{\nu \in N} \|u\|_{E;\nu}} \|L(u)\|_{F;\mu} \leq 1,$$

which gives the desired inequality with $c = 1/\eta$. This still holds if $\sup_{\nu \in N} \|u\|_{E;\nu}$ is zero, in which case $\|L(u)\|_{F;\mu} = 0$, otherwise $v = 2u/\|L(u)\|_{F;\mu}$ would not satisfy the derived property for v.

Conversely, the stated inequality and linearity imply that

$$\|Lv - Lu\|_{F;\mu} \leq c \sup_{\nu \in N} \|v - u\|_{E;\nu},$$

which implies continuity. □

1.8. Continuous extension and restriction

We wish to show that continuous extension from Ω to $\overline{\Omega}$ is an isomorphism in the spaces of uniformly continuous functions $\mathbf{C}_b(\Omega; E)$ (bounded functions) and $\mathbf{C}_D(\Omega; E)$ (functions supported in D). Recall that an **isomorphism** from a separated semi-normed space to another is a continuous linear bijection whose inverse mapping is also linear and continuous.

More precisely, we will show that restriction from $\overline{\Omega}$ to Ω is an isomorphism. This result is equivalent because this mapping is the inverse of continuous extension.

THEOREM 1.26.– *Let $\Omega \subset \mathbb{R}^d$, E be a separated semi-normed space, $\mathcal{N}_E$ the set indexing its semi-norms and D a compact subset of $\mathbb{R}^d$. Then:*

(a) *For every $f \in \mathbf{C}_b(\overline{\Omega}; E)$ and $\nu \in \mathcal{N}_E$,*

$$\|f|_\Omega\|_{\mathbf{C}_b(\Omega;E);\nu} = \|f\|_{\mathbf{C}_b(\overline{\Omega};E);\nu}.$$

(b) *If E is a Neumann space, then the mapping $f \mapsto f|_\Omega$ is an isomorphism from $\mathbf{C}_b(\overline{\Omega}; E)$ onto $\mathbf{C}_b(\Omega; E)$ and from $\mathbf{C}_D(\overline{\Omega}; E)$ onto $\mathbf{C}_D(\Omega; E)$.* ▮

Proof. (a) Since, by Definition 1.4 (b), $\mathbf{C}_b(\Omega; E)$ is endowed with the semi-norms of $\mathcal{C}_b(\Omega; E)$ (Definition 1.3 (b)),

$$\|f\|_{\mathbf{C}_b(\overline{\Omega};E);\nu} = \sup_{x\in\overline{\Omega}} \|f(x)\|_{E;\nu} \geq \sup_{x\in\Omega} \|f(x)\|_{E;\nu} = \|f|_\Omega\|_{\mathbf{C}_b(\Omega;E);\nu}. \tag{1.9}$$

Conversely, let $a = \|f\|_{\mathbf{C}_b(\overline{\Omega};E);\nu}$, $\epsilon > 0$ and $x \in \overline{\Omega}$ such that

$$\|f(x)\|_{E;\nu} \geq a - \epsilon.$$

Since f is uniformly continuous, by Definition 1.2 (b), there exists $\eta > 0$ such that, for every $y \in \overline{\Omega}$ satisfying $|y - x| \leq \eta$,

$$\|f(y) - f(x)\|_{E;\nu} \leq \epsilon. \tag{1.10}$$

But by the characterization of the closure $\overline{\Omega}$ from Theorem A.11, there exists $y \in \Omega$ such that $|y - x| \leq \eta$. Therefore, (1.10) implies that

$$\|f|_\Omega\|_{\mathbf{C}_b(\Omega;E);\nu} \geq \|f(y)\|_{E;\nu} \geq a - 2\epsilon.$$

This holds for every $\epsilon > 0$, which gives the reverse inequality in (1.9), which must therefore be an equality.

(b) Here, E is assumed to be a Neumann space.

Isomorphism from $\mathbf{C}_b(\overline{\Omega}; E)$ *onto* $\mathbf{C}_b(\Omega; E)$. Every function g in $\mathbf{C}_b(\Omega; E)$ has an extension $\overline{g}$ that is uniformly continuous on $\overline{\Omega}$ (Theorem A.38, which may be applied because Ω is sequentially dense in $\overline{\Omega}$ by Theorem A.25 (b)). This function $\overline{g}$ is bounded because the inequality $\|g(x)\|_{E;\nu} \leq b_\nu$ in Ω may be continuously extended to $\overline{\Omega}$ (Theorem A.37 (a) and (b)). Therefore, $\overline{g} \in \mathbf{C}_b(\overline{\Omega}; E)$ and $\overline{g}|_\Omega = g$. Thus, the mapping $f \mapsto f|_\Omega$ is surjective. The equality (a) then shows that it is bijective and bicontinuous.

Isomorphism from $\mathbf{C}_D(\overline{\Omega}; E)$ *onto* $\mathbf{C}_D(\Omega; E)$. This isomorphism follows from the previous one because, when D is closed, as is the case here, the support of g is included in D if and only if the support of $\overline{g}$ is included in D (by Definition 1.5 of the support). □

Identification. If we identify f with its restriction in Theorem 1.26, we obtain the following topological equality:

$$\text{If } E \text{ is a Neumann space, } \mathbf{C}_b(\overline{\Omega}; E) \approxeq \mathbf{C}_b(\Omega; E). \tag{1.11}$$

Some care is required with identifications, as it is explained in the section **Dangerous identifications** of Volume 1 [73, § 14.6, pp. 240–243]. □

1.9. Separation and permutation of variables

We can separate the variables of continuous functions as follows.

THEOREM 1.27.– *Let $\Omega_1 \subset \mathbb{R}^{d_1}$, $\Omega_2 \subset \mathbb{R}^{d_2}$ and E be a separated semi-normed space. For any function defined on $\Omega_1 \times \Omega_2$, let $\underline{f}$ be the function obtained by separating the variables, i.e.*

$$(\underline{f}(x_1))(x_2) \stackrel{\text{def}}{=} f(x_1, x_2).$$

Then:

(a) *The mapping $f \mapsto \underline{f}$ is an isomorphism from $\mathcal{C}(\Omega_1 \times \Omega_2; E)$ onto $\mathcal{C}(\Omega_1; \mathcal{C}(\Omega_2; E))$ and from $\mathcal{C}_b(\Omega_1 \times \Omega_2; E)$ onto $\mathcal{C}_b(\Omega_1; \mathcal{C}_b(\Omega_2; E))$.*

(b) *The mapping $f \mapsto \underline{f}$ is linear, continuous and injective from $\mathbf{C}_b(\Omega_1 \times \Omega_2; E)$ into $\mathbf{C}_b(\Omega_1; \mathbf{C}_b(\Omega_2; E))$.*

If Ω_1 is compact, or if Ω_1 is bounded and E is a Neumann space, then this mapping is an isomorphism.

(c) *For any two compact subsets D_1 of $\mathbb{R}^{d_1}$ and D_2 of $\mathbb{R}^{d_2}$, the mapping $f \mapsto \underline{f}$ is an isomorphism from $\mathbf{C}_{D_1 \times D_2}(\Omega_1 \times \Omega_2; E)$ onto $\mathbf{C}_{D_1}(\Omega_1; \mathbf{C}_{D_2}(\Omega_2; E))$.* ▮

Proof. Let $\{\| \|_{E;\nu} : \nu \in \mathcal{N}_E\}$ be the family of semi-norms of E.

(a) 1. Isomorphism in $\mathcal{C}(\Omega_1 \times \Omega_2; E)$. We will proceed in three steps:

Separation of variables. Let $f \in \mathcal{C}(\Omega_1 \times \Omega_2; E)$. For every $x_1 \in \Omega_1$, $\underline{f}(x_1)$ belongs to $\mathcal{C}(\Omega_2; E)$ because $x_2 \mapsto f(x_1, x_2)$ is continuous by restriction. Let $(x_1^n)_{n\in\mathbb{N}}$ be a sequence in Ω_1 that converges to x_1. By Definition 1.3 (a) of the semi-norms of $\mathcal{C}(\Omega; E)$, for any compact set K_2 included in Ω_2 and $\nu \in \mathcal{N}_E$,

$$\|\underline{f}(x_1^n) - \underline{f}(x_1)\|_{\mathcal{C}(\Omega_2;E);K_2,\nu} = \sup_{x_2 \in K_2} \|f(x_1^n, x_2) - f(x_1, x_2)\|_{E;\nu}. \tag{1.12}$$

Let $Q_1 = \{x_1^n\}_{n\in\mathbb{N}} \cup \{x_1\}$. The set $Q_1 \times K_2$ is compact in $\mathbb{R}^{d_1+d_2}$ by the Borel–Lebesgue theorem (Theorem A.23 (b)) because it is closed and bounded, so f is uniformly continuous on this set by Heine's theorem (Theorem A.34). Therefore, for every $\epsilon > 0$, there exists $\eta > 0$ such that $x_2 \in K_2$ and $|(x_1^n, x_2) - (x_1, x_2)| \leq \eta$ together imply that $\|f(x_1^n, x_2) - f(x_1, x_2)\|_{E;\nu} \leq \epsilon$.

By (1.12), $|x_1^n - x_1| \leq \eta$ therefore implies that $\|\underline{f}(x_1^n) - \underline{f}(x_1)\|_{\mathcal{C}(\Omega_2;E);K_2,\nu} \leq \epsilon$. This proves that $\underline{f}$ is sequentially continuous and hence continuous (Theorem A.29, since $\mathbb{R}^{d_1}$ is metrizable) from Ω_1 into $\mathcal{C}(\Omega_2, E)$. In other words, $\underline{f} \in \mathcal{C}(\Omega_1; \mathcal{C}(\Omega_2; E))$.

Regrouping the variables. Let $h \in \mathcal{C}(\Omega_1; \mathcal{C}(\Omega_2; E))$ and denote by f the function obtained by regrouping its variables. Let $((x_1^n, x_2^n))_{n\in\mathbb{N}}$ be a sequence in $\Omega_1 \times \Omega_2$ converging to a point (x_1, x_2) in $\Omega_1 \times \Omega_2$. The set $Q_2 = \{x_2^n\}_{n\in\mathbb{N}} \cup \{x_2\}$ is compact and included in Ω_2. Therefore, by (1.12) (since $h = \underline{f}$), for every $\nu \in \mathcal{N}_E$,

$$\|f(x_1^n, x_2^n) - f(x_1, x_2^n)\|_{E;\nu} \leq \|h(x_1^n) - h(x_1)\|_{\mathcal{C}(\Omega_2;E);Q_2,\nu} \to 0,$$

since h is continuous and hence sequentially continuous (Theorem A.29 again). Moreover,

$$\|f(x_1, x_2^n) - f(x_1, x_2)\|_{E;\nu} = \|(h(x_1))(x_2^n) - (h(x_1))(x_2)\|_{E;\nu} \to 0,$$

since $h(x_1) \in \mathcal{C}(\Omega_2; E)$. These two properties imply that

$$\|f(x_1^n, x_2^n) - f(x_1, x_2)\|_{E;\nu} \to 0.$$

This shows that f is sequentially continuous and hence continuous (Theorem A.29 again, since $\mathbb{R}^{d_1+d_2}$ is metrizable). In other words, $f \in \mathcal{C}(\Omega_1 \times \Omega_2; E)$.

Isomorphism. Let K_1 be a compact set included in Ω_1, K_2 a compact set included in Ω_2 and $\nu \in \mathcal{N}_E$. Once again by Definition 1.3 (a) of the semi-norms of $\mathcal{C}(\Omega; E)$,

$$\|\underline{f}\|_{\mathcal{C}(\Omega_1;\mathcal{C}(\Omega_2;E));K_1,K_2,\nu} = \sup_{x_1\in K_1} \|\underline{f}(x_1)\|_{\mathcal{C}(\Omega_2;E);K_2,\nu} =$$

$$= \sup_{x_1\in K_1} \sup_{x_2\in K_2} \|(\underline{f}(x_1))(x_2)\|_{E;\nu} = \sup_{(x_1,x_2)\in K_1\times K_2} \|f(x_1, x_2)\|_{E;\nu} =$$

$$= \|f\|_{\mathcal{C}(\Omega_1\times\Omega_2;E);K_1\times K_2,\nu}.$$

Hence, $f \mapsto \underline{f}$ and its inverse mapping, which are both linear, are continuous by the characterization of continuous linear mappings from Theorem 1.25. This shows that $f \mapsto \underline{f}$ is an isomorphism from $\mathcal{C}(\Omega_1 \times \Omega_2; E)$ onto $\mathcal{C}(\Omega_1; \mathcal{C}(\Omega_2; E))$.

2. Isomorphism in $\mathcal{C}_b(\Omega_1 \times \Omega_2; E)$. A function $f \in \mathcal{C}(\Omega_1 \times \Omega_2; E)$ is bounded if and only if $\underline{f}$ is bounded from Ω_1 into $\mathcal{C}_b(\Omega_2; E)$, since, by Definition 1.3 (b) of the semi-norms of $\mathcal{C}_b(\Omega; E)$, for every $\nu \in \mathcal{N}_E$,

$$\sup_{x_1 \in \Omega_1} \|\underline{f}(x_1)\|_{\mathcal{C}_b(\Omega_2;E);\nu} = \sup_{x_1 \in \Omega_1} \sup_{x_2 \in \Omega_2} \|(\underline{f}(x_1))(x_2)\|_{E;\nu}$$
$$= \sup_{(x_1,x_2) \in \Omega_1 \times \Omega_2} \|f(x_1, x_2)\|_{E;\nu}.$$

By point **1**, the mapping $f \mapsto \underline{f}$ is therefore a bijection from $\mathcal{C}_b(\Omega_1 \times \Omega_2; E)$ onto $\mathcal{C}_b(\Omega_1; \mathcal{C}_b(\Omega_2; E))$. This is an isomorphism because the above equality can be written as

$$\|\underline{f}\|_{\mathcal{C}_b(\Omega_1;\mathcal{C}_b(\Omega_2;E));\nu} = \|f\|_{\mathcal{C}_b(\Omega_1 \times \Omega_2;E);\nu}. \qquad (1.13)$$

(b) 1. Continuity and injectivity in $\mathbf{C}_b(\Omega_1 \times \Omega_2; E)$. Let $f \in \mathbf{C}_b(\Omega_1 \times \Omega_2; E)$. For every $x_1 \in \Omega_1$, $\underline{f}(x_1)$ is uniformly continuous (since $x_2 \mapsto f(x_1, x_2)$ is uniformly continuous by restriction) and, for every $y_1 \in \Omega_1$ and $\nu \in \mathcal{N}_E$,

$$\|\underline{f}(y_1) - \underline{f}(x_1)\|_{\mathcal{C}_b(\Omega_2;E);\nu} = \sup_{x_2 \in \Omega_2} \|f(y_1, x_2) - f(x_1, x_2)\|_{E;\nu}. \qquad (1.14)$$

By Definition 1.2 (b) of uniform continuity, for every $\epsilon > 0$, there exists $\eta > 0$ such that $|(y_1, x_2) - (x_1, x_2)| \leq \eta$ implies $\|f(y_1, x_2) - f(x_1, x_2)\|_{E;\nu} \leq \epsilon$. By (1.14), $|y_1 - x_1| \leq \eta$ therefore implies that $\|\underline{f}(y_1) - \underline{f}(x_1)\|_{\mathcal{C}_b(\Omega_2;E);\nu} \leq \epsilon$. This shows that $\underline{f}$ is uniformly continuous from Ω_1 into $\mathbf{C}_b(\Omega_2; E)$, since, by Definition 1.4 (b), the latter is endowed with the semi-norms of $\mathcal{C}_b(\Omega; E)$. Hence, $\underline{f} \in \mathbf{C}_b(\Omega_1; \mathbf{C}_b(\Omega_2; E))$.

Thus, once again by the characterization of continuous linear mappings from Theorem 1.25, the equality (1.13) shows that the mapping $f \mapsto \underline{f}$ is continuous from $\mathbf{C}_b(\Omega_1 \times \Omega_2; E)$ into $\mathbf{C}_b(\Omega_1; \mathbf{C}_b(\Omega_2; E))$.

2. Isomorphism in $\mathbf{C}_b(\Omega_1 \times \Omega_2; E)$. Let us study each case separately.

Case where Ω_1 is compact. Let $h \in \mathbf{C}_b(\Omega_1; \mathbf{C}_b(\Omega_2; E))$ and $f \in \mathcal{C}(\Omega_1 \times \Omega_2; E)$ the function obtained by regrouping its variables with (c).

Let us first prove by contradiction that f is uniformly continuous. Suppose not. Then there exists $\nu \in \mathcal{N}_E$, $a > 0$ and for every $n \in \mathbb{N}$, points (x_1^n, x_2^n) and (y_1^n, y_2^n) in $\Omega_1 \times \Omega_2$ such that:

$$|(x_1^n, x_2^n) - (y_1^n, y_2^n)| \leq \frac{1}{n}, \quad \|f(x_1^n, x_2^n) - f(y_1^n, y_2^n)\|_{E;\nu} \geq a. \qquad (1.15)$$

We assumed that Ω_1 is compact, which is equivalent to being sequentially compact in $\mathbb{R}^{d_1}$ by the Borel–Lebesgue theorem (Theorem A.23 (b)), so there exists $x_1 \in \Omega_1$ and some subsequence (relabeled as the original sequence!) such that $x_1^n \to x_1$ as $n \to \infty$. This implies that $y_1^n \to x_1$. Therefore, together with (1.14) (since $h = \underline{f}$),

$$\|f(x_1^n, x_2^n) - f(x_1, x_2^n)\|_{E;\nu} \leq \|h(x_1^n) - h(x_1)\|_{C_b(\Omega_2;E);\nu} \to 0, \tag{1.16}$$

$$\|f(x_1, x_2^n) - f(x_1, y_2^n)\|_{E;\nu} = \|(h(x_1))(x_2^n) - (h(x_1))(y_2^n)\|_{E;\nu} \to 0, \tag{1.17}$$

$$\|f(x_1, y_2^n) - f(y_1^n, y_2^n)\|_{E;\nu} \leq \|h(x_1) - h(y_1^n)\|_{C_b(\Omega_2;E);\nu} \to 0, \tag{1.18}$$

respectively, since $h \in \mathbf{C}_b(\Omega_1; \mathbf{C}_b(\Omega_2; E))$, $h(x_1) \in \mathbf{C}_b(\Omega_2; E)$, and, once again, $h \in \mathbf{C}_b(\Omega_1; \mathbf{C}_b(\Omega_2; E))$. These three properties imply that

$$\|f(x_1^n, x_2^n) - f(y_1^n, y_2^n)\|_{E;\nu} \to 0, \tag{1.19}$$

which contradicts (1.15). Hence, f must be uniformly continuous.

Thus, $f \in \mathbf{C}_b(\Omega_1 \times \Omega_2; E)$, which shows that the mapping $f \mapsto \underline{f}$ is a bijection from $\mathbf{C}_b(\Omega_1 \times \Omega_2; E)$ onto $\mathbf{C}_b(\Omega_1; \mathbf{C}_b(\Omega_2; E))$. The equality (1.13) then shows that it is in fact an isomorphism of these spaces.

Case where Ω_1 is bounded and E is a Neumann space. Let $h \in \mathbf{C}_b(\Omega_1; \mathbf{C}_b(\Omega_2; E))$. Since E is a Neumann space, $\mathbf{C}_b(\Omega_2; E)$ is also a Neumann space (Theorem 1.12), so, by Theorem 1.26 (b), h has an extension $Ph \in \mathbf{C}_b(\overline{\Omega_1}; \mathbf{C}_b(\Omega_2; E))$.

Since the domain Ω_1 is bounded, $\overline{\Omega_1}$ is closed and bounded and hence compact in $\mathbb{R}^{d_1}$, so the previous case provides k in $\mathbf{C}_b(\overline{\Omega_1} \times \Omega_2; E)$ such that $\underline{k} = Ph$. The restriction f of k to $\Omega_1 \times \Omega_2$ belongs to $\mathbf{C}_b(\Omega_1 \times \Omega_2; E)$ and satisfies $\underline{f} = h$. Therefore, $f \mapsto \underline{f}$ is once again an isomorphism from $\mathbf{C}_b(\Omega_1 \times \Omega_2; E)$ onto $\mathbf{C}_b(\Omega_1; \mathbf{C}_b(\Omega_2; E))$.

(c) Isomorphism in $\mathbf{C}_{D_1 \times D_2}(\Omega_1 \times \Omega_2; E)$. We will proceed in three steps:

Separation of variables. If $f \in \mathbf{C}_{D_1 \times D_2}(\Omega_1 \times \Omega_2; E)$, then the support of $\underline{f}$ is included in D_1 and, for every $x_1 \in \Omega_1$, the support of $\underline{f}(x_1)$ is included in D_2 for all x_1. Hence, $\underline{f} \in \mathbf{C}_{D_1}(\Omega_1; \mathbf{C}_{D_2}(\Omega_2; E))$ by case (b).

Regrouping the variables. Let $h \in \mathbf{C}_{D_1}(\Omega_1; \mathbf{C}_{D_2}(\Omega_2; E))$ and $f \in \mathcal{C}(\Omega_1 \times \Omega_2; E)$ the function obtained by regrouping its variables with (c). Then the support of f is included in $D_1 \times D_2$.

Let us prove by contradiction that f is uniformly continuous. Suppose not. Then there exists $\nu \in \mathcal{N}_E$, $a > 0$ and, for every $n \in \mathbb{N}$, points (x_1^n, x_2^n) and (y_1^n, y_2^n) in $\Omega_1 \times \Omega_2$ satisfying (1.15). For each n, either x_1^n or y_1^n belongs to D_1, otherwise $f(x_1^n, x_2^n) = 0_E = f(y_1^n, y_2^n)$, which contradicts (1.15). Switching them if necessary, we may assume that x_1^n belongs to D_1. Since the latter is compact, there exists

$x_1 \in \Omega_1$ and a subsequence such that $x_1^n \to x_1$, as above. Once again, convergences (1.16)–(1.18) and hence (1.19) are satisfied, which contradicts (1.15). Therefore, f must indeed be uniformly continuous.

Thus, $f \in \mathbf{C}_{D_1 \times D_2}(\Omega_1 \times \Omega_2; E)$, and the mapping $f \mapsto \underline{f}$ is a bijection from $\mathbf{C}_{D_1 \times D_2}(\Omega_1 \times \Omega_2; E)$ onto $\mathbf{C}_{D_1}(\Omega_1; \mathbf{C}_{D_2}(\Omega_2; E))$.

Isomorphism. The equality (1.13) then shows that it is an isomorphism of these spaces, since $\mathbf{C}_D(\Omega; E)$ is endowed with the semi-norms of $\mathcal{C}_b(\Omega; E)$ by Definition 1.7 (b).

□

Let us permute the variables of functions of continuous functions.

THEOREM 1.28.– *Let $\Omega_1 \subset \mathbb{R}^{d_1}$, $\Omega_2 \subset \mathbb{R}^{d_2}$ and E be a separated semi-normed space. Let $\check{}$ denote permutation of the variables of functions of functions, i.e.*

$$(\check{f}(x_2))(x_1) \stackrel{\text{def}}{=} (f(x_1))(x_2).$$

Then:

(a) *The mapping $f \mapsto \check{f}$ is an isomorphism from $\mathcal{C}(\Omega_1; \mathcal{C}(\Omega_2; E))$ onto $\mathcal{C}(\Omega_2; \mathcal{C}(\Omega_1; E))$ and from $\mathcal{C}_b(\Omega_1; \mathcal{C}(\Omega_2; E))$ onto $\mathcal{C}_b(\Omega_2; \mathcal{C}_b(\Omega_1; E))$.*

(b) *For any two compact subsets D_1 of $\mathbb{R}^{d_1}$ and D_2 of $\mathbb{R}^{d_2}$, the mapping $f \mapsto \check{f}$ is an isomorphism from $\mathbf{C}_{D_1}(\Omega_1; \mathbf{C}_{D_2}(\Omega_2; E))$ onto $\mathbf{C}_{D_2}(\Omega_2; \mathbf{C}_{D_1}(\Omega_1; E))$.*

(c) *If Ω_1 and Ω_2 are compact, or if Ω_1 and Ω_2 are bounded and E is a Neumann space, then the mapping $f \mapsto \check{f}$ is an isomorphism from $\mathbf{C}_b(\Omega_1; \mathbf{C}_b(\Omega_2; E))$ onto $\mathbf{C}_b(\Omega_2; \mathbf{C}_b(\Omega_1; E))$.* ▮

Proof. Let $f \in \mathcal{C}(\Omega_1; \mathcal{C}(\Omega_2; E))$. Regrouping its variables yields $g \in \mathcal{C}(\Omega_1 \times \Omega_2; E)$ (Theorem 1.27 (a)). Permuting its variables, i.e. writing $\widehat{g}(x_2, x_1) = g(x_1, x_2)$, yields $\widehat{g} \in \mathcal{C}(\Omega_2 \times \Omega_1; E)$. Finally, separating the variables of $\widehat{g}$ yields $\check{f} \in \mathcal{C}(\Omega_2; \mathcal{C}(\Omega_1; E))$ (Theorem 1.27 (a) once again).

Similarly, the other properties can be deduced from the properties in Theorem 1.27.

□

1.10. Sequential compactness in $C_b(\Omega; E)$

Let us characterize the relatively sequentially compact sets in $\mathbf{C}_b(\Omega; E)$ (a subset $\mathcal{F}$ of a separated semi-normed space F is said to be *relatively sequentially compact* if every sequence in $\mathcal{F}$ has a convergent subsequence in F). This is a variant of the **Arzelà-Ascoli theorem**[12].

THEOREM 1.29.– *Let $\mathcal{F} \subset \mathbf{C}_b(\Omega; E)$, where Ω is a bounded subset of $\mathbb{R}^d$ and E is a separated semi-normed space with a family of semi-norms $\{\| \ \|_{E;\nu} : \nu \in \mathcal{N}_E\}$. Suppose that, for every $\nu \in \mathcal{N}_E$ and $\epsilon > 0$, there exists $\eta > 0$ such that*

$$\sup_{y\in\Omega,\ x\in\Omega,\ |y-x|\leq\eta} \|f(y) - f(x)\|_{E;\nu} \leq \epsilon \tag{1.20}$$

and, for all $x \in \Omega$,

every sequence in $\{f(x) : f \in \mathcal{F}\}$ has a convergent subsequence in E.

Then

every sequence in $\mathcal{F}$ has a convergent subsequence in $\mathbf{C}_b(\Omega; E)$. ▮

Notes. The hypothesis (1.20) means that $\mathcal{F}$ is uniformly equicontinuous [Vol. 1, Definition 10.5]. □

Optimality of Theorem 1.29. The conditions of this theorem are necessary and sufficient for every sequence of $\mathcal{F}$ to have a convergent subsequence [SIMON, 75, forthcoming]. □

Utility of Theorem 1.29. We will use this theorem to show that every bounded sequence in $\mathcal{C}^{m+1}_K(\Omega)$ has a convergence subsequence in $\mathcal{C}^m_K(\Omega)$ (Theorem 2.25) and, using this fact, that every bounded sequence in $\mathcal{D}(\Omega)$ has a convergent subsequence [Vol. 3], a result that will in turn be useful for the Schwartz kernel theorem. □

Compact versus sequentially compact The definitions of both notions of compactness in a semi-normed space are recalled in the Appendix (Definitions A.17 and A.18). In spaces of continuous functions, we only use sequential compactness, not compactness.

Note that there are separated semi-normed spaces where *compact* is neither stronger nor weaker than *sequentially compact* [Vol. 1, Properties (2.6) and (2.7), p. 43]. If this holds in a given space E, then it also holds in $\mathbf{C}_b(\Omega; E)$. □

---

12 **History of the Arzelà-Ascoli theorem.** In 1883 [4, pp. 545–549], Giulio ASCOLI gave a construction with the essence of Theorem 1.29 in one dimension, namely $d = 1$. Cesare ARZELÀ showed the converse in 1889 [2]. Both of their manuscripts use geometric language that makes our statement difficult to recognize. In 1895 [3, pp. 56–60], Cesare ARZELÀ stated Theorem 1.29 in one dimension. Maurice FRÉCHET extended the result in 1906 to functions defined on a subset of a metrizable space in his thesis [38].

Proof of Theorem 1.29. Let $(f_n)_{n\in\mathbb{N}}$ be a sequence in $\mathcal{F}$ and $(x_k)_{k\in\mathbb{N}}$ a sequentially dense sequence in Ω. Such a sequence is guaranteed to exist because the space $\mathbb{R}^d$ is sequentially separable (Theorem A.23 (a)), so the set Ω is also sequentially separable (Theorem A.25 (c)).

Construction of a convergent subsequence $(f_{n,n})_{n\in\mathbb{N}}$ *in* $\{x_k\}_{k\in\mathbb{N}}$. For every k, the sequence $(f_n(x_k))_{n\in\mathbb{N}}$ has a convergent subsequence $(f_{k,n}(x_k))_{n\in\mathbb{N}}$ by the hypotheses. Write $f(x_k)$ for its limit. We can extract nested subsequences, that is, such that $(f_{k+1,n}(x_k))_{n\in\mathbb{N}}$ is subsequence of $(f_{k,n}(x_k))_{n\in\mathbb{N}}$ for every k. As $n\to\infty$, for every k, the diagonal sequence[13] then satisfies

$$f_{n,n}(x_k)\to f(x_k). \tag{1.21}$$

Convergence of $f_{n,n}$ *on the whole of* Ω. Let $x\in\Omega$. The sequence $(f_{n,n}(x))_{n\in\mathbb{N}}$ has a convergent subsequence, say

$$g_n(x)\to f(x). \tag{1.22}$$

Thus, f is defined on the set $X=\{x_k\}_{k\in\mathbb{N}}\cup\{x\}$.

Let $\nu\in\mathcal{N}_E$, $\epsilon>0$, and $\eta>0$ be such that the equicontinuity hypothesis (1.20) is satisfied. Then $|x_k-x|\le\eta$ implies $\|g_n(x_k)-g_n(x)\|_{E;\nu}\le\epsilon$. Hence, taking the limit in n,

$$\|f(x_k)-f(x)\|_{E;\nu}\le\epsilon.$$

Therefore, f is continuous at the point x by Definition 1.2 (a). By the hypotheses, the sequence $(x_k)_{k\in\mathbb{N}}$ has a subsequence such that $x_k\to x$. For this subsequence, $f(x_k)\to f(x)$. Hence, $f(x)$ does not depend on the choice of subsequence $(g_n(x))_{n\in\mathbb{N}}$ satisfying (1.22). This implies that the entire sequence $(f_{n,n}(x))_{n\in\mathbb{N}}$ must converge. Thus, for every $x\in\Omega$,

$$f_{n,n}(x)\to f(x). \tag{1.23}$$

Uniform convergence. Let us show by contradiction that

$$f_{n,n}\to f \ \text{ in } \mathbf{C}_b(\Omega;E). \tag{1.24}$$

If not, there exists $\nu\in\mathcal{N}_E$, $\epsilon>0$, and a subsequence $(g_n)_{n\in\mathbb{N}}$ in $(f_{n,n})_{n\in\mathbb{N}}$ such that, for every $n\in\mathbb{N}$, there exists $y_n\in\Omega$ satisfying

$$\|g_n(y_n)-f(y_n)\|_{E;\nu}\ge 3\epsilon. \tag{1.25}$$

13 **History of extracting a diagonal subsequence.** The diagonal procedure for extracting a subsequence was introduced by Georg CANTOR (see his works [17]).

Since the set Ω is bounded, the sequence $(y_n)_{n\in\mathbb{N}}$ has a subsequence that converges to some limit y by the Bolzano–Weierstrass theorem (Theorem A.23 (c)). Extracting the corresponding subsequences yields (1.25) together with $y_n \to y$.

Let $n_0 \in \mathbb{N}$ be such that $n \geq n_0$ implies $|y_n - y| \leq \eta$, where η satisfies the equicontinuity hypothesis (1.20). For every $n \geq n_0$ and $m \in \mathbb{N}$, this gives

$$\|g_m(y_n) - g_m(y)\|_{E;\nu} \leq \epsilon, \tag{1.26}$$

which in turn, taking the limit in m together with (1.23), implies that

$$\|f(y_n) - f(y)\|_{E;\nu} \leq \epsilon. \tag{1.27}$$

The inequalities (1.25), (1.26) for $m = n$ and (1.27) now imply that

$$\|g_n(y) - f(y)\|_{E;\nu} \geq \epsilon.$$

This contradicts the convergence $g_n(y) \to f(y)$ from (1.23). Therefore, (1.24) must be satisfied, and $(f_n)_{n\in\mathbb{N}}$ must indeed have a convergent subsequence in $\mathbf{C}_\mathrm{b}(\Omega; E)$. □

Chapter 2

Differentiable Functions

This chapter and Chapter 3 are dedicated to properties of differential functions taking values in a semi-normed space that will be useful later.

In addition to the finite increment theorem and Schwarz's theorem (Theorems 2.5 and 2.12), we study the links between differentiability and the existence of partial derivatives (Theorems 2.10, 2.11 and 2.13). We then define the spaces $\mathcal{C}^m(\Omega; E)$, $\mathbf{C}^m(\Omega; E)$, $\mathcal{K}^m(\Omega; E)$ and variants of these spaces (§ 2.5), before comparing them (§ 2.6) and studying their metrizability (Theorem 2.19), their filtering properties (Theorem 2.22) and their sequential completeness properties (§ 2.8). Finally, we study the spaces $\mathcal{C}^m(\overline{\Omega}; E)$ defined on the closure of the open set Ω.

These topics are all classical, except for the construction of the topology of $\mathcal{K}^m(\Omega; E)$ in Definition 2.14 using the semi-norms $\|f\|_{\mathcal{K}^m(\Omega;E);q,\nu} = \sup_{0\leq|\beta|\leq m} \sup_{x\in\Omega} q(x)\|\partial^\beta f(x)\|_{E;\nu}$ indexed by $q \in \mathcal{C}^+(\Omega)$ and $\nu \in \mathcal{N}_E$, which provides an advantageous replacement for the *inductive limit topology* of the $\mathcal{C}^m_K(\Omega; E)$.

2.1. Differentiability

Given a semi-normed space E endowed with semi-norms $\{\| \,\|_{E;\nu} : \nu \in \mathcal{N}_E\}$, the space $E^d = \{(u_1, \ldots, u_d) : u_i \in E, \forall i\}$ is endowed with the semi-norms, indexed by $\nu \in \mathcal{N}_E$,

$$\|u\|_{E^d;\nu} \stackrel{\text{def}}{=} (\|u_1\|^2_{E;\nu} + \cdots + \|u_d\|^2_{E;\nu})^{1/2}. \tag{2.1}$$

For $z \in \mathbb{R}^d$ and $u \in E^d$, we denote

$$z \cdot u \stackrel{\text{def}}{=} z_1 u_1 + \cdots + z_d u_d.$$

Then

$$\|z \cdot u\|_{E;\nu} \leq |z|\|u\|_{E^d;\nu}. \tag{2.2}$$

Indeed, $\|z \cdot u\|_{E;\nu} \leq \sum_i |z_i|\|u_i\|_{E;\nu} \leq (\sum_i z_i^2)^{1/2}(\sum_i \|u_i\|^2_{E;\nu})^{1/2} = |z|\|u\|_{E^d;\nu}$ by the Cauchy–Schwarz inequality in $\mathbb{R}^d$ (the inequality (A.1) on p. 232). Finally, we denote by $\mathbf{e}_i$ the ith basis vector of $\mathbb{R}^d$, namely

$$(\mathbf{e}_i)_i = 1, \quad (\mathbf{e}_i)_j = 0 \text{ if } j \neq i.$$

Let us introduce the notion of differentiability for a function taking values in a semi-normed space.

DEFINITION 2.1.– *Let f be a function from an open subset Ω of $\mathbb{R}^d$ into a separated semi-normed space E with a family of semi-norms $\{\| \ \|_{E;\nu} : \nu \in \mathcal{N}_E\}$.*

(a) *We say that f is **differentiable at the point** x of Ω if there exists an element of E^d, denoted $\nabla f(x)$ and called the **gradient** of f at x, such that, for every $\nu \in \mathcal{N}_E$ and $\epsilon > 0$, there exists $\eta > 0$ such that, if $z \in \mathbb{R}^d$, $|z| \leq \eta$, and $x + z \in \Omega$, then*

$$\|f(x+z) - f(x) - z \cdot \nabla f(x)\|_{E;\nu} \leq \epsilon\,|z|. \tag{2.3}$$

*We say that f is **differentiable** if it is differentiable at every point of Ω, and we say that f is **continuously differentiable** if ∇f is also continuous from Ω into E^d.*

*We say that f is m **times differentiable**, where $m \in \mathbb{N}$, if it has successive gradients ∇f, $\nabla^2 f$, ..., $\nabla^m f$ (which are elements of E^d, E^{d^2}, ..., E^{d^m} respectively), and we say that f is m **times continuously differentiable** if its successive gradients are also continuous. This definition is extended to $m = 0$ by writing $\nabla^0 f \overset{\text{def}}{=} f$.*

*We say that f is **infinitely differentiable** if it is m times differentiable for every $m \in \mathbb{N}$.*

(b) *When $d = 1$, the differentiability of f at the point x reduces to the existence of an element of E, denoted as $f'(x)$ and called the **derivative** at x, such that, for every $\nu \in \mathcal{N}_E$ and $\epsilon > 0$, there exists $\eta > 0$ such that if $t \in \mathbb{R}$, $|t| \leq \eta$ and $x + t \in \Omega$, then*

$$\|f(x+t) - f(x) - tf'(x)\|_{E;\nu} \leq \epsilon\,|t|. \tag{2.4}$$

In this case, the gradient has a single component $f'(x) = \nabla f(x)$. ▮

The derivative[1] is often denoted as $\mathrm{d}f/\mathrm{d}x$ instead of f', especially when we wish to specify the variable with respect to which we are differentiating.

Let us show that the gradient is unique.

---

1 **History of the notion of derivative of a real function.** EUCLID, in his *Elements* [33, Book III, p. 16], was already studying the tangents of curves. Pierre de FERMAT, in 1636, found the tangent for a curve of equation $y = x^m$ with a calculation foreshadowing the derivative. In 1671, Isaac NEWTON introduced the *fluxion* of a function $y = f(x)$, which he denoted $\dot{y}$ [60, p. 76]. Gottfried von LEIBNIZ developed

THEOREM 2.2.– *The gradient of a function from an open subset of $\mathbb{R}^d$ into a separated semi-normed space is unique at every point where the function is differentiable.* ▌

Proof. Let us reuse the notation from Definition 2.1. Suppose that f is differentiable at the point x. Let $D = {}_1\nabla f(x) - {}_2\nabla f(x)$ be the difference of two possible gradients, where $\nu \in \mathcal{N}_E$, $\epsilon > 0$ and η_1 and η_2 both satisfy (2.3) for ${}_1\nabla f(x)$ and ${}_2\nabla f(x)$, respectively. Since Ω is open, there exists $\zeta > 0$ such that if $|z| \leq \zeta$, then $x + z \in \Omega$. If $|z| \leq \eta$, where $\eta = \inf\{\eta_1, \eta_2, \zeta\}$, then

$$\|z \cdot D\|_{E;\nu} \leq \sum_{j=1}^{2} \|f(x+z) - f(x) - z \cdot {}_j\nabla f(x)\|_{E;\nu} \leq 2\epsilon\,|z|.$$

For $z = \eta \mathbf{e}_i$, after dividing by η, it follows that $\|D_i\|_{E;\nu} \leq 2\epsilon$. This holds for every $\epsilon > 0$, so $\|D_i\|_{E;\nu} = 0$. This in turn holds for every $\nu \in \mathcal{N}_E$, so $D_i = 0_E$ because E is separated (Definition 1.1). Finally, this holds for every $i \in [\![1, d]\!]$, so ${}_1\nabla f(x) = {}_2\nabla f(x)$. □

Utility of assuming that Ω is open. This hypothesis guarantees the uniqueness of the gradient at every point where it exists. For example, if Ω is just a single point, then every function would be differentiable and any element of E could be taken as the gradient. However, the notion of differentiability can be meaningfully extended to the closure of an open set while preserving the uniqueness of the gradient (see § 2.9, p 52). □

Let us give an upper bound for the growth of a function in the neighborhood of a point where it is differentiable.

THEOREM 2.3.– *Let f be a function from an open subset Ω of $\mathbb{R}^d$ into a separated semi-normed space E that is differentiable at some point x. Then, for every semi-norm $\|\ \|_{E;\nu}$ of E, there exists $\eta > 0$ and $c \geq 0$ such that if $|z| \leq \eta$ and $x + z \in \Omega$, then*

$$\|f(x+z) - f(x)\|_{E;\nu} \leq c\,|z|. \ ▌$$

Proof. Taking $\epsilon = 1$ in the inequality (2.3) gives us $\eta > 0$ such that if $|z| \leq \eta$ and $x + z \in \Omega$, then $\|f(x+z) - f(x) - z \cdot \nabla f(x)\|_{E;\nu} \leq |z|$, which implies the desired inequality with $c = 1 + \|\nabla f(x)\|_{E^d;\nu}$ by the inequality (2.2). □

---

infinitesimal calculus in 1675 [54]. The notion of derivative was made rigorous in 1821 by Augustin CAUCHY [20, p. 22].
History of the notation. The notation $\mathrm{d}y/\mathrm{d}x$ was introduced by Gottfried VON LEIBNIZ in 1675 [54]. Joseph Louis LAGRANGE instead wrote $f'x$ in 1772 [51].
The symbol ∇ was introduced by Sir William Rowan HAMILTON in 1847 by flipping over the Greek letter Δ, already used in a similar context (to denote the *Laplacian*); this symbol was named *nabla* by Peter Guthrie TAIT on the advice of William Robertson SMITH in 1870, referencing an Ancient Greek harp of similar shape ($\nu\acute{\alpha}\beta\lambda\alpha$).

Let us observe that the differentiability of a function taking values in a space E^ℓ is equivalent to the differentiability of each of its components.

THEOREM 2.4.– *Let $f = (f_1, \dots, f_\ell)$ be a function from an open subset Ω of $\mathbb{R}^d$ into E^ℓ, where E is a separated semi-normed space and $\ell \in \mathbb{N}$.*

Then f is m times differentiable (from Ω into E^ℓ) if and only if each f_j is m times differentiable (from Ω into E). Likewise, the function f is m times continuously differentiable if and only if each f_j is m times continuously differentiable. ▌

Proof. Let $j \in [\![1, \ell]\!]$, $k \in [\![0, m-1]\!]$, $x \in \Omega$, $z \in \mathbb{R}^d$ such that $x + z \in \Omega$, and

$$D_j \stackrel{\text{def}}{=} \|\nabla^k f_j(x+z) - \nabla^k f_j(x) - z \cdot \nabla(\nabla^k f_j)(x)\|_{E^{dk};\nu}.$$

If f_j is m times differentiable, then, by the definition of this property (Definition 2.1), for every $\epsilon > 0$, there exists $\eta_j > 0$ such that $|z| \le \eta_j$ implies $D_j \le (\epsilon/\sqrt{\ell})|z|$. If this holds for each j and if $|z| \le \inf_{1 \le j \le \ell} \eta_j$, then, by the definition of the semi-norms of the Cartesian powers of E (p. 31),

$$\|\nabla^k f(x+z) - \nabla^k f(x) - z \cdot \nabla(\nabla^k f)(x)\|_{E^{\ell dk};\nu} = \Big(\sum_{j=1}^{\ell} D_j^2\Big)^{1/2} \le \epsilon\,|z|,$$

which proves that $\nabla^k f$ is differentiable. Hence, f is m times differentiable.

The converse also holds, since this inequality implies that $D_j \le \epsilon\,|z|$.

Furthermore, f and its successive gradients $\nabla^k f$ are continuous if and only if the f_j and the $\nabla^k f_j$ are continuous, since

$$\|\nabla^k f(y) - \nabla^k f(x)\|_{E^{\ell dk};\nu} = \Big(\sum_{j=1}^{\ell} \|\nabla^k f_j(y) - \nabla^k f_j(x)\|^2_{E^{dk};\nu}\Big)^{1/2}. \quad \square$$

2.2. Finite increment theorem

Let us state the **finite increment theorem** for a function in terms of its gradient on a **line segment** $[x, x+z] = \{x + tz : 0 \le t \le 1\}$.

THEOREM 2.5.– *Let f be a differentiable function from an open subset Ω of $\mathbb{R}^d$ into a separated semi-normed space E with a family of semi-norms $\{\|\ \|_{E;\nu} : \nu \in \mathcal{N}_E\}$.*

Then, for every $x \in \Omega$ and $z \in \mathbb{R}^d$ such that $[x, x+z] \subset \Omega$ and for every $\nu \in \mathcal{N}_E$:

(a) $\|f(x+z)-f(x)\|_{E;\nu} \le |z| \sup_{0\le t\le 1} \|\nabla f(x+tz)\|_{E^d;\nu}$;

(b) $\|f(x+z)-f(x)-z\cdot\nabla f(x)\|_{E;\nu} \le |z| \sup_{0\le t\le 1} \|\nabla f(x+tz)-\nabla f(x)\|_{E^d;\nu}$. ∎

Remark. The upper bounds on the right-hand side of the inequalities (a) and (b) are not necessarily finite. □

Proof. (a) Let us proceed in two steps.

1. *Decomposition of* $[0,1]$ *into elementary intervals.* Let $\epsilon > 0$. By Definition 2.1 of the differentiability of f at the point $x+tz$, there exists $\eta_t > 0$ such that if $|s| \le \eta_t$ and $x+tz+sz \in \Omega$, then

$$\|f(x+tz+sz)-f(x+tz)-sz\cdot\nabla f(x+tz)\|_{E;\nu} \le \epsilon\,|s|.$$

Writing

$$g(t) \stackrel{\text{def}}{=} f(x+tz) \text{ and } \delta(t) \stackrel{\text{def}}{=} z\cdot\nabla f(x+tz),$$

it follows that

$$\|g(t+s)-g(t)-s\delta(t)\|_{E;\nu} \le \epsilon\,|s|. \tag{2.5}$$

The open sets $\mathcal{O}_t = (t-\eta_t, t+\eta_t)$ form a covering of the compact set $[0,1]$ as t ranges over this set, so there exists a finite subcovering $\{\mathcal{O}_t\}_{t\in\mathcal{T}}$. Therefore, there is t_i and a_i such that

$$\begin{gathered}0 \le t_1 \le t_2 \le \cdots \le t_n \le 1, \quad 0 = a_1 \le a_2 \le \cdots \le a_n \le a_{n+1} = 1,\\ a_i \le t_i \le a_{i+1}, \quad |a_i - t_i| \le \eta_{t_i}, \quad |a_{i+1}-t_i| \le \eta_{t_i}.\end{gathered} \tag{2.6}$$

Concretely, these parameters can be determined by choosing t_1 to be the element (or one of the elements) of $\{t\in\mathcal{T} : t-\eta_t < 0\}$ that maximizes $t+\eta_t$, then, for each $i \ge 1$, choosing t_{i+1} to be the element (or one of the elements) of $\{t\in\mathcal{T} : t-\eta_t < t_i+\eta_{t_i}\}$ that maximizes $t+\eta_t$, stopping when $t_n+\eta_{t_n} \ge 1$, which necessarily occurs. Then $t_{i+1} \ge t_i$, and $\mathcal{O}_{t_{i+1}} \cap \mathcal{O}_{t_i}$ is non-empty, so this intersection contains a point a_{i+1} such that $t_i \le a_{i+1} \le t_{i+1}$; to conclude, we simply pick $a_1 = 0$ and $a_{n+1} = 1$.

2. *Estimating the growth.* Suppose that $M = \sup_{0\le t\le 1} \|\delta(t)\|_{F;\mu}$ is finite; if not, the stated inequality is trivial. If $|s| \le \eta_t$, then (2.5) gives

$$\|g(t+s)-g(t)\|_{F;\mu} \le (M+\epsilon)|s|.$$

For $t = t_i$, and $s = a_i - t_i$ and $s = a_{i+1}-t_i$, respectively, it follows that

$$\begin{aligned}\|g(a_i)-g(t_i)\|_{F;\mu} &\le (M+\epsilon)(t_i-a_i),\\ \|g(a_{i+1})-g(t_i)\|_{F;\mu} &\le (M+\epsilon)(a_{i+1}-t_i),\end{aligned}$$

and therefore

$$\|g(a_{i+1}) - g(a_i)\|_{F;\mu} \leq (M + \epsilon)(a_{i+1} - a_i).$$

Summing over i gives

$$\|g(1) - g(0)\|_{F;\mu} \leq \sum_{1 \leq i \leq n} \|g(a_{i+1}) - g(a_i)\|_{F;\mu} \leq M + \epsilon.$$

Since this inequality holds for every $\epsilon > 0$, it also holds for $\epsilon = 0$, which gives the desired inequality.

(b) The inequality (b) follows by applying the inequality (a) to the function $f - L$, where $L(y) = y \cdot \nabla f(x)$. To see this, observe that $\nabla L = \nabla f(x)$ because we have $L(y + z) - L(y) - z \cdot \nabla f(x) = 0$ for every $z \in \mathbb{R}^d$. □

Before we consider functions with zero gradient below, let us show that a **locally constant** function, namely a function that is constant on a ball around any given point of its domain, is constant on every connected component of this domain.

THEOREM 2.6.– *Let f be a function from an open subset Ω of $\mathbb{R}^d$ into a separated semi-normed space E such that, for every $x \in \Omega$, there exists a ball $B(x, r)$ included in Ω on which f is constant.*

Then f is constant on every connected component of Ω. ▮

Proof. Given $c \in E$, the set $X_c = \{x \in \Omega : f(x) = c\}$ is open by the hypotheses. The set $Y_c = \{x \in \Omega : f(x) \neq c\}$ is also open as the union of the sets X_d for $d \neq c$. If ω is a connected component of Ω and $a \in \omega$, then the open sets $X_{f(a)}$ and $Y_{f(a)}$ cover ω, so one of these sets must be equal to it and the other must be empty. Since $X_{f(a)}$ is not empty, it must be equal to ω. Therefore, f is constant on ω. □

Let us show that a function whose partial derivatives are zero is constant on every connected component of its domain.

THEOREM 2.7.– *Let $f \in \mathcal{C}(\Omega; E)$, where Ω is an open subset of $\mathbb{R}^d$ and E is a separated semi-normed space. Suppose that*

$$\nabla f = 0.$$

Then f is constant on every connected component of Ω. ▮

Proof. Let $x \in \Omega$ and $r > 0$ be such that the ball $B(x, r)$ is included in Ω. For every $y \in B(x, r)$, the inequality (a) from the finite increment theorem (Theorem 2.5) gives $\|f(y) - f(x)\|_{E;\nu} = 0$ for every semi-norm $\| \ \|_{E;\nu}$ of E. Therefore, $f(y) = f(x)$.

Thus, f is locally constant. Hence, it is constant on each connected component of Ω by Theorem 2.6. □

2.3. Partial derivatives

Let us define the notion of partial derivative[2].

DEFINITION 2.8.– *Let f be a function from an open subset Ω of $\mathbb{R}^d$ into a separated semi-normed space E with a family of semi-norms $\{\| \ \|_{E;\nu} : \nu \in \mathcal{N}_E\}$. We say that f has a* ***partial derivative*** *$\partial_i f(x) \in E$ at the point x of Ω, where $i \in [\![1, d]\!]$, if the mapping $x_i \mapsto f(x)$ is differentiable at the point x_i with derivative $\partial_i f(x)$.*

In other words, f has the partial derivative $\partial_i f(x)$ if, for every $\nu \in \mathcal{N}_E$ and every $\epsilon > 0$, there exists $\eta > 0$ such that if $t \in \mathbb{R}$, $|t| \leq \eta$ and $x + t\mathbf{e}_i \in \Omega$, then

$$\|f(x + t\mathbf{e}_i) - f(x) - t\partial_i f(x)\|_{E;\nu} \leq \epsilon\,|t|. \tag{2.7}$$

■

Clarification. More precisely, f has the partial derivative $\partial_i f(x)$ at the point x if the function

$$s \mapsto f(x_1, \ldots, x_{i-1}, s, x_{i+1}, \ldots, x_d),$$

defined on the open subset $\{x_i : x \in \Omega\}$ of $\mathbb{R}$, has the derivative $\partial_i f(x)$ at the point x_i. □

We often write $\partial f/\partial x_i$ instead of $\partial_i f$, especially when we wish to specify the variable with respect to which we are differentiating.

Let us show that every differentiable function has partial derivatives.

THEOREM 2.9.– *If a function f from an open subset Ω of $\mathbb{R}^d$ into a separated semi-normed space is differentiable at the point x of Ω, then, for all $i \in [\![1, d]\!]$, it has the unique partial derivative*

$$\partial_i f(x) = (\nabla f(x))_i$$

at this point, and

$$\nabla f(x) = (\partial_1 f(x), \ldots, \partial_d f(x)).$$

When $d = 1$, the unique partial derivative $\partial_1 f(x)$ coincides with the derivative $f'(x)$.

■

2 **History of partial derivatives.** Partial derivatives first appeared in 1747 in the works of Alexis Claude CLAIRAUT and Jean le Rond D'ALEMBERT [27], and in 1755 in the works of Leonhard EULER [34]. The symbol ∂ was introduced by Nicolas de Caritat DE CONDORCET, in 1770 [25].

Proof. By taking $z = t\mathbf{e}_i$ in the inequality (2.3) characterizing the derivative from Definition 2.1 (a), we obtain

$$\|f(x + t\mathbf{e}_i) - f(x) - t\nabla_i f(x)\|_{E;\nu} \leq \epsilon\,|t|,$$

which is the inequality (2.7) characterizing the partial derivative, with

$$\partial_i f(x) = \nabla_i f(x).$$

If $d = 1$, the inequality (2.4) characterizing the derivative from Definition 2.1 (b) implies the characterization (2.7) of the unique partial derivative with $\partial_1 f(x) = f'(x)$. □

Conversely, let us show that every function with continuous partial derivatives is continuously differentiable.

THEOREM 2.10.– *Every function f from an open subset of $\mathbb{R}^d$ into a separated semi-normed space with continuous partial derivatives $\partial_i f$ for every $i \in [\![1, d]\!]$ is continuous and continuously differentiable.* ▌

Proof. Let f be a function from an open subset Ω of $\mathbb{R}^d$ into a separated semi-normed space E with continuous partial derivatives. We will show that it is differentiable by finding an upper bound for the variation in each of the d dimensions, followed by an upper bound for the multidirectional variation.

One-directional variation. Given $x \in \Omega$ and $z \in \mathbb{R}^d$ such that the line segment $[x, x + z]$ is included in Ω, $i \in [\![1, d]\!]$, and $t \in [0, 1]$, we write

$$\mathbf{y}_i \stackrel{\text{def}}{=} (x_1 + z_1, \ldots, x_{i-1} + z_{i-1}, x_i, \ldots, x_d)$$

and

$$g_i(t) \stackrel{\text{def}}{=} f(\mathbf{y}_i + tz_i\mathbf{e}_i) - tz_i\partial_i f(x).$$

Definition 2.8 of the partial derivatives shows that the function $t \mapsto f(\mathbf{y}_i + tz_i\mathbf{e}_i)$ is differentiable, so g_i is differentiable and

$$g_i'(t) = z_i\partial_i f(\mathbf{y}_i + tz_i\mathbf{e}_i) - z_i\partial_i f(x).$$

The finite increment theorem (Theorem 2.5 (a)) then implies that, for every semi-norm $\|\ \|_{E;\nu}$ of E,

$$\|g_i(1) - g_i(0)\|_{E;\nu} \leq \sup_{0 \leq t \leq 1} \|g_i'(t)\|_{E;\nu} \leq$$
$$\leq |z_i| \sup_{0 \leq t \leq 1} \|\partial_i f(\mathbf{y}_i + tz_i\mathbf{e}_i) - \partial_i f(x)\|_{E;\nu} \qquad (2.8)$$

(since in one dimension, in this case t, we have $\nabla g_i = g_i'$).

Multidirectional variation. Write

$$D(z) \overset{\text{def}}{=} f(x+z) - f(x) - \sum_{i=1}^{d} z_i \partial_i f(x).$$

By the definitions of $\mathbf{y}_i$ and g_i, we have $\mathbf{y}_1 = x$, $\mathbf{y}_d = x + z$ and $\mathbf{y}_i + z_i \mathbf{e}_i = \mathbf{y}_{i+1}$, and thus

$$D(z) = \sum_{i=1}^{d} f(\mathbf{y}_i + z_i \mathbf{e}_i) - f(\mathbf{y}_i) - z_i \partial_i f(x) = \sum_{i=1}^{d} g_i(1) - g_i(0).$$

Since the functions $\partial_i f$ are continuous by the hypotheses, for every $\epsilon > 0$, there exists $\eta > 0$ such that if $\zeta \in \mathbb{R}^d$ and $|\zeta| \leq \eta$, then $\|\partial_i f(x+\zeta) - \partial_i f(x)\|_{E;\nu} \leq \epsilon/d$. Thus, if $|z| \leq \eta$, the inequality (2.8) implies

$$\|D(z)\|_{E;\nu} \leq \sum_{i=1}^{d} \|g_i(1) - g_i(0)\|_{E;\nu} \leq \frac{\epsilon}{d} \sum_{i=1}^{d} |z_i| \leq \epsilon\, |z|.$$

By Definition 2.1 of differentiability, this shows that f is differentiable at the point x and $\nabla f(x) = (\partial_1 f(x), \dots, \partial_d f(x))$. □

Utility of continuity of the partial derivatives. In Theorem 2.10, we cannot omit the hypothesis that the partial derivatives are continuous[3]. For example, the function defined on $\mathbb{R}^2$ by

$$f(x_1, x_2) = x_1 x_2 / (x_1^2 + x_2^2) \text{ if } x \neq (0,0) \text{ and } f(0,0) = 0 \tag{2.9}$$

has partial derivatives at every point but is not differentiable or even continuous at $(0,0)$. □

Let us observe that a differentiable function is m times differentiable if and only if its partial derivatives are $m-1$ times differentiable.

THEOREM 2.11.– *Let f be a differentiable function from an open subset of $\mathbb{R}^d$ into a separated semi-normed space, and let $m \in \mathbb{N}$.*

Then f is m times differentiable if and only if, for every $i \in [\![1, d]\!]$, its partial derivatives $\partial_i f$ are $m-1$ times differentiable. It is m times continuously differentiable if and only if the $\partial_i f$ are $m-1$ times continuously differentiable. ▌

3 **History of the insufficiency of partial derivatives.** In his remarkable *Cours d'Analyse de l'École Royale Polytechnique* in 1821, Augustin CAUCHY "effortlessly" showed [20, p. 46] that if a real function of two real variables is continuous in each variable separately, then it is continuous in both together. Unfortunately, this statement is false.
But this was not shown until 1870, when Carl Johannes THOMAE [79, p. 15] considered the function defined by $f(x_1, x_2) = \sin(4 \arctan(x_1/x_2))$ for non-zero x_1 and x_2 and $f(x_1, 0) = f(0, x_2) = 0$. The more straightforward counterexample of (2.9) was given in 1872 by Hermann Amandus SCHWARZ [69, p. 220].

Proof. Let f be a differentiable function from an open subset Ω of $\mathbb{R}^d$ into a separated semi-normed space E. By Theorem 2.9, $\partial_i f = (\nabla f)_i$ for every $i \in [\![1, d]\!]$.

By Definition 2.1, f is m times differentiable if and only if ∇f is $m-1$ times differentiable from Ω into E^d. This is equivalent to saying that each of its components $(\nabla f)_i$, namely $\partial_i f$, is $m-1$ times differentiable from Ω into E by Theorem 2.4. The same argument works if we replace differentiable by continuously differentiable. □

2.4. Higher order partial derivatives

Let us show that successive partial derivatives commute. This is known as **Schwarz's theorem**[4].

THEOREM 2.12.– *Let f be a twice differentiable function from an open subset of $\mathbb{R}^d$ into a separated semi-normed space. Then, for every i and j in $[\![1, d]\!]$,*

$$\partial_j \partial_i f = \partial_i \partial_j f. \blacksquare$$

Proof. Preliminary estimate. Let f be a twice differentiable function from an open subset Ω of $\mathbb{R}^d$ into a separated semi-normed space E. We will proceed by estimating

$$D(y) \stackrel{\text{def}}{=} f(x+y+z) - f(x+y) - f(x+z) + f(x) - \sum_{i=1}^{d} \sum_{j=1}^{d} y_i z_j \partial_j \partial_i f(x),$$

where $x \in \Omega$ and where $y \in \mathbb{R}^d$ and $z \in \mathbb{R}^d$ are sufficiently small. The derivative of this expression is

$$\partial_i D(y) = \partial_i f(x+y+z) - \partial_i f(x+y) - \sum_{j=1}^{d} z_j \partial_j \partial_i f(x),$$

and so

$$\begin{aligned} \partial_i D(y) = \Big(\partial_i f(x+y+z) - \partial_i f(x) - \sum_{j=1}^{d} (y+z)_j \partial_j \partial_i f(x)\Big) - \\ - \Big(\partial_i f(x+y) - \partial_i f(x) - \sum_{j=1}^{d} y_j \partial_j \partial_i f(x)\Big). \end{aligned} \tag{2.10}$$

4 **History of Schwarz's theorem (Theorem 2.12).** Karl WEIERSTRASS showed in an unpublished lecture from 1861 that the second-order partial derivatives of a function f commute at any point x where they are continuous; in other words, $\partial^2 f/\partial x_1 \partial x_2(x) = \partial^2 f/\partial x_2 \partial x_1(x)$.
Hermann Amandus SCHWARZ showed in 1873 that this result still holds whenever just one of the two terms is continuous.

By Definition 2.1 of differentiability, given that Ω is open, for every semi-norm $\|\ \|_{E;\nu}$ of E and all $\epsilon > 0$, there exists $\eta > 0$ such that if $|y| \leq \eta$, then $x + y \in \Omega$ and

$$\left\| \partial_i f(x+y) - \partial_i f(x) - \sum_{j=1}^{d} y_j \partial_j \partial_i f(x) \right\|_{E;\nu} \leq \epsilon\, |y|.$$

If $|y| \leq \eta/2$ and $|z| \leq \eta/2$, this inequality is also satisfied when y is replaced by $y+z$, so (2.10) gives

$$\|\partial_i D(y)\|_{E;\nu} \leq \epsilon(|y| + |y+z|) \leq 2\epsilon(|y| + |z|).$$

Since $D(0) = 0_E$, the finite increment theorem (Theorem 2.5 (a)) gives us

$$\|D(y)\|_{E;\nu} \leq |y| \sup_{0 \leq t \leq 1} \|\partial_i D(ty)\|_{E;\nu} \leq 2\epsilon(|y| + |z|)^2.$$

Conclusion. Now write $D_z(y)$ instead of $D(y)$. Then

$$D_z(y) - D_y(z) = \sum_{i=1}^{d} \sum_{j=1}^{d} -y_i z_j \partial_j \partial_i f(x) + z_i y_j \partial_j \partial_i f(x) =$$

$$= -\sum_{i=1}^{d} \sum_{j=1}^{d} y_i z_j \big(\partial_j \partial_i f(x) - \partial_i \partial_j f(x)\big).$$

Therefore,

$$\left\| \sum_{i=1}^{d} \sum_{j=1}^{d} y_i z_j \big(\partial_j \partial_i f(x) - \partial_i \partial_j f(x)\big) \right\|_{E;\nu} \leq \|D_z(y)\|_{E;\nu} + \|D_y(z)\|_{E;\nu} \leq$$

$$\leq 4\epsilon(|y| + |z|)^2.$$

For $y = \eta \mathbf{e}_i/2$ and $z = \eta \mathbf{e}_j/2$, after dividing by $\eta^2/4$, it follows that

$$\|\partial_j \partial_i f(x) - \partial_i \partial_j f(x)\|_{E;\nu} \leq 4\epsilon.$$

This holds for every $\epsilon > 0$, so the left-hand side is zero. This also holds for every semi-norm of E. Therefore, since this space is separated, $\partial_j \partial_i f(x) - \partial_i \partial_j f(x) = 0_E$. □

Since the **successive partial derivatives** commute, we can reorder them as follows for any $\beta \in \mathbb{N}^{*d}$:

$$\partial^\beta f \stackrel{\text{def}}{=} \partial_1^{\beta_1} \cdots \partial_d^{\beta_d} f, \quad \partial^0 f \stackrel{\text{def}}{=} f.$$

We also denote $|\beta| = \beta_1 + \cdots + \beta_d$.

Let us show that every m times differentiable function has partial derivatives of orders lower than or equal to m, and that the converse holds if the partial derivatives are continuous.

THEOREM 2.13.– *Let f be a function from an open subset of $\mathbb{R}^d$ into a separated semi-normed space and let $m \in \mathbb{N}$. Then:*

(a) *If f is m times differentiable, it has partial derivatives $\partial^\beta f$ for all $\beta \in \mathbb{N}^{*d}$ such that $0 \leq |\beta| \leq m$.*

(b) *The following three properties are equivalent:*
— *f is m times continuously differentiable;*
— *f has continuous partial derivatives $\partial^\beta f$ for all $\beta \in \mathbb{N}^{*d}$ such that $0 \leq |\beta| \leq m$;*
— *f has continuous partial derivatives $\partial^\beta f$ for all $\beta \in \mathbb{N}^{*d}$ such that $|\beta| = m$.* ▮

Proof. Let Ω be an open subset of $\mathbb{R}^d$ and E a separated semi-normed space as in the statement of the theorem.

(a) We will show that this property holds by induction on m. Assume thus that it holds for $m-1$ and that f is m times differentiable. Each $\partial_i f$ is $m-1$ times differentiable by Theorem 2.11, so, by the property for $m-1$, it has partial derivatives $\partial^\alpha \partial_i f$ for all $\alpha \in \mathbb{N}^{*d}$ such that $0 \leq |\alpha| \leq m-1$. Since every derivative $\partial^\beta f$ is of this form for $1 \leq |\beta| \leq m$, this shows that property (a) then holds for m. In addition, the case $m = 1$ holds by Theorem 2.9. Hence, by induction, it holds for every $m \in \mathbb{N}$.

(b) If f is m times continuously differentiable, we can show in the same way as above that it has continuous partial derivatives for every $\beta \in \mathbb{N}^{*d}$ such that $0 \leq |\beta| \leq m$, and so this is certainly true for every $\beta \in \mathbb{N}^{*d}$ such that $|\beta| = m$.

Let us show the converse, again by induction on m. Assume thus that it holds for $m-1$ and that f has continuous partial derivatives $\partial^\beta f$ for all $\beta \in \mathbb{N}^{*d}$ such that $|\beta| = m$. Each $\partial_i f$ has continuous partial derivatives $\partial^\alpha \partial_i f$ for all α such that $|\alpha| = m-1$, so it is $m-1$ times continuously differentiable by the converse for $m-1$. Furthermore, f is differentiable by Theorem 2.10, so it is m times continuously differentiable by Theorem 2.11. This proves the converse for m. Since the case $m = 1$ holds by Theorem 2.10, the converse holds for every $m \in \mathbb{N}$ by induction. □

2.5. Spaces $\mathcal{C}^m(\Omega; E)$, $\mathcal{C}_b^m(\Omega; E)$, $\mathcal{C}_K^m(\Omega; E)$, $\mathrm{C}_b^m(\Omega; E)$ and $\mathcal{K}^m(\Omega; E)$

Let us define the space of m times differentiable functions.

DEFINITION 2.14.– *Let $m \in \mathbb{N}^*$, Ω an open subset of $\mathbb{R}^d$ and E a separated semi-normed space with a family of semi-norms $\{\| \ \|_{E;\nu} : \nu \in \mathcal{N}_E\}$.*

(a) *We denote by $\mathcal{C}^m(\Omega; E)$ the vector space of m times continuously differentiable functions from Ω into E with semi-norms, indexed by the compact sets $K \subset \Omega$ and $\nu \in \mathcal{N}_E$,*

$$\|f\|_{\mathcal{C}^m(\Omega;E);K,\nu} \stackrel{\text{def}}{=} \sup_{0\leq|\beta|\leq m} \sup_{x\in K} \|\partial^\beta f(x)\|_{E;\nu}.$$

(b) *We denote*

$$\mathcal{C}^m_{\rm b}(\Omega; E) \stackrel{\text{def}}{=} \{f \in \mathcal{C}^m(\Omega; E) : \partial^\beta f \text{ is bounded}, \forall\beta,\ 0 \leq |\beta| \leq m\}$$

endowed with the semi-norms, indexed by $\nu \in \mathcal{N}_E$,

$$\|f\|_{\mathcal{C}^m_{\rm b}(\Omega;E);\nu} \stackrel{\text{def}}{=} \sup_{0\leq|\beta|\leq m} \sup_{x\in \Omega} \|\partial^\beta f(x)\|_{E;\nu}.$$

(c) *Given a compact set $K \subset \Omega$, we denote*

$$\mathcal{C}^m_K(\Omega; E) \stackrel{\text{def}}{=} \{f \in \mathcal{C}^m(\Omega; E) : \operatorname{supp} f \subset K\},$$

endowed with the semi-norms of $\mathcal{C}^m_{\rm b}(\Omega; E)$.

(d) *We denote*

$$\mathbf{C}^m_{\rm b}(\Omega; E) \stackrel{\text{def}}{=}$$
$$\{f \in \mathcal{C}^m(\Omega; E) : \partial^\beta f \text{ is uniformly continuous and bounded}, \forall\beta,\ 0 \leq |\beta| \leq m\},$$

endowed with the semi-norms of $\mathcal{C}^m_{\rm b}(\Omega; E)$.

(e) *We denote*

$$\mathcal{K}^m(\Omega; E) \stackrel{\text{def}}{=} \{f \in \mathcal{C}^m(\Omega; E) : \operatorname{supp} f \text{ is compact}\}$$

endowed with the semi-norms, indexed by $q \in \mathcal{C}^+(\Omega)$ and $\nu \in \mathcal{N}_E$,

$$\|f\|_{\mathcal{K}^m(\Omega;E);q,\nu} \stackrel{\text{def}}{=} \sup_{0\leq|\beta|\leq m} \sup_{x\in \Omega} q(x)\|\partial^\beta f(x)\|_{E;\nu}. \blacksquare$$

Justification. In (a), the mapping $f \mapsto \sup_{0\leq|\beta|\leq m} \sup_{x\in K} \|\partial^\beta f(x)\|_{E;\nu}$ is a semi-norm on $\mathcal{C}^m(\Omega; E)$ because the upper envelope of semi-norms is itself a semi-norm whenever it is everywhere finite (Theorem A.6). This is the case here because, first, for each x, the mapping $f \mapsto \|\partial^\beta f(x)\|_{E;\nu}$ is a semi-norm, and second, for each f, we have $\sup_{0\leq|\beta|\leq m} \sup_{x\in K} \|\partial^\beta f(x)\|_{E;\nu} < \infty$ because $\partial^\beta f$ is continuous (Theorem 2.13 (b)) and hence bounded on the compact set K (Theorem A.34).

It can be shown in the same way that semi-norms are defined in (b) and (e). □

Benefit of the semi-norms of $\mathcal{K}^m(\Omega; E)$. The semi-norms from Definition 2.14 (e), which are newly introduced here, allow us to avoid the usual use of the *inductive limit topology* of the $\mathcal{C}_K^m(\Omega; E)$. They generate this topology [SIMON, 75, forthcoming], whose less straightforward definition therefore becomes unnecessary. □

Let us now consider spaces of infinitely differentiable functions.

DEFINITION 2.15.– *Let Ω be an open subset of $\mathbb{R}^d$ and E a separated semi-normed space with a family of semi-norms $\{\| \ \|_{E;\nu} : \nu \in \mathcal{N}_E\}$.*

(a) *We denote by $\mathcal{C}^\infty(\Omega; E)$ the vector space of infinitely differentiable functions from Ω into E endowed with the semi-norms, indexed by $m \in \mathbb{N}^*$, the compact sets $K \subset \Omega$ and $\nu \in \mathcal{N}_E$,*

$$\|f\|_{\mathcal{C}^\infty(\Omega;E);m,K,\nu} \overset{\text{def}}{=} \sup_{0 \le |\beta| \le m} \sup_{x \in K} \|\partial^\beta f(x)\|_{E;\nu}.$$

(b) *We denote*

$$\mathcal{C}_{\mathrm{b}}^\infty(\Omega; E) \overset{\text{def}}{=} \{f \in \mathcal{C}^\infty(\Omega; E) : \partial^\beta f \text{ is bounded, } \forall \beta \in \mathbb{N}^{*d}\}$$

endowed with the semi-norms, indexed by $m \in \mathbb{N}^*$ and $\nu \in \mathcal{N}_E$,

$$\|f\|_{\mathcal{C}_{\mathrm{b}}^\infty(\Omega;E);m,\nu} \overset{\text{def}}{=} \sup_{0 \le |\beta| \le m} \sup_{x \in \Omega} \|\partial^\beta f(x)\|_{E;\nu}.$$

(c) *Given a compact set $K \subset \Omega$, we denote*

$$\mathcal{C}_K^\infty(\Omega; E) \overset{\text{def}}{=} \{f \in \mathcal{C}^\infty(\Omega; E) : \operatorname{supp} f \subset K\}$$

endowed with the semi-norms of $\mathcal{C}_{\mathrm{b}}^\infty(\Omega; E)$.

(d) *We denote*

$$\mathbf{C}_{\mathrm{b}}^\infty(\Omega; E) \overset{\text{def}}{=} \{f \in \mathcal{C}^\infty(\Omega; E) : \partial^\beta f \text{ is uniformly continuous and bounded } \forall \beta \in \mathbb{N}^{*d}\}$$

endowed with the semi-norms of $\mathcal{C}_{\mathrm{b}}^\infty(\Omega; E)$.

(e) *We denote*

$$\mathcal{K}^\infty(\Omega; E) \overset{\text{def}}{=} \{f \in \mathcal{C}^\infty(\Omega; E) : \operatorname{supp} f \text{ is compact}\}$$

endowed with the semi-norms, indexed by $m \in \mathbb{N}^*$, $q \in \mathcal{C}^+(\Omega)$ and $\nu \in \mathcal{N}_E$,

$$\|f\|_{\mathcal{K}^\infty(\Omega;E);m,q,\nu} \overset{\text{def}}{=} \sup_{0 \le |\beta| \le m} \sup_{x \in \Omega} q(x) \|\partial^\beta f(x)\|_{E;\nu}. \ \blacksquare$$

Justification. It can be shown that semi-norms are indeed defined by the same reasoning as in Definition 2.14 (for (a), we need to observe that $\partial^\beta f$ is differentiable and hence continuous, see Theorem 2.2). □

Existence of infinitely differentiable functions with compact support. It is worth noting that the vector space $\mathcal{K}^\infty(\Omega)$ is not trivially just the zero function, as we will see in Theorems 3.19 and 3.21. □

The space $\mathcal{D}(\Omega)$. We will denote by $\mathcal{D}(\Omega)$ the space $\mathcal{K}^\infty(\Omega)$ endowed with the family of semi-norms, also indexed by $q \in \mathcal{C}^+(\Omega)$,

$$\|\varphi\|_{\mathcal{D}(\Omega);q} \stackrel{\text{def}}{=} \sup_{x\in\Omega} \sup_{0\le|\beta|\le q(x)} q(x)|\partial^\beta\varphi(x)|.$$

This family dominates the family from Definition 2.15 (e), which implies that $\mathcal{D}(\Omega) \subsetneq \mathcal{K}^\infty(\Omega)$. The space endowed with these semi-norms will play a key role in the construction of distributions in Volume 3. □

2.6. Comparison and metrizability of spaces of differentiable functions

Let us state a few elementary properties of these spaces.

THEOREM 2.16.– *Let $0 \le m \le \infty$, Ω an open subset of $\mathbb{R}^d$, E a separated semi-normed space and K a compact set included in Ω. Then:*

(a) *The spaces $\mathcal{C}^m(\Omega; E)$, $\mathcal{C}^m_{\mathrm{b}}(\Omega; E)$, $\mathcal{C}^m_K(\Omega; E)$, $\mathbf{C}^m_{\mathrm{b}}(\Omega; E)$ and $\mathcal{K}^m(\Omega; E)$ are separated semi-normed spaces.*

(b) *The spaces $\mathcal{C}^m_K(\Omega; E)$ and $\mathbf{C}^m_{\mathrm{b}}(\Omega; E)$ are closed vector subspaces of $\mathcal{C}^m_{\mathrm{b}}(\Omega; E)$.*

(c) $$\mathcal{C}^m_K(\Omega; E) \subsetneq \mathcal{K}^m(\Omega; E) \subsetneq \mathbf{C}^m_{\mathrm{b}}(\Omega; E) \subsetneq \mathcal{C}^m_{\mathrm{b}}(\Omega; E) \subsetneq \mathcal{C}^m(\Omega; E).$$ ▮

Proof. This can be shown in the same way as for $m = 0$ in Theorems 1.10 (for $\mathcal{C}(\Omega; E)$, $\mathcal{C}_{\mathrm{b}}(\Omega; E)$, $\mathcal{C}_K(\Omega; E)$ and $\mathbf{C}_{\mathrm{b}}(\Omega; E)$) and 1.19 (for $\mathcal{K}(\Omega; E)$). □

Let us show that these spaces become smaller as m increases.

THEOREM 2.17.– *Let $0 \le m' \le m \le \infty$, Ω an open subset of $\mathbb{R}^d$, E a separated semi-normed space and K a compact set included in Ω.*

Then $\mathcal{C}^m(\Omega; E) \subsetneq \mathcal{C}^{m'}(\Omega; E)$, $\mathcal{C}^m_{\mathrm{b}}(\Omega; E) \subsetneq \mathcal{C}^{m'}_{\mathrm{b}}(\Omega; E)$, $\mathcal{C}^m_K(\Omega; E) \subsetneq \mathcal{C}^{m'}_K(\Omega; E)$, $\mathbf{C}^m_{\mathrm{b}}(\Omega; E) \subsetneq \mathbf{C}^{m'}_{\mathrm{b}}(\Omega; E)$ and $\mathcal{K}^m(\Omega; E) \subsetneq \mathcal{K}^{m'}(\Omega; E)$. ▮

Proof. This follows from the definitions of these spaces (Definitions 2.14 and 2.15). □

Let us observe that the topologies of $\mathcal{C}^m(\Omega; E)$, $\mathcal{C}^m_{\mathrm{b}}(\Omega; E)$, $\mathbf{C}^m_{\mathrm{b}}(\Omega; E)$ and $\mathcal{K}^m(\Omega; E)$ coincide on $\mathcal{C}^m_K(\Omega; E)$.

THEOREM 2.18.– *Let $0 \le m \le \infty$, Ω an open subset of $\mathbb{R}^d$, E a separated semi-normed space and K a compact set included in Ω.*

Then the topologies of $\mathcal{C}^m(\Omega; E)$, $\mathcal{C}^m_{\mathrm{b}}(\Omega; E)$, $\mathcal{C}^m_K(\Omega; E)$, $\mathbf{C}^m_{\mathrm{b}}(\Omega; E)$ and $\mathcal{K}^m(\Omega; E)$ coincide on $\mathcal{C}^m_K(\Omega; E)$. ▌

Proof. Let $\{\| \ \|_{E;\nu} : \nu \in \mathcal{N}_E\}$ be the family of semi-norms of E. Since we have $\mathcal{C}^m_K(\Omega; E) \underset{\to}{\subset} \mathcal{K}^m(\Omega; E) \underset{\to}{\subset} \mathbf{C}^m_{\mathrm{b}}(\Omega; E) \underset{\to}{\subset} \mathcal{C}^m_{\mathrm{b}}(\Omega; E) \underset{\to}{\subset} \mathcal{C}^m(\Omega; E)$ by Theorem 2.16 (c), it suffices to show that the family of semi-norms of $\mathcal{C}^m(\Omega; E)$ dominates the family of $\mathcal{C}^m_K(\Omega; E)$ on the latter space. The space $\mathcal{C}^m_K(\Omega; E)$ being endowed with the semi-norms of $\mathcal{C}^m_{\mathrm{b}}(\Omega; E)$ by Definition 2.14 (c), this follows from the fact that for every $f \in \mathcal{C}^m_K(\Omega; E)$ and $\nu \in \mathcal{N}_E$,

$$\|f\|_{\mathcal{C}^m_{\mathrm{b}}(\Omega;E);\nu} = \sup_{x \in \Omega} \|f(x)\|_{E;\nu} = \sup_{x \in K} \|f(x)\|_{E;\nu} = \|f\|_{\mathcal{C}^m(\Omega;E);K,\nu}. \ \square$$

Let us state some metrizability properties.

THEOREM 2.19.– *Let $0 \le m \le \infty$, Ω an open subset of $\mathbb{R}^d$, E a separated semi-normed space and K a compact set included in Ω. Then:*

(a) *If E is metrizable, then $\mathcal{C}^m(\Omega; E)$, $\mathcal{C}^m_{\mathrm{b}}(\Omega; E)$, $\mathcal{C}^m_K(\Omega; E)$ and $\mathbf{C}^m_{\mathrm{b}}(\Omega; E)$ are metrizable.*

(b) *If E is a normed space and $m < \infty$, then $\mathcal{C}^m_{\mathrm{b}}(\Omega; E)$, $\mathcal{C}^m_K(\Omega; E)$ and $\mathbf{C}^m_{\mathrm{b}}(\Omega; E)$ are normed spaces.* ▌

Proof. (a) By Definition 1.13 of metrizability, if E is metrizable, then we can endow it with a family of semi-norms indexed by a countable set $\mathcal{N}_E$.

Hence, $\mathcal{C}^m_{\mathrm{b}}(\Omega; E)$ is metrizable if $m < \infty$ because its semi-norms are indexed by $\mathcal{N}_E$ in this case (Definition 2.14 (b)). Similarly, $\mathcal{C}^\infty_{\mathrm{b}}(\Omega; E)$ is metrizable because its semi-norms are indexed by $\mathbb{N}^* \times \mathcal{N}_E$ (Definition 2.15 (b)), which is itself countable (Theorem A.2). Furthermore, $\mathbf{C}^m_{\mathrm{b}}(\Omega; E)$ and $\mathcal{C}^m_K(\Omega; E)$ are also metrizable for $m \in [\![1, \infty]\!]$, since they are endowed with the semi-norms of $\mathcal{C}^m_{\mathrm{b}}(\Omega; E)$ (Definitions 2.14 (c) and (d) and 2.15 (c) and (d)).

Now consider $\mathcal{C}^m(\Omega; E)$. Every compact set $K \subset \Omega$ is included in one of the compact sets $K_n = \overline{\Omega^n_{1/n}}$, which are included in Ω and indexed by $n \in \mathbb{N}$ by Lemma 1.15. The family of semi-norms of $\mathcal{C}^m(\Omega; E)$ (Definition 2.14 (a)) is therefore equivalent to the subfamily associated with the K_n, which is indexed by $\mathbb{N} \times \mathcal{N}_E$ if $m < \infty$ and by $\mathbb{N}^* \times \mathbb{N} \times \mathcal{N}_E$ if $m = \infty$. Since $\mathcal{N}_E$ is countable, these sets are also countable (Theorem A.2), and so $\mathcal{C}^m(\Omega; E)$ is metrizable.

(b) If E is a normed space, then $\mathcal{N}_E$ is just a single element, so the same is true for the families of semi-norms of $\mathcal{C}^m_{\rm b}(\Omega; E)$, $\mathcal{C}^m_K(\Omega; E)$ and $\mathbf{C}^m_{\rm b}(\Omega; E)$, which are therefore also normed spaces. □

Let us state a few facts about extending functions with compact support by 0_E.

Theorem 2.20.– *Let f be a function from Ω into E, where Ω is an open subset of $\mathbb{R}^d$ and E is a separated semi-normed space, and let $\widetilde{f}$ be its extension by 0_E on $\mathbb{R}^d \setminus \Omega$. Additionally, let $0 \leq m \leq \infty$, and suppose that K is a compact set included in Ω.*

(a) *If $f \in \mathcal{C}^m_K(\Omega; E)$, then $\widetilde{f} \in \mathcal{C}^m_K(\mathbb{R}^d; E)$.*

(b) *If $f \in \mathcal{K}^m(\Omega; E)$, then $\widetilde{f} \in \mathcal{K}^m(\mathbb{R}^d; E)$.* ■

Proof. (a) Let $f \in \mathcal{C}^m_K(\Omega; E)$, where $m < \infty$. By the definition of this space (Definition 2.14 (c)), $\widetilde{f}$ is m times continuously differentiable on Ω, since it is equal to f on this set. The same is true on the open set $\mathbb{R}^d \setminus K$, since $\widetilde{f}$ is zero on this set, and hence on the whole of $\mathbb{R}^d$. By the definition of the support (Definition 1.5), the support of $\widetilde{f}$ is included in K, like the support of f. Therefore, $\widetilde{f} \in \mathcal{C}^m_K(\mathbb{R}^d; E)$.

If $f \in \mathcal{C}^\infty_K(\Omega; E)$, $\widetilde{f}$ is therefore m times continuously differentiable for every $m \in \mathbb{N}^*$, and its support is included in K. Thus, $\widetilde{f} \in \mathcal{C}^\infty_K(\mathbb{R}^d; E)$ by Definition 2.15 (c) of this space.

(b) This follows from (a), since a function belongs to $\mathcal{K}^m(\Omega; E)$ if and only if it belongs to one of the $\mathcal{C}^m_K(\Omega; E)$ (Theorem 1.18). □

2.7. Filtering properties of spaces of differentiable functions

Let us define the notion of a filtering family of semi-norms.

DEFINITION 2.21.– *A family $\{\| \ \|_\nu : \nu \in \mathcal{N}\}$ of semi-norms on a vector space E is said to be **filtering** if, for every finite subset N of $\mathcal{N}$, there exists $\mu \in \mathcal{N}$ such that, for every $u \in E$,*

$$\sup_{\nu \in N} \|u\|_\nu \leq \|u\|_\mu. \ \blacksquare$$

Recall that some authors incorporate this definition directly into the definition of a semi-normed space (see the remark *Caution*, p. 2).

Let us state some filtering properties of spaces of differentiable functions.

THEOREM 2.22.– *Let $0 \leq m \leq \infty$, Ω an open subset of $\mathbb{R}^d$, E a separated semi-normed space and K a compact set included in Ω.*

If the family of semi-norms of E is filtering, then the families of semi-norms of $\mathcal{C}^m(\Omega; E)$, $\mathcal{C}^m_{\mathrm{b}}(\Omega; E)$, $\mathcal{C}^m_K(\Omega; E)$, $\mathbf{C}^m_{\mathrm{b}}(\Omega; E)$ and $\mathcal{K}^m(\Omega; E)$ are also filtering.

In particular, the families of semi-norms of $\mathcal{C}^m(\Omega)$, $\mathcal{C}^m_{\mathrm{b}}(\Omega)$, $\mathcal{C}^m_K(\Omega)$, $\mathbf{C}^m_{\mathrm{b}}(\Omega)$ and $\mathcal{K}^m(\Omega)$ are filtering. $\blacksquare$

Proof. Let $\{\| \ \|_{E;\nu} : \nu \in \mathcal{N}_E\}$ be the family of semi-norms of E, and suppose that it is filtering. In other words (Definition 2.21), suppose that, given finitely many elements $\nu_1, \dots, \nu_n$ of $\mathcal{N}_E$, there exists $\nu \in \mathcal{N}_E$ such that $\|u\|_{E;\nu} \geq \|u\|_{E;\nu_i}$ for every $u \in E$ and $i \in [\![1, n]\!]$.

The space $\mathcal{C}^m(\Omega; E)$ for finite m. Let $K_1, \dots, K_n$ be compact subsets of $\mathbb{R}^d$ that are included in Ω. Their union $K = K_1 \cup \cdots \cup K_n$ is also compact because, in $\mathbb{R}^d$, compactness is equivalent to being closed and bounded by the Borel–Lebesgue theorem (Theorem A.23 (b)) and both of these properties are preserved by finite unions. For every $f \in \mathcal{C}^m(\Omega; E)$, by the definition of the semi-norms of this space (Definition 2.14 (a)), $\|f\|_{\mathcal{C}^m(\Omega;E);K,\nu} = \sup_{0 \leq |\beta| \leq m} \sup_{x \in K} \|\partial^\beta f(x)\|_{E;\nu}$, which is greater than the semi-norms $\|f\|_{\mathcal{C}^m(\Omega;E);K_i,\nu_i}$ for $i \in [\![1, n]\!]$.

The space $\mathcal{C}^\infty(\Omega; E)$. Suppose further that $m_1, \dots, m_n$ are integers, and let m be their maximum. For every $f \in \mathcal{C}^\infty(\Omega; E)$, by the definition of the semi-norms of this space (Definition 2.15 (a)), $\|f\|_{\mathcal{C}^\infty(\Omega;E);m,K,\nu} = \sup_{0 \leq |\beta| \leq m} \sup_{x \in K} \|\partial^\beta f(x)\|_{E;\nu}$, which is greater than the semi-norms $\|f\|_{\mathcal{C}^\infty(\Omega;E);m_i,K_i,\nu_i}$ for $i \in [\![1, n]\!]$.

The space $\mathcal{C}^m_{\mathrm{b}}(\Omega; E)$. By the definition of the semi-norms of $\mathcal{C}^m_{\mathrm{b}}(\Omega; E)$ (Definitions 2.14 (b) and 2.15 (b)), simply proceed in the same way as above after replacing the K_i and K by Ω.

The spaces $\mathcal{C}^m_K(\Omega; E)$ *and* $\mathbf{C}^m_{\mathrm{b}}(\Omega; E)$. The families of semi-norms of these spaces are filtering because they coincide with the family of semi-norms of $\mathcal{C}^m_{\mathrm{b}}(\Omega; E)$ by Definitions 2.14 (c) and (d) and 2.15 (c) and (d).

The space $\mathcal{K}^m(\Omega; E)$ *for finite* m. Suppose now that $q_1, \ldots, q_n$ are functions in $\mathcal{C}^+(\Omega)$. Then $q = q_1 + \cdots + q_n$ belongs to $\mathcal{C}^+(\Omega)$ and, for every $f \in \mathcal{K}^m(\Omega; E)$, by the definition of the semi-norms of this space (Definition 2.14 (e)), we have $\|f\|_{\mathcal{K}^m(\Omega;E);q,\nu} = \sup_{0\le|\beta|\le m} \sup_{x\in\Omega} q(x)\|\partial^\beta f(x)\|_{E;\nu}$, which is greater than the semi-norms $\|f\|_{\mathcal{K}^m(\Omega;E);q_i,\nu_i}$ for $i \in [\![1, n]\!]$.

The space $\mathcal{K}^\infty(\Omega; E)$. Let $m_1, \ldots, m_n$ be integers with maximum m, let $q_1, \ldots, q_n$ be functions in $\mathcal{C}^+(\Omega)$ and let $q = q_1 + \cdots + q_n$. For every $f \in \mathcal{K}^\infty(\Omega; E)$, by Definition 2.15 (e), $\|f\|_{\mathcal{K}^\infty(\Omega;E);m,q,\nu} = \sup_{0\le|\beta|\le m} \sup_{x\in\Omega} q(x)\|\partial^\beta f(x)\|_{E;\nu}$, which is greater than the semi-norms $\|f\|_{\mathcal{K}^\infty(\Omega;E);m_i,q_i,\nu_i}$ for $i \in [\![1, n]\!]$.

Real-valued spaces. The family of semi-norms of $\mathbb{R}$ is just its norm and therefore is filtering, so the above results hold whenever $E = \mathbb{R}$. □

2.8. Sequential completeness of spaces of differentiable functions

Let us state a differentiation property for limits that will be useful for our proofs of sequential completeness and compactness later.

THEOREM 2.23.– *Let* f_n, f *and* g_i *be functions in* $\mathcal{C}(\Omega; E)$, *where* Ω *is an open subset of* $\mathbb{R}^d$ *and* E *is a separated semi-normed space. Suppose that for every* $n \in \mathbb{N}$, $n \to \infty$, *and every* $i \in [\![1, d]\!]$:

$$f_n \in \mathcal{C}^1(\Omega; E),$$
$$f_n \to f \text{ in } \mathcal{C}(\Omega; E),$$
$$\partial_i f_n \to g_i \text{ in } \mathcal{C}(\Omega; E).$$

Then

$$f \in \mathcal{C}^1(\Omega; E),\ \partial_i f = g_i \text{ and } f_n \to f \text{ in } \mathcal{C}^1(\Omega; E).$$ ■

Proof. Let $\{\|\ \|_{E;\nu} : \nu \in \mathcal{N}_E\}$ be the family of semi-norms of E. Given $x \in \Omega$, choose h sufficiently small that the line segment $[x, x+h]$ is also included in Ω. Then the finite increment theorem (Theorem 2.5 (b)) implies that, for every $\nu \in \mathcal{N}_E$,

$$\|f_n(x+h) - f_n(x) - h \cdot \nabla f_n(x)\|_{E;\nu} \le |h| \sup_{0\le t\le 1} \|\nabla f_n(x+th) - \nabla f_n(x)\|_{E^d;\nu}.$$

Taking the limit, we obtain the same inequality on f except with g instead of ∇f. This implies that f is differentiable and $\partial_i f = g_i$. Hence, $f \in \mathcal{C}^1(\Omega; E)$ and $f_n \to f$ in this space. □

Let us state some sequential completeness properties of $\mathcal{C}^m(\Omega; E)$, $\mathcal{C}_{\rm b}^m(\Omega; E)$, $\mathcal{C}_K^m(\Omega; E)$, $\mathbf{C}_{\rm b}^m(\Omega; E)$ and $\mathcal{K}^m(\Omega; E)$.

THEOREM 2.24.– *Let $0 \leq m \leq \infty$, Ω an open subset of $\mathbb{R}^d$, E a separated semi-normed space and K a compact set included in Ω. Then:*

(a) *If E is a Neumann space, then $\mathcal{C}^m(\Omega; E)$, $\mathcal{C}_{\rm b}^m(\Omega; E)$, $\mathcal{C}_K^m(\Omega; E)$ and $\mathbf{C}_{\rm b}^m(\Omega; E)$ are Neumann spaces.*

(b) *If E is a Fréchet space, then $\mathcal{C}^m(\Omega; E)$, $\mathcal{C}_{\rm b}^m(\Omega; E)$, $\mathcal{C}_K^m(\Omega; E)$ and $\mathbf{C}_{\rm b}^m(\Omega; E)$ are Fréchet spaces.*

(c) *If E is a Banach space and $m < \infty$, then $\mathcal{C}_{\rm b}^m(\Omega; E)$, $\mathcal{C}_K^m(\Omega; E)$ and $\mathbf{C}_{\rm b}^m(\Omega; E)$ are Banach spaces.*

(d) *If E is a Banach space, then $\mathcal{K}^m(\Omega; E)$ is a Neumann space.*

In particular:

(e) *The spaces $\mathcal{C}^m(\Omega)$, $\mathcal{C}_{\rm b}^m(\Omega)$, $\mathcal{C}_K^m(\Omega)$, $\mathbf{C}_{\rm b}^m(\Omega)$ and $\mathcal{K}^m(\Omega)$ are Neumann spaces.*

(f) *The spaces $\mathcal{C}^m(\Omega)$, $\mathcal{C}_{\rm b}^m(\Omega)$, $\mathcal{C}_K^m(\Omega)$ and $\mathbf{C}_{\rm b}^m(\Omega)$ are Fréchet spaces.*

(g) *The spaces $\mathcal{C}_{\rm b}^m(\Omega)$, $\mathcal{C}_K^m(\Omega)$ and $\mathbf{C}_{\rm b}^m(\Omega)$ are Banach spaces if $m < \infty$.* ▮

Proof. (a) *Sequential completeness of $\mathcal{C}^m(\Omega; E)$. Case $m < \infty$.* Let $(f_n)_{n\in\mathbb{N}}$ be a Cauchy sequence in $\mathcal{C}^m(\Omega; E)$. For $0 \leq |\beta| \leq m$, $(\partial^\beta f_n)_{n\in\mathbb{N}}$ is a Cauchy sequence in $\mathcal{C}(\Omega; E)$, which is sequentially complete (Theorem 1.12), so it has a limit f_β. Theorem 2.23 shows that $\partial_i f_\beta = f_{\beta+\mathbf{e}_i}$ for $|\beta| \leq m - 1$, which by iterating shows that $f_\beta = \partial^\beta f$ for every $|\beta| \leq m$, where $f = f_0$. Hence, $f \in \mathcal{C}^m(\Omega; E)$, and $f_n \to f$ in this space, which is therefore sequentially complete.

Case $m = \infty$. Let $(f_n)_{n\in\mathbb{N}}$ be a Cauchy sequence in $\mathcal{C}^\infty(\Omega; E)$. For every $m \in \mathbb{N}$, this sequence is Cauchy in $\mathcal{C}^m(\Omega; E)$, so it has a limit f that cannot depend on m, since the sequence must have the same limit in $\mathcal{C}(\Omega; E)$. Hence, $f_n \to f$ in $\mathcal{C}^\infty(\Omega; E)$.

Sequential completeness of $\mathcal{C}_{\rm b}^m(\Omega; E)$. Case $m < \infty$. Let $(f_n)_{n\in\mathbb{N}}$ be a Cauchy sequence in $\mathcal{C}_{\rm b}^m(\Omega; E)$. This sequence is Cauchy in $\mathcal{C}^m(\Omega; E)$, so it has a limit f in this space. For every $|\beta| \leq m$, $\partial^\beta f_n \to \partial^\beta f$ in $\mathcal{C}(\Omega; E)$. This convergence also occurs in $\mathcal{C}_{\rm b}(\Omega; E)$, since this is a sequentially complete space (Theorem 1.12) in which $(\partial^\beta f_n)_{n\in\mathbb{N}}$ is Cauchy. Hence, $f_n \to f$ in $\mathcal{C}_{\rm b}^m(\Omega; E)$.

Case $m = \infty$. Argue in the same way as for $\mathcal{C}^\infty(\Omega; E)$.

Sequential completeness of $\mathcal{C}_K^m(\Omega; E)$ *and* $\mathbf{C}_{\mathrm{b}}^m(\Omega; E)$. This follows from the fact that these spaces are closed subspaces of $\mathcal{C}_{\mathrm{b}}^m(\Omega; E)$ (Theorem 2.16 (b)) and the fact that every closed subspace of a sequentially complete space is sequentially complete (Theorem A.27).

(b) *Case where* E *is a Fréchet space.* In this case, $\mathcal{C}^m(\Omega; E)$, $\mathcal{C}_{\mathrm{b}}^m(\Omega; E)$, $\mathcal{C}_K^m(\Omega; E)$ and $\mathbf{C}_{\mathrm{b}}^m(\Omega; E)$ are metrizable (Theorem 2.19 (a)). Together with (a), this implies that they are Fréchet spaces.

(c) *Case where* E *is a Banach space. Properties of* $\mathcal{C}_{\mathrm{b}}^m(\Omega; E)$, $\mathcal{C}_K^m(\Omega; E)$ *and* $\mathbf{C}_{\mathrm{b}}^m(\Omega; E)$ *for* $m < \infty$. In this case, these spaces are normed spaces (part (b) of Theorem 2.19). Together with (a), this implies that they are Banach spaces.

(d) *Sequential completeness of* $\mathcal{K}^m(\Omega; E)$. *Case* $m < \infty$. Let $(f_n)_{n\in\mathbb{N}}$ be a Cauchy sequence in $\mathcal{K}^m(\Omega; E)$. This sequence is Cauchy in $\mathcal{C}^m(\Omega; E)$, so it has a limit f in this space. For every $|\beta| \le m$, $\partial^\beta f_n \to \partial^\beta f$ in $\mathcal{C}(\Omega; E)$. This convergence also occurs in $\mathcal{K}(\Omega; E)$, since this is a sequentially complete space (Theorem 1.21 because E is now a Banach space) in which $(\partial^\beta f_n)_{n\in\mathbb{N}}$ is Cauchy. Hence, $f_n \to f$ in $\mathcal{K}^m(\Omega; E)$.

Case $m = \infty$. Proceed in the same way as (a) for $\mathcal{C}^\infty(\Omega; E)$.

(e), (f), *and* (g). These properties follow from (b) and (c), since $\mathbb{R}$ is a Banach space (Theorem A.23 (a)). □

Let us state a sequential compactness property in $\mathcal{C}_K^m(\Omega)$.

THEOREM 2.25.– *Let* $m \in \mathbb{N}^*$, Ω *an open subset of* $\mathbb{R}^d$ *and* K *a compact subset of* Ω.

Then every bounded sequence in $\mathcal{C}_K^{m+1}(\Omega)$ *has a convergent subsequence in* $\mathcal{C}_K^m(\Omega)$. ▌

Proof. Reduce to the case $\Omega = \mathbb{R}^d$ by extending each function by 0. We will then argue by induction on m.

Case $m = 0$. Let $(f_n)_{n\in\mathbb{N}}$ be a bounded sequence in $\mathcal{C}_K^1(\mathbb{R}^d)$. By the finite increment theorem (Theorem 2.5 (a)), for every $h \in \mathbb{R}^d$,

$$\sup_{x\in\mathbb{R}^d} |f_n(x+h) - f_n(x)| \le |h| \sup_{x\in\mathbb{R}^d} |\nabla f_n(x)|. \tag{2.11}$$

Let ω be a bounded open set containing K. By Ascoli's theorem (Theorem 1.29), there exists a subsequence $(f_{\sigma(n)}|_\omega)_{n\in\mathbb{N}}$ of $(f_n|_\omega)_{n\in\mathbb{N}}$ that converges to some function f_ω

in $\mathcal{C}_b(\omega)$. Indeed:
— $(f_n|_\omega)_{n\in\mathbb{N}}$ is uniformly equicontinuous by (2.11);
— for all x, every sequence in $\{f_n(x) : n \in \mathbb{N}\}$ contains a convergent subsequence by the Bolzano–Weierstrass theorem (Theorem A.23 (c) because this set is bounded in $\mathbb{R}$).

Then $f_{\sigma(n)}|_\omega \to f_\omega$ in $\mathcal{C}_K(\omega)$ because this space is closed in $\mathcal{C}_b(\omega)$ by Theorem 1.10 (b). Therefore, $f_{\sigma(n)} \to \widetilde{f_\omega}$ in $\mathcal{C}_K(\mathbb{R}^d)$, where $\tilde{}$ denotes extension by 0. This proves the result for $m = 0$.

General case. Suppose that the result holds for $m-1$ and let $(f_n)_{n\in\mathbb{N}}$ be a bounded sequence in $\mathcal{C}_K^{m+1}(\mathbb{R}^d)$. Then $(f_n)_{n\in\mathbb{N}}$ and the $(\partial_i f_n)_{n\in\mathbb{N}}$ are bounded in $\mathcal{C}_K^m(\mathbb{R}^d)$. By the induction hypothesis, we can therefore successively choose subsequences such that

$$f_n \to f \text{ and } \partial_i f_n \to f_i \text{ in } \mathcal{C}_K^{m-1}(\mathbb{R}^d).$$

Therefore, by Theorem 2.23, $f_i = \partial_i f$. Hence, $f_n \to f$ in $\mathcal{C}_K^m(\mathbb{R}^d)$, which proves the result for m. Since we already showed the result for the case $m = 0$, it holds for all m by induction. □

2.9. The space $\mathcal{C}^m(\overline{\Omega}; E)$ and the set $\mathcal{C}^m(\Omega; U)$

Let us define the space of "differentiable" functions on the closure of a bounded open set. This space will occasionally be useful later.

DEFINITION 2.26.– *Let E be a separated semi-normed space, $\{\| \ \|_{E;\nu} : \nu \in \mathcal{N}_E\}$ its semi-norms, $m \in \mathbb{N}^*$, and*

$$\Omega \text{ a bounded open subset of } \mathbb{R}^d.$$

We denote

$$\mathcal{C}^m(\overline{\Omega}; E) \overset{\text{def}}{=} \left\{ \begin{array}{l} f \in \mathcal{C}(\overline{\Omega}; E) : f|_\Omega \in \mathcal{C}^m(\Omega; E) \text{ and, for every } |\beta| \leq m, \\ \partial^\beta(f|_\Omega) \text{ has an extension in } \mathcal{C}(\overline{\Omega}; E), \text{ denoted } \partial^\beta f \end{array} \right\}$$

endowed with the semi-norms, indexed by $\nu \in \mathcal{N}_E$,

$$\|f\|_{\mathcal{C}^m(\overline{\Omega};E);\nu} \overset{\text{def}}{=} \|f|_\Omega\|_{\mathcal{C}_b^m(\Omega;E);\nu} = \sup_{0\leq|\beta|\leq m} \sup_{x\in\Omega} \|\partial^\beta f(x)\|_{E;\nu}. \blacksquare$$

Justification. The semi-norms $\|f|_\Omega\|_{\mathcal{C}_b^m(\Omega;E);\nu}$ are well defined because $f|_\Omega$ and its derivatives of order $\leq m$ are bounded on Ω. Indeed, by the hypotheses, they have continuous extensions on $\overline{\Omega}$ that are bounded by Theorem A.34 because $\overline{\Omega}$ is compact by the Borel–Lebesgue theorem (Theorem A.23 (b)) as a bounded set like Ω.

The notation $\partial^\beta f$ can be used to denote the extension because it is unique for any given $f \in \mathcal{C}^m(\overline{\Omega}; E)$ by the continuous extension theorem (Theorem A.37 (a)). □

The space $\mathcal{C}^1([a,b];E)$. In particular, $\mathcal{C}^1([a,b];E)$ is the space of functions that are continuous on $[a,b]$, continuously differentiable on (a,b) and whose derivatives have a continuous extension on $[a,b]$. □

Let us characterize $\mathcal{C}^m(\overline{\Omega};E)$ when Ω is bounded and E is sequentially complete.

THEOREM 2.27.– *Let $m \in \mathbb{N}^*$,*

Ω a bounded open subset of $\mathbb{R}^d$ and E a Neumann space.

Then

$$\mathcal{C}^m(\overline{\Omega};E) = \{f \in \mathcal{C}(\overline{\Omega};E) : f|_{\Omega} \in \mathbf{C}_{\mathrm{b}}^m(\Omega;E)\}. \blacksquare$$

Proof. If $f \in \mathcal{C}^m(\overline{\Omega};E)$, then itself and the $\partial^\beta f$ for $|\beta| \leq m$ are uniformly continuous and bounded on the compact set $\overline{\Omega}$ by Heine's theorem (Theorem A.34), since they are continuous by Definition 2.26. Therefore, so are their restrictions to Ω, namely $f|_{\Omega}$ and $\partial^\beta(f|_{\Omega})$, and hence $f|_{\Omega} \in \mathbf{C}_{\mathrm{b}}^m(\Omega;E)$.

Conversely, if $f|_{\Omega} \in \mathbf{C}_{\mathrm{b}}^m(\Omega;E)$, that is, if $f|_{\Omega}$ and the $\partial^\beta(f|_{\Omega})$ for $|\beta| \leq m$ are uniformly continuous and bounded, then they have unique continuous extensions to $\overline{\Omega}$ (Theorem A.38, which we can apply because E is a Neumann space and Ω is sequentially dense in $\overline{\Omega}$, see Theorem A.25 (b)). Therefore, $f \in \mathcal{C}^m(\overline{\Omega};E)$ by Definition 2.26. □

Caution when defining $\mathcal{C}^m(\Lambda;E)$ for arbitrary Λ. Instead of Definition 2.26, some authors choose the alternative definition that, for every $\Lambda \subset \mathbb{R}^d$,

$$\mathcal{C}^m(\Lambda;E) \overset{\mathrm{def}}{=} \{f \in \mathcal{C}(\Lambda;E) : f \textit{ has an extension } \overline{f} \in \mathcal{C}^m(\mathbb{R}^d;E)\}.$$

This is equivalent to Definition 2.26 when Λ is the closure $\overline{\Omega}$ of an open set Ω. More generally, it is satisfactory whenever Λ is included in the closure of its interior, since this guarantees that the derivatives of f are uniquely defined on Λ.

However, **this definition is not satisfactory without any hypotheses on** Λ, in which case the derivatives are not necessarily uniquely defined on Λ. Indeed, a function $f \in \mathcal{C}(\Lambda;E)$ might have several extensions $\overline{f} \in \mathcal{C}^m(\mathbb{R}^d;E)$ (which is not a problem) for which the restrictions to Λ of the derivatives $\partial^\beta \overline{f}$ do not coincide (which is a problem). For example, if $\Lambda = \{x \in \mathbb{R}^2 : x_2 = 0\}$ and $f \in \mathcal{C}^1(\Lambda)$, the derivative $\partial_2 f$ is not uniquely defined (any continuous function works). □

Let us show that a differentiable function on an interval with a point removed is differentiable on the entire interval if the function and its derivative can be continuously extended to this interval.

THEOREM 2.28.– *Let f and g be two functions in $\mathcal{C}((a,b);E)$, where (a,b) is an open interval of $\mathbb{R}$ and E is a separated semi-normed space. In addition, let $c \in (a,b)$ and $X = (a,c) \cup (c,b)$. If*

$$f|_X \in \mathcal{C}^1(X;E) \ \textit{and} \ (f|_X)' = g|_X,$$

then

$$f \in \mathcal{C}^1((a,b);E) \ \textit{and} \ f' = g. \blacksquare$$

Proof. We need to show that f is differentiable at the point c with $f'(c) = g(c)$.

Let $u \in (c, c+r)$ and $z > 0$, $z \leq r$, where $r = \inf\{b-c, c-a\}/2$, and let $\| \ \|_{E;\nu}$ be a semi-norm of E. Since $\nabla f = f' = g$ on (c, b), the finite increment theorem (Theorem 2.5 (b)) gives

$$\|f(u+z) - f(u) - zg(u)\|_{E;\nu} \leq |z| \sup_{u \leq x \leq u+z} \|g(x) - g(u)\|_{E;\nu}. \tag{2.12}$$

Since f and g are continuous, taking the limit $u \to c$ gives us this inequality for $u = c$ and $z > 0$.

Now, by choosing $u \in (c-r, c)$ and $z < 0$, $|z| \leq r$ and letting $u \to c$, we obtain (2.12) for $u = c$ and $z < 0$. Thus, for every z such that $|z| \leq r$,

$$\|f(c+z) - f(c) - zg(c)\|_{E;\nu} \leq |z| \sup_{|x-c| \leq |z|} \|g(x) - g(c)\|_{E;\nu}.$$

By the definition of continuity (Definition 1.2), for every $\epsilon > 0$, there exists $\eta > 0$ such that $|x - c| \leq \eta$ implies $\|g(x) - g(c)\|_{E;\nu} \leq \epsilon$, and so

$$\|f(c+z) - f(c) - zg(c)\|_{E;\nu} \leq \epsilon\, |z|.$$

By the characterization of differentiability (2.4) from Definition 2.1 (b), this shows that f is differentiable at the point c and $f'(c) = g(c)$. □

Now consider the set of functions taking values in a subset of a separated semi-normed space.

DEFINITION 2.29.– *Given $\Omega \subset \mathbb{R}^d$, $m \in [\![0, \infty]\!]$, E a semi-normed space and $U \subset E$, we denote:*

$$\mathcal{C}^m(\Omega; U) \stackrel{\text{def}}{=} \{f \in \mathcal{C}^m(\Omega; E) : f(x) \in U,\ \forall x \in \Omega\},$$

$$\mathbf{C}^m_{\mathrm{b}}(\Omega; U) \stackrel{\text{def}}{=} \{f \in \mathbf{C}^m_{\mathrm{b}}(\Omega; E) : f(x) \in U,\ \forall x \in \Omega\}.\ \blacksquare$$

The sets $\mathcal{C}^m(\Omega; U)$ and $\mathbf{C}^m_{\mathrm{b}}(\Omega; U)$ are not vector spaces if U is not a vector space. Accordingly, we will not endow them with a topology, which we reserve for vector spaces.

Chapter 3

Differentiating Composite Functions and Others

In this chapter, we study the derivative of the image Lf of a differentiable function f under a linear mapping L (§ 3.1), the image $T(f_1, \ldots, f_d)$ of several functions under a multilinear mapping T (§ 3.2 to 3.4), the image $S \circ f$ under a differentiable mapping S (§ 3.7) and the image $f \circ T$ under a change of variables T (Theorem 3.12). We also differentiate with respect to a separated variable (Theorem 3.14) and a translation vector (Theorem 3.18).

None of our results are surprising, except for the generalization of differentiability to mappings S from one semi-normed space into another (Definition 3.15), which is a new result, together with Theorem 3.17 on the derivative of $S \circ f$, which makes use of it.

3.1. Image under a linear mapping

If f is a function taking values in a vector space E and L is a linear mapping from E into a vector space, we denote $Lf \stackrel{\text{def}}{=} L \circ f$, namely

$$(Lf)(x) \stackrel{\text{def}}{=} L(f(x)). \tag{3.1}$$

Let us show that differentiation commutes with continuous linear mappings.

THEOREM 3.1.– *Let f be an m times differentiable function from an open subset Ω of $\mathbb{R}^d$ into a separated semi-normed space E, where $1 \leq m \leq \infty$, and let L be a continuous linear mapping from E into a separated semi-normed space F.*

Then Lf is m times differentiable and, for every $i \in [\![1, d]\!]$ and $\beta \in \mathbb{N}^{*d}$ such that $|\beta| \leq m$,

$$\partial_i(Lf) = L(\partial_i f) \text{ and } \partial^\beta(Lf) = L(\partial^\beta f). \blacksquare$$

Proof. *Case* $m = 1$. Let $\{\| \|_{E;\nu} : \nu \in \mathcal{N}_E\}$ and $\{\| \|_{F;\mu} : \mu \in \mathcal{N}_F\}$ be the families of semi-norms of E and F. By the characterization of continuous linear mappings from Theorem 1.25, for every $\mu \in \mathcal{N}_F$, there exists a finite subset N of $\mathcal{N}_E$ and $c \geq 0$ such that, for every $u \in E$,

$$\|L(u)\|_{F;\mu} \leq c \sup_{\nu \in N} \|u\|_{E;\nu}. \tag{3.2}$$

Let $x \in \Omega$ and $\epsilon > 0$. By Definition 2.1 of differentiability, for every $\nu \in \mathcal{N}_E$, there exists $\eta_\nu > 0$ such that if $|z| \leq \eta_\nu$ and $x + z \in \Omega$, then

$$\|f(x+z) - f(x) - z \cdot \nabla f(x)\|_{E;\nu} \leq \frac{\epsilon}{c}. \tag{3.3}$$

Since L is linear, using the notation from (3.1), we have

$$L\big(f(x+z) - f(x) - z \cdot \nabla f(x)\big) = (Lf)(x+z) - (Lf)(x) - \sum_{i=1}^{d} z_i (L\nabla_i f)(x).$$

Let $\mu \in \mathcal{N}_F$ and $\eta = \inf_{\nu \in N} \eta_\nu$. Then $\eta > 0$ and, if $|z| \leq \eta$ and $x + z \in \Omega$, the inequalities (3.2) and (3.3) give

$$\Big\|(Lf)(x+z) - (Lf)(x) - \sum_{i=1}^{d} z_i (L\nabla_i f)(x)\Big\|_{F;\mu} \leq \epsilon\, |z|.$$

This proves that f is differentiable at the point x and $\nabla_i(Lf) = L(\nabla_i f)$. In other words, $\partial_i(Lf) = L(\partial_i f)$, since $\nabla_i = \partial_i$ (Theorem 2.9).

Case $m \geq 2$. Let us show by induction that Lf is m times differentiable whenever f is m times differentiable. Thus, suppose that this holds for $m-1$ and that f is m times differentiable. The $\partial_i f$ are $m-1$ times differentiable by Theorem 2.11, so the induction hypothesis for $m-1$ implies that the same is true for the $L(\partial_i f)$, namely the $\partial_i(Lf)$, as we saw just above. Therefore, Lf is m times differentiable, again by Theorem 2.11. This proves the induction hypothesis for m. Since it holds for $m = 1$, it holds for every $m \in \mathbb{N}$ by induction, and hence also for $m = \infty$.

The equality $\partial^\beta(Lf) = L(\partial^\beta f)$ for $|\beta| \leq m$ can also be shown by induction on m, since ∂^β is of the form $\partial^\alpha \partial_i$, where $|\alpha| \leq m-1$, so the equalities for 1 and $m-1$ successively imply $\partial^\alpha \partial_i(Lf) = \partial^\alpha(L(\partial_i f)) = L(\partial^\alpha \partial_i f)$. □

Let us state some continuity properties of the image under a linear mapping. We denote by $\mathcal{L}(E; F)$ the space of continuous linear mappings from E into F.

THEOREM 3.2.– *Let $L \in \mathcal{L}(E; F)$, where E and F are separated semi-normed spaces, Ω is an open subset of $\mathbb{R}^d$, $0 \leq m \leq \infty$ and D is a compact subset of $\mathbb{R}^d$.*

Then the mapping $f \mapsto Lf$ is continuous and linear from $\mathcal{C}^m(\Omega; E)$ into $\mathcal{C}^m(\Omega; F)$, from $\mathcal{C}_b^m(\Omega; E)$ into $\mathcal{C}_b^m(\Omega; F)$, from $\mathcal{K}^m(\Omega; E)$ into $\mathcal{K}^m(\Omega; F)$ and from $\mathbf{C}_D(\Omega; E)$ into $\mathbf{C}_D(\Omega; F)$. ∎

Proof. Continuity of Lf. Let $\{\| \ \|_{E;\nu} : \nu \in \mathcal{N}_E\}$ and $\{\| \ \|_{F;\mu} : \mu \in \mathcal{N}_F\}$ be the families of semi-norms of E and F. By the characterization of continuous linear mappings from Theorem 1.25, for every $\mu \in \mathcal{N}_F$, there exists a finite subset N_μ of $\mathcal{N}_E$ and $c_\mu \in \mathbb{R}$ such that, for every $u \in E$,

$$\|Lu\|_{F;\mu} \leq c_\mu \sup_{\nu \in N_\mu} \|u\|_{E;\nu}.$$

If $f \in \mathcal{C}(\Omega; E)$, $x \in \Omega$ and $\epsilon > 0$, then, by Definition 1.2 (a) of continuity, for every $\nu \in \mathcal{N}_E$, there exists $\zeta_\nu > 0$ such that if $y \in \Omega$ and $|y - x| \leq \zeta_\nu$, then $\|f(y) - f(x)\|_{E;\nu} \leq \epsilon/c_\mu$. If $|y - x| \leq \inf_{\nu \in N_\mu} \zeta_\nu$, then

$$\|L(f(y)) - L(f(x))\|_{F;\mu} \leq c_\mu \sup_{\nu \in N_\mu} \|f(y) - f(x)\|_{E;\nu} \leq \epsilon, \tag{3.4}$$

which shows that $Lf \in \mathcal{C}(\Omega; F)$.

Continuity of L on $\mathcal{C}^m(\Omega; E)$. Let $f \in \mathcal{C}^m(\Omega; E)$. Its derivatives $\partial^\beta f$ are continuous for every β such that $|\beta| \leq m$, so their images $L(\partial^\beta f)$ are continuous, as we saw just above. Since $L(\partial^\beta f) = \partial^\beta(Lf)$ by Theorem 3.1, it follows that $Lf \in \mathcal{C}^m(\Omega; F)$. The function $f \mapsto Lf$ being linear, we simply need to check that it is continuous.

If $m < \infty$, then, by Definition 2.14 (a) of the semi-norms of $\mathcal{C}^m(\Omega; E)$, for every compact set K included in Ω and every $\mu \in \mathcal{N}_F$, since $\partial^\beta(Lf) = L(\partial^\beta f)$,

$$\begin{aligned} \|Lf\|_{\mathcal{C}^m(\Omega;F);K,\mu} &= \sup_{|\beta| \leq m} \sup_{x \in K} \|\partial^\beta(Lf)(x)\|_{F;\mu} = \\ &\leq c_\mu \max_{\nu \in N_\mu} \sup_{|\beta| \leq m} \sup_{x \in K} \|\partial^\beta f(x)\|_{E;\nu} = c_\mu \sup_{\nu \in N_\mu} \|f\|_{\mathcal{C}^m(\Omega;E);K,\nu}. \end{aligned} \tag{3.5}$$

Again by the characterization from Theorem 1.25, this shows that L is continuous from $\mathcal{C}^m(\Omega; E)$ into $\mathcal{C}^m(\Omega; F)$.

The continuity for $m = \infty$ follows from this, since, by Definition 2.15 (a) of the semi-norms of $\mathcal{C}^\infty(\Omega; E)$, for every $n \in \mathbb{N}^*$, $\|f\|_{\mathcal{C}^\infty(\Omega;F);n,K,\mu} = \|f\|_{\mathcal{C}^n(\Omega;F);K,\mu}$.

Continuity of L on $\mathcal{C}_b^m(\Omega; E)$. If $f \in \mathcal{C}_b^m(\Omega; E)$, then $Lf \in \mathcal{C}_b^m(\Omega; F)$, since its derivatives $\partial^\beta(Lf)$ are continuous for $|\beta| \leq m$, as we saw just above, and are also bounded because, for every $\mu \in \mathcal{N}_F$, since once again $\partial^\beta(Lf) = L(\partial^\beta f)$,

$$\sup_{x \in \Omega} \|\partial^\beta(Lf)(x)\|_{F;\mu} \leq c_\mu \sup_{\nu \in N_\mu} \sup_{x \in \Omega} \|\partial^\beta f(x)\|_{E;\nu} < \infty.$$

The continuity of L from $\mathcal{C}_{\mathrm{b}}^m(\Omega; E)$ into $\mathcal{C}_{\mathrm{b}}^m(\Omega; F)$ follows by considering the upper bound on Ω instead of on K in (3.5), by Definition 2.14 (b) of the semi-norms of $\mathcal{C}_{\mathrm{b}}^m(\Omega; E)$ or by Definition 2.15 (b) if $m = \infty$.

Continuity of L on $\mathcal{K}^m(\Omega; E)$. If $f \in \mathcal{K}^m(\Omega; E)$, then Lf belongs to $\mathcal{K}^m(\Omega; F)$ because its derivatives $\partial^\beta(Lf)$ are continuous for $|\beta| \leq m$ as above and its support is compact, since its support is included in the support of f.

The continuity of L from $\mathcal{K}^m(\Omega; E)$ into $\mathcal{K}^m(\Omega; F)$ is obtained by multiplying by $q(x)$ the semi-norms at the point x in (3.5), by Definition 2.14 (e) of the semi-norms of $\mathcal{K}^m(\Omega; E)$ or Definition 2.15 (e) if $m = \infty$.

Continuity of L on $\mathbf{C}_D(\Omega; E)$. If $f \in \mathbf{C}_D(\Omega; E)$, then Lf belongs to $\mathbf{C}_D(\Omega; F)$, since it is uniformly continuous (this follows from the first inequality of (3.4)) and its support is included in the support of f and hence in the support of D.

The continuity of L from $\mathbf{C}_D(\Omega; E)$ into $\mathbf{C}_D(\Omega; F)$ follows from its continuity from $\mathcal{C}_{\mathrm{b}}(\Omega; E)$ into $\mathcal{C}_{\mathrm{b}}(\Omega; F)$, since $\mathbf{C}_D(\Omega; E)$ is endowed with the semi-norms of $\mathcal{C}_{\mathrm{b}}(\Omega; E)$ by Definition 1.7 (b). □

Let us observe that the image under a semi-norm preserves various notions of continuity (however, it does not preserve differentiability in general).

THEOREM 3.3.– *Let $\Omega \subset \mathbb{R}^d$, E a separated semi-normed space and $\| \ \|_{E;\nu}$ one of its semi-norms.*

Then the mapping $f \mapsto \|f\|_{E;\nu}$ is uniformly continuous from $\mathcal{C}(\Omega; E)$ into $\mathcal{C}(\Omega)$ and from $\mathbf{C}_{\mathrm{b}}(\Omega; E)$ into $\mathbf{C}_{\mathrm{b}}(\Omega)$. ▌

Proof. Let $f \in \mathcal{C}(\Omega; E)$. The function $\|f\|_{E;\nu}$ is continuous because, every semi-norm being a contraction (Theorem A.5),

$$\big| \|f(y)\|_{E;\nu} - \|f(x)\|_{E;\nu} \big| \leq \|f(y) - f(x)\|_{E;\nu}.$$

If in addition $g \in \mathcal{C}(\Omega; E)$, then $\big| \|f(x)\|_{E;\nu} - \|g(x)\|_{E;\nu} \big| \leq \|f(x) - g(x)\|_{E;\nu}$ for the same reason, so, by Definition 1.3 of the semi-norms of $\mathcal{C}(\Omega; E)$, for every compact set $K \subset \Omega$,

$$\begin{aligned} \big\| \|f\|_{E;\nu} - \|g\|_{E;\nu} \big\|_{\mathcal{C}(\Omega);K} &= \sup_{x \in K} \big| \|f(x)\|_{E;\nu} - \|g(x)\|_{E;\nu} \big| \\ &\leq \sup_{x \in K} \|f(x) - g(x)\|_{E;\nu} = \|f - g\|_{\mathcal{C}(\Omega;E);K,\nu}. \end{aligned}$$

Hence, the mapping $f \mapsto \|f\|_{E;\nu}$ is uniformly continuous from $\mathcal{C}(\Omega; E)$ into $\mathcal{C}(\Omega)$.

The continuity from $\mathbf{C}_{\mathrm{b}}(\Omega; E)$ into $\mathbf{C}_{\mathrm{b}}(\Omega)$ follows similarly from Definition 1.4 of their semi-norms. □

Extension of Theorem 3.3. This result also holds if we replace the space $\mathcal{C}(\Omega; E)$ by the spaces $\mathcal{C}_b(\Omega; E)$, $\mathcal{C}_K(\Omega; E)$ and $\mathcal{K}(\Omega; E)$, but we will not need this later. □

3.2. Image under a multilinear mapping: Leibniz rule

We denote by $\mathcal{L}^\ell(E_1 \times \cdots \times E_\ell; F)$ the vector space of continuous multilinear mappings from a Cartesian product $E_1 \times \cdots \times E_\ell$ of separated semi-normed spaces into a separated semi-normed space F.

The product $E_1 \times \cdots \times E_\ell = \{(u_1, \ldots, u_\ell) : u_i \in E_i, \forall i\}$ is endowed with the semi-norms, indexed by $\nu_1 \in \mathcal{N}_{E_1}, \ldots, \nu_\ell \in \mathcal{N}_{E_\ell}$,

$$\|(u_1, \ldots, u_\ell)\|_{E_1 \times \cdots \times E_\ell; \nu_1, \ldots, \nu_\ell} \stackrel{\text{def}}{=} \|u_1\|_{E_1;\nu_1} + \cdots + \|u_\ell\|_{E_\ell;\nu_\ell}.$$

These semi-norms make it a separated semi-normed space, which allows us to define the notion of continuity for mappings on it.

Let us show that the image of a differentiable function under a continuous multilinear mapping is continuous and differentiable, and let us state the generalized **Leibniz rule**.

THEOREM 3.4.– *Let f_1, f_2, ..., f_ℓ be differentiable functions from an open subset Ω of $\mathbb{R}^d$ into the separated semi-normed spaces $E_1, \ldots, E_\ell$, respectively, and let T be a continuous multilinear mapping from $E_1 \times \cdots \times E_\ell$ into a separated semi-normed space F.*

Then $T(f_1, \ldots, f_\ell)$ is differentiable from Ω into F and, for every $i \in [\![1, d]\!]$,

$$\partial_i (T(f_1, \ldots, f_\ell)) = \sum_{j=1}^{\ell} T(f_1, \ldots, \partial_i f_j, \ldots, f_\ell). \blacksquare$$

Proof. Let $\{\| \ \|_{E_j;\nu_j} : \nu_j \in \mathcal{N}_{E_j}\}$ and $\{\| \ \|_{F;\mu} : \mu \in \mathcal{N}_F\}$ be the families of semi-norms of E_j and F. Suppose that $x \in \Omega$ and $z \in \mathbb{R}^d$ satisfy $x + z \in \Omega$. Decompose

$$T\big(f_1(x+z), \ldots, f_\ell(x+z)\big) - T\big(f_1(x), \ldots, f_\ell(x)\big) - \\ - \sum_{i=1}^{d} z_i \sum_{j=1}^{\ell} T\big(f_1(x), \ldots, \partial_i f_j(x), \ldots, f_\ell(x)\big) = \sum_{i=1}^{d} \sum_{j=1}^{\ell} A_{ij} + z_i B_{ij},$$

where

$$A_{ij} \stackrel{\text{def}}{=} T\big(f_1(x+z), \ldots, f_{j-1}(x+z), \\ , f_j(x+z) - f_j(x) - z_i \partial_i f_j(x), f_{j+1}(x), \ldots, f_\ell(x)\big)$$

and

$$B_{ij} \stackrel{\text{def}}{=} T(f_1(x+z),\ldots,f_{j-1}(x+z),\partial_i f_j(x),f_{j+1}(x),\ldots,f_\ell(x)) - \\ - T((f_1(x),\ldots,f_{j-1}(x),\partial_i f_j(x),f_{j+1}(x),\ldots,f_\ell(x)).$$

Let $\mu \in \mathcal{N}_F$. By the characterization of continuous linear mappings from Theorem A.40, there exists finite subsets N_1 of $\mathcal{N}_{E_1}$, ..., N_ℓ of $\mathcal{N}_{E_\ell}$ and $c \geq 0$ such that, for every $(u_1,\ldots,u_\ell) \in E_1 \times \cdots \times E_\ell$,

$$\|T(u_1,\ldots,u_\ell)\|_{F;\mu} \leq c \sup_{\nu_1 \in N_1} \|u_1\|_{E_1;\nu_1} \cdots \sup_{\nu_\ell \in N_\ell} \|u_\ell\|_{E_\ell;\nu_\ell}. \tag{3.6}$$

By Theorem 2.3, every f_j is locally bounded. In other words, for every $\nu_j \in \mathcal{N}_{E_j}$, there exists $\eta_{\nu_j} > 0$ and $\gamma_{\nu_j} > 0$ such that if $|z| \leq \eta_{\nu_j}$, then

$$\|f_j(x+z)\|_{E_j;\nu_j} \leq \gamma_{\nu_j}.$$

Let $\epsilon > 0$ and $\gamma'_j = \prod_{1\leq k\leq \ell,\, k\neq j} \sup_{\nu_k \in N_k} \gamma_{\nu_k}$. By Definition 2.1 of differentiability, for every $\nu_j \in \mathcal{N}_{E_j}$, there exists $\eta'_{\nu_j} > 0$ such that, if $|z| \leq \eta'_{\nu_j}$,

$$\|f_j(x+z) - f_j(x) - z \cdot \nabla f_j(x)\|_{E_j;\nu_j} \leq \frac{\epsilon}{2cd\ell\gamma'_{\nu_j}}|z|. \tag{3.7}$$

Let $\eta = \inf_{1\leq j\leq \ell,\, \nu_j \in N_j} \inf\{\eta_{n_{\nu_j}}, \eta'_{n_{\nu_j}}\}$. If $|z| \leq \eta$, then the inequalities (3.6) and (3.7) imply, since $z \cdot \nabla f_j(x) = \sum_{i=1}^d z_i \partial_i f_j(x)$ by Theorem 2.9, that, for every j,

$$\Big\|\sum_{i=1}^d A_{ij}\Big\|_{F;\mu} \leq \frac{\epsilon}{2\ell}|z|.$$

Moreover, since the f_j and T are continuous, there exists $\eta'' > 0$ such that, if $|z| \leq \eta''$, then, for every i and j,

$$\|B_{ij}\|_{F;\mu} \leq \frac{\epsilon}{2d\ell}.$$

Thus, if $|z| \leq \inf\{\eta, \eta''\}$,

$$\Big\|\sum_{i=1}^d \sum_{j=1}^\ell A_{ij} + z_i B_{ij}\Big\|_{F;\mu} \leq \epsilon\,|z|.$$

In light of the decomposition above, this proves that $T(f_1,\ldots,f_\ell)$ is differentiable at the point x and $\partial_j(T(f_1,\ldots,f_\ell))(x) = \sum_{i=1}^\ell T(f_1,\ldots,\partial_j f_i,\ldots,f_\ell)(x)$. □

Let us state a few differentiability properties for arbitrary orders.

THEOREM 3.5.– *Let $T \in \mathcal{L}^\ell(E_1 \times \cdots \times E_\ell; F)$, where E_1, ..., E_ℓ and F are separated semi-normed spaces, Ω is an open subset of $\mathbb{R}^d$ and $0 \leq m \leq \infty$.*

(a) *If* $f_1 \in \mathcal{C}^m(\Omega; E_1), \ldots, f_\ell \in \mathcal{C}^m(\Omega; E_\ell)$, *then* $T(f_1, \ldots, f_\ell) \in \mathcal{C}^m(\Omega; F)$.

(b) *If* $f_1 \in \mathbf{C}^m_\mathrm{b}(\Omega; E_1), \ldots, f_\ell \in \mathbf{C}^m_\mathrm{b}(\Omega; E_\ell)$, *then* $T(f_1, \ldots, f_\ell) \in \mathbf{C}^m_\mathrm{b}(\Omega; F)$. ▮

Proof. (a) If all of the f_j are continuous, then so are $(f_1, \ldots, f_\ell)$ and hence so is the composite function $T(f_1, \ldots, f_\ell)$ by Theorem A.35. This proves the stated property for $m = 0$.

For $m \geq 1$, we will argue by induction. Suppose that the property holds for $m - 1$ and that $f_j \in \mathcal{C}^m(\Omega; E_j)$ for each j. By Theorem 3.4, the function $T(f_1, \ldots, f_\ell)$ is differentiable and

$$\partial_i(T(f_1, \ldots, f_\ell)) = \sum_{j=1}^{\ell} T(f_1, \ldots, \partial_i f_j, \ldots, f_\ell).$$

Since $\partial_i f_j \in \mathcal{C}^{m-1}(\Omega; E_j)$, the induction hypothesis for $m - 1$ shows that the right-hand side, and therefore the left-hand side, belongs to $\mathcal{C}^{m-1}(\Omega; F)$. Theorem 2.11 then gives $T(f_1, \ldots, f_\ell) \in \mathcal{C}^m(\Omega; F)$, which shows the induction hypothesis for m. Since it holds for $m = 0$, it holds for all $m \in \mathbb{N}^*$ by induction, and therefore also for $m = \infty$.

(b) If the functions f_j are all uniformly continuous and bounded, then the same is true for $(f_1, \ldots, f_\ell)$. Moreover, T is uniformly continuous and bounded on the set $U = (f_1, \ldots, f_\ell)(\Omega)$, like on any other bounded set (Theorem A.41). The composite function $T(f_1, \ldots, f_\ell)$ is therefore uniformly continuous (Theorem A.35). It is bounded because $(T(f_1, \ldots, f_\ell))(\Omega) = T(U)$ is bounded. This proves the stated property for $m = 0$.

The general statement can be deduced by induction on m in the same way as (a). □

Image of a non-bounded uniformly continuous function. The analogous version of Theorem 3.5 (b) does not hold in the space $\mathbf{C}(\Omega; E)$: the image of a uniformly continuous function under a continuous multilinear mapping is **not uniformly continuous** in general.

For example, if the f_j are linear functions and T is a multilinear mapping, then $T(f_1, \ldots, f_\ell)$ is a multilinear mapping and is therefore not uniformly continuous for $\ell \geq 2$ whenever it is non-zero [Vol. 1, Property (7.4), p. 132]. □

Let us consider a special case of the product by a real function that we will frequently use later, as well as the **Leibniz rule**[1].

1 **History of the Leibniz rule.** Gottfried VON LEIBNIZ gave the expression of the derivative of a product in the form $\mathrm{d}\,xv = x\,\mathrm{d}v + v\,\mathrm{d}x$ in 1675 [54, p. 467].

THEOREM 3.6.– *Let $f_1 \in \mathcal{C}^1(\Omega; E)$ and $f_2 \in \mathcal{C}^1(\Omega)$, where Ω is an open subset of $\mathbb{R}^d$ and E is a separated semi-normed space.*

Then $f_1 f_2 \in \mathcal{C}^1(\Omega; E)$ and, for every $i \in [\![1, d]\!]$,

$$\partial_i(f_1 f_2) = f_1\, \partial_i f_2 + f_2\, \partial_i f_1. \blacksquare$$

Proof. Theorem 3.4 for $T(u_1, u_2) = u_1 u_2$ implies differentiability and the expression of the partial derivatives, since the product is bilinear and continuous from $E \times \mathbb{R}$ into E (Theorem A.30). Theorem 3.5 (a) implies that the product belongs to $\mathcal{C}^1(\Omega; E)$. □

Let us state the Leibniz rule to calculate derivatives of arbitrary order for a bilinear mapping.

THEOREM 3.7.– *Let $f_1 \in \mathcal{C}^m(\Omega; E_1)$ and $f_2 \in \mathcal{C}^m(\Omega; E_2)$, where Ω is an open subset of $\mathbb{R}^d$, E_1 and E_2 are separated semi-normed spaces and $m \in \mathbb{N}$. Additionally, let $T \in \mathcal{L}^2(E_1 \times E_2; F)$, where F is a separated semi-normed space.*

Then $T(f_1, f_2) \in \mathcal{C}^m(\Omega; F)$ and, for every $\beta \in \mathbb{N}^{*d}$ such that $|\beta| \leq m$,

$$\partial^\beta(T(f_1, f_2)) = \sum_{0 \leq \sigma \leq \beta} C_\beta^\sigma T(\partial^{\beta-\sigma} f_1, \partial^\sigma f_2),$$

where $C_\beta^\sigma = \beta!/(\sigma!\,(\beta-\sigma)!)$, $\beta! = \beta_1! \ldots \beta_d!$, $n! = 1 \times 2 \times \cdots \times n$ and $\sigma \leq \beta$ means that $\sigma_i \leq \beta_i$ for each i. $\blacksquare$

Proof. The fact that $T(f_1, f_2)$ belongs to $\mathcal{C}^m(\Omega; F)$ follows from Theorem 3.5 (a). Let us calculate its derivatives.

Case of a single variable. Denote by $f^{(m)}$ the mth derivative of f. We need to show that

$$(T(f_1, f_2))^{(m)} = \sum_{0 \leq i \leq m} C_m^i T\big(f_1^{(m-i)}, f_2^{(i)}\big). \tag{3.8}$$

For $m = 1$, this is the formula from Theorem 3.4, which can be stated as

$$(T(f_1, f_2))^{(1)} = T\big(f_1^{(1)}, f_2\big) + T\big(f_1, f_2^{(1)}\big),$$

since $C_1^0 = C_1^1 = 1$. The equality (3.8) follows by induction on m because, if it holds for $m-1$, it also holds for m. Indeed, writing $j = i+1$,

$$\begin{aligned}(T(f_1, f_2))^{(m)} &= ((T(f_1, f_2))^{(m-1)})^{(1)} \\ &= \sum_{0 \le i \le m-1} C_{m-1}^i (T(f_1^{(m-1-i)}, f_2^{(i)}))^{(1)} \\ &= \sum_{0 \le i \le m-1} C_{m-1}^i (T(f_1^{(m-i)}, f_2^{(i)}) + T(f_1^{(m-1-i)}, f_2^{(i+1)})) \\ &= \sum_{0 \le i \le m-1} C_{m-1}^i T(f_1^{(m-i)}, f_2^{(i)}) + \sum_{1 \le j \le m} C_{m-1}^{j-1} T(f_1^{(m-j)}, f_2^{(j)}),\end{aligned}$$

which gives (3.8) because $C_{m-1}^i + C_{m-1}^{i-1} = C_m^i$ for $1 \le i \le m$ and $C_m^0 = C_m^m = 1$.

Case of d variables. The stated formula is true for $d = 1$ by (3.8). The general formula follows by induction on d because, if it holds for $d-1$, it also holds for d. Indeed, writing $\beta_* = (\beta_2, \ldots, \beta_d)$ and $\partial_*^{\beta_*} = \partial_2^{\beta_2} \cdots \partial_d^{\beta_d}$,

$$\begin{aligned}\partial^\beta (T(f_1, f_2)) &= \partial_1^{\beta_1} (\partial_*^{\beta_*} T(f_1, f_2))) \\ &= \sum_{0 \le \sigma_* \le \beta_*} C_{\beta_*}^{\sigma_*} \partial_1^{\beta_1} (T(\partial_*^{\beta_* - \sigma_*} f_1, \partial_*^{\sigma_*} f_2)) \\ &= \sum_{0 \le \sigma_1 \le \beta_1} \sum_{0 \le \sigma_* \le \beta_*} C_{\beta_1}^{\sigma_1} C_{\beta_*}^{\sigma_*} T(\partial_1^{\beta_1 - \sigma_1} \partial_*^{\beta_* - \sigma_*} f_1, \partial_1^{\sigma_1} \partial_*^{\sigma_*} f_2) \\ &= \sum_{0 \le \sigma \le \beta} C_\beta^\sigma T(\partial^{\beta - \sigma} f_1, \partial^\sigma f_2). \ \square\end{aligned}$$

3.3. Dual formula of the Leibniz rule

Let us state a dual formula of the Leibniz rule.

THEOREM 3.8.– *Let $f_1 \in \mathcal{C}^m(\Omega)$ and $f_2 \in \mathcal{C}^m(\Omega)$, where Ω is an open subset of $\mathbb{R}^d$ and $m \in \mathbb{N}$. Then, for every $\beta \in \mathbb{N}^{*d}$ such that $|\beta| \le m$,*

$$f_1 \partial^\beta f_2 = \sum_{\sigma \in \mathbb{N}^{*d} :\, 0 \le \sigma \le \beta} (-1)^{|\beta - \sigma|} C_\beta^\sigma \partial^\sigma (\partial^{\beta - \sigma} f_1 \, f_2),$$

where $C_\beta^\sigma = \beta!/(\sigma!(\beta-\sigma)!)$, $\beta! = \beta_1! \ldots \beta_d!$, $n! = 1 \times 2 \times \cdots \times n$ and $\sigma \le \beta$ means that $\sigma_i \le \beta_i$ for every i. ▌

Proof. Observe that $\partial^{\beta-\sigma} f_1 f_2 \in \mathcal{C}^{|\sigma|}(\Omega)$ (Theorem 3.5 (a)).

Case of a single variable. Denote by $f^{(m)}$ the mth derivative of f. We need to show that

$$f_1 f_2^{(m)} = \sum_{0 \le i \le m} (-1)^{m-i} C_m^i \big(f_1^{(m-i)} f_2\big)^{(i)}. \tag{3.9}$$

For $m = 1$, since $C_1^0 = C_1^1 = 1$, this equality can be written as

$$f_1 f_2^{(1)} = -f_1^{(1)} f_2 + (f_1 f_2)^{(1)},$$

which follows from the Leibniz rule in Theorem 3.6.

The equality (3.9) follows by induction on m because if it holds for $m-1$, it also holds for m. Indeed (writing $j = i+1$),

$$\begin{aligned}
f_1 f_2^{(m)} &= f_1 \big(f_2^{(1)}\big)^{(m-1)} = \\
&= \sum_{0 \le i \le m-1} (-1)^{m-1-i} C_{m-1}^i \big(f_1^{(m-1-i)} f_2^{(1)}\big)^{(i)} = \\
&= \sum_{0 \le i \le m-1} (-1)^{m-1-i} C_{m-1}^i \Big(\big(f_1^{(m-1-i)} f_2\big)^{(i+1)} - \big(f_1^{(m-i)} f_2\big)^{(i)} \Big) = \\
&= \sum_{1 \le j \le m} (-1)^{m-j} C_{m-1}^{j-1} \big(f_1^{(m-j)} f_2\big)^{(j)} + \sum_{0 \le i \le m-1} (-1)^{m-i} C_{m-1}^i \big(f_1^{(m-i)} f_2\big)^{(i)},
\end{aligned}$$

which gives (3.9) because $C_{m-1}^i + C_{m-1}^{i-1} = C_{m-1}^i$ for $1 \le i \le m$ and $C_m^0 = C_m^m = 1$ for all m.

Case of d variables. The stated formula is true for $d = 1$ by (3.9). The general formula follows by induction on d because if it holds for $d-1$, it also holds for d. Indeed, writing $\beta_* = (\beta_2, \ldots, \beta_d)$ and $\partial_*^{\beta_*} = \partial_2^{\beta_2} \cdots \partial_d^{\beta_d}$,

$$\begin{aligned}
f_1 \partial^\beta f_2 &= f_1 \partial_1^{\beta_1} \partial_*^{\beta_*} f_2 = \\
&= \sum_{0 \le \sigma_1 \le \beta_1} (-1)^{\beta_1 - \sigma_1} C_{\beta_1}^{\sigma_1} \partial_1^{\sigma_1} \big(\partial_1^{\beta_1 - \sigma_1} f_1 \, \partial_*^{\beta_*} f_2\big) = \\
&= \sum_{0 \le \sigma_1 \le \beta_1} \sum_{0 \le \sigma_* \le \beta_*} (-1)^{\beta_1 - \sigma_1 + |\beta_* - \sigma_*|} C_{\beta_1}^{\sigma_1} C_{\beta_*}^{\sigma_*} \partial_*^{\sigma_*} \partial_1^{\sigma_1} \big(\partial_*^{\beta_* - \sigma_*} \partial_1^{\beta_1 - \sigma_1} f_1 \, f_2\big) = \\
&= \sum_{0 \le \sigma \le \beta} (-1)^{|\beta - \sigma|} C_\beta^\sigma \partial^\sigma \big(\partial^{\beta - \sigma} f_1 \, f_2\big),
\end{aligned}$$

because $\partial_*^{\sigma_*} \partial_1^{\sigma_1} = \partial_1^{\sigma_1} \partial_*^{\sigma_*} = \partial^\beta$ by Schwarz's theorem (Theorem 2.12). $\square$

3.4. Continuity of the image under a multilinear mapping

Let us begin by stating a few inequalities satisfied by the image under a multilinear mapping.

THEOREM 3.9.– *Let $T \in \mathcal{L}^\ell(E_1 \times \cdots \times E_\ell; F)$, where $E_1, \ldots, E_\ell$ and F are separated semi-normed spaces, and let $\{\| \, \|_{E_j;\nu_j} : \nu_j \in \mathcal{N}_{E_j}\}$ and $\{\| \, \|_{F;\mu} : \mu \in \mathcal{N}_F\}$ be the families of semi-norms of E_j and F.*

Let $\mu \in \mathcal{N}_F$, $c \geq 0$ and, for every $j \in [\![1, \ell]\!]$, N_j a finite subset of $\mathcal{N}_{E_j}$ such that: for every $u_1 \in E_1, \ldots, u_\ell \in E_\ell$,

$$\|T(u_1, \ldots, u_\ell)\|_{F;\mu} \leq c \prod_{j=1}^{\ell} \sup_{\nu_j \in N_j} \|u_j\|_{E_j;\nu_j}. \tag{3.10}$$

Finally, let $f_1 \in \mathcal{C}^m(\Omega; E_1), \ldots, f_\ell \in \mathcal{C}^m(\Omega; E_\ell)$, where Ω is an open subset of $\mathbb{R}^d$ and $m < \infty$. Then $T(f_1, \ldots, f_\ell) \in \mathcal{C}^m(\Omega; F)$ and:

(a) *For every $\beta \in \mathbb{N}^{*d}$ such that $|\beta| \leq m$ and $x \in \Omega$,*

$$\|\partial^\beta(T(f_1, \ldots, f_\ell))(x)\|_{F;\mu} \leq c\ell^{|\beta|} \prod_{j=1}^{\ell} \sup_{\nu_j \in N_j} \sup_{|\alpha| \leq |\beta|} \|\partial^\alpha f_j(x)\|_{E_j;\nu_j}.$$

(b) *For every compact set $K \subset \Omega$,*

$$\|T(f_1, \ldots, f_\ell)\|_{\mathcal{C}^m(\Omega;F);K,\mu} \leq c\ell^m \prod_{j=1}^{\ell} \sup_{\nu_j \in N_j} \|f_j\|_{\mathcal{C}^m(\Omega;E_j);K,\nu_j}.$$

(c) *If $f_j \in \mathcal{C}^m_{\mathrm{b}}(\Omega; E_j)$ for every $j \in [\![1, d]\!]$, then $T(f_1, \ldots, f_\ell) \in \mathcal{C}^m_{\mathrm{b}}(\Omega; F)$ and*

$$\|T(f_1, \ldots, f_\ell)\|_{\mathcal{C}^m_{\mathrm{b}}(\Omega;F);\mu} \leq c\ell^m \prod_{j=1}^{\ell} \sup_{\nu_j \in N_j} \|f_j\|_{\mathcal{C}^m_{\mathrm{b}}(\Omega;E_j);\nu_j}.$$

(d) *If $f_1 \in \mathcal{K}^m(\Omega; E_1)$, then $T(f_1, \ldots, f_\ell) \in \mathcal{K}^m(\Omega; F)$ and, for every $q \in \mathcal{C}^+(\Omega)$ and every compact set $K \subset \Omega$ outside of which $\|\partial^\alpha f_1\|_{E_1;\nu_1} = 0$ for every $|\alpha| \leq m$ and $\nu_1 \in N_1$,*

$$\|T(f_1, \ldots, f_\ell)\|_{\mathcal{K}^m(\Omega;F);q,\mu} \leq$$

$$\leq c\ell^m \sup_{\nu_1 \in N_1} \|f_1\|_{\mathcal{K}^m(\Omega;E_1);q,\nu_1} \prod_{j=2}^{\ell} \sup_{\nu_j \in N_j} \|f_j\|_{\mathcal{C}^m(\Omega;E_j);K,\nu_j}. \blacksquare$$

Simplification of (d). The statement (d) could be simplified by restricting to compact sets K that contain the support of f_1. We do not restrict ourselves to this simple case because some K smaller than the support will be useful for the proof of Theorem 3.11 (d) (see (3.12)). □

In particular, the product of real functions satisfies the following inequality.

THEOREM 3.10.– *Let $f_1 \in C_b^m(\Omega)$ and $f_2 \in C_b^m(\Omega)$, where Ω is an open subset of $\mathbb{R}^d$ and $m < \infty$. Then $f_1 f_2 \in C_b^m(\Omega)$ and*

$$\|f_1 f_2\|_{C_b^m(\Omega)} \leq 2^m \|f_1\|_{C_b^m(\Omega)} \|f_2\|_{C_b^m(\Omega)}. \ \blacksquare$$

Proof. This is a special case of Theorem 3.9 (c) (with $\ell = 2$ and $c = 1$). □

Proof of Theorem 3.9. We proved that $T(f_1, \ldots, f_\ell) \in \mathcal{C}^m(\Omega; F)$ in Theorem 3.5.

(a) The stated inequality is immediate for $m = 0$ because $\partial^0 g = g$.

For $m \geq 1$, we will argue by induction on m by assuming that the inequality holds for $m - 1$. Every derivative of order $|\beta| = m$ is of the form $\partial^\beta = \partial^\alpha \partial_i$, with $|\alpha| = |\beta| - 1$ and $1 \leq i \leq d$. The expression of $\partial_i(T(f_1, \ldots, f_\ell))$ from Theorem 3.4 gives

$$\partial^\alpha \partial_i (T(f_1, \ldots, f_\ell)) = \sum_{k=1}^{\ell} \partial^\alpha (T(f_1, \ldots, \partial_i f_k, \ldots, f_\ell)). \tag{3.11}$$

The inequality for $m - 1$ implies

$$\|\partial^\alpha (T(f_1, \ldots, \partial_i f_k, \ldots, f_\ell))(x)\|_{F;\mu} \leq c\ell^{|\alpha|} \prod_{j=1}^{\ell} \sup_{\nu_j \in N_j} \sup_{|\sigma| \leq |\alpha|+1} \|\partial^\sigma f_j(x)\|_{E_j;\nu_j}.$$

This implies the inequality for m, since, by (3.11), $\|\partial^\alpha \partial_i (T(f_1, \ldots, f_\ell))(x)\|_{F;\mu}$ is less than ℓ times the right-hand side.

(b) For $m < \infty$ and K a compact set included in Ω, Definition 2.14 (a) of the semi-norms of $\mathcal{C}^m(\Omega; E)$ and inequality (a) together imply

$$\|T(f_1, \ldots, f_\ell)\|_{\mathcal{C}^m(\Omega;F);K,\mu} = \sup_{|\beta| \leq m} \sup_{x \in K} \|\partial^\beta (T(f_1, \ldots, f_\ell))(x)\|_{F;\mu} \leq$$

$$\leq c\ell^m \prod_{j=1}^{\ell} \sup_{\nu_j \in N_j} \sup_{|\alpha| \leq m} \sup_{x \in K} \|\partial^\alpha f_j(x)\|_{E_j;\nu_j} = c\ell^m \prod_{j=1}^{\ell} \sup_{\nu_j \in N_j} \|f_j\|_{\mathcal{C}^m(\Omega;E_j);K,\nu_j}.$$

(c) By Definition 2.14 (b) of the semi-norms of $\mathcal{C}_b^m(\Omega; E)$, this case can be shown by repeating the proof of (b) with an upper bound on Ω instead of on K.

(d) Suppose that $f_1 \in \mathcal{K}^m(\Omega; E_1)$. Since the support of $T(f_1, \dots, f_\ell)$ is included in the support of f_1, which is compact, it is also compact. Therefore,

$$T(f_1, \dots, f_\ell) \in \mathcal{K}^m(\Omega; F).$$

For $q \in \mathcal{C}^+(\Omega)$, if the $\|\partial^\alpha f_1\|_{E_1;\nu_1}$ are zero outside of a compact set K that is included in Ω, then Definitions 2.14 (e) and (a) of the semi-norms of $\mathcal{K}^m(\Omega; E)$ and $\mathcal{C}^m(\Omega; E)$ and the inequality (a) together imply

$$\|T(f_1, \dots, f_\ell)\|_{\mathcal{K}^m(\Omega;F);q,\mu} = \sup_{|\beta|\le m} \sup_{x\in\Omega} q(x) \, \|\partial^\beta(T(f_1, \dots, f_\ell))(x)\|_{F;\mu} \le$$

$$\le c\ell^m \sup_{\nu_1\in N_1} \sup_{|\alpha|\le m} \sup_{x\in\Omega} q(x) \|\partial^\alpha f_1(x)\|_{E_1;\nu_1} \prod_{j=2}^{\ell} \sup_{\nu_j\in N_j} \sup_{|\alpha|\le m} \sup_{x\in K} \|\partial^\alpha f_j(x)\|_{E_j;\nu_j}$$

$$= c\ell^m \sup_{\nu_1\in N_1} \|f_1\|_{\mathcal{K}^m(\Omega;E_1);q,\nu_1} \prod_{j=2}^{\ell} \sup_{\nu_j\in N_j} \|f_j\|_{\mathcal{C}^m(\Omega;E_j);K,\nu_j}. \quad \square$$

Next, let us consider the continuity properties of the image under a multilinear mapping.

THEOREM 3.11.– *Let $T \in \mathcal{L}^\ell(E_1 \times \cdots \times E_\ell; F)$, where $E_1, \dots, E_\ell$, and F are separated semi-normed spaces, Ω is an open subset of $\mathbb{R}^d$ and $0 \le m \le \infty$. Then:*

(a) *The mapping $(f_1, \dots, f_\ell) \mapsto T(f_1, \dots, f_\ell)$ is continuous from the product space $\mathcal{C}^m(\Omega; E_1) \times \cdots \times \mathcal{C}^m(\Omega; E_\ell)$ into $\mathcal{C}^m(\Omega; F)$.*

(b) *The mapping $(f_1, \dots, f_\ell) \mapsto T(f_1, \dots, f_\ell)$ is continuous from the product space $\mathcal{C}^m_{\mathrm{b}}(\Omega; E_1) \times \cdots \times \mathcal{C}^m_{\mathrm{b}}(\Omega; E_\ell)$ into $\mathcal{C}^m_{\mathrm{b}}(\Omega; F)$.*

(c) *The mapping $(f_1, \dots, f_\ell) \mapsto T(f_1, \dots, f_\ell)$ is sequentially continuous from the product space $\mathcal{K}^m(\Omega; E_1) \times \mathcal{C}^m(\Omega; E_2) \times \cdots \times \mathcal{C}^m(\Omega; E_\ell)$ into $\mathcal{K}^m(\Omega; F)$.*

(d) *Given $f_1 \in \mathcal{K}^m(\Omega; E_1)$, $f_2 \in \mathcal{C}^m(\Omega; E_2)$, …, $f_d \in \mathcal{C}^m(\Omega; E_d)$, for $j \ge 2$, the mapping $f_j \mapsto T(f_1, \dots, f_\ell)$ is continuous from $\mathcal{C}^m(\Omega; E_j)$ into $\mathcal{K}^m(\Omega; F)$.* ▮

Proof. (a) *Continuity for the spaces $\mathcal{C}^m$.* By the characterization of multilinear continuous applications (Theorem A.40), for every $\mu \in \mathcal{N}_F$, there exist finite subsets N_1 of $\mathcal{N}_{E_1}$, …, N_ℓ of $\mathcal{N}_{E_\ell}$ and $c \in \mathbb{R}$ such that the inequality (3.10) from Theorem 3.9 is satisfied. The inequality (b) of this theorem then implies the desired continuity for $m < \infty$.

The continuity for $m = \infty$ then follows because, by Definitions 2.15 (a) and 2.14 (a) of the semi-norms of $\mathcal{C}^\infty$ and $\mathcal{C}^m$, $\| \; \|_{\mathcal{C}^\infty(\Omega;F);m,K,\mu} = \| \; \|_{\mathcal{C}^m(\Omega;F);K,\mu}$.

(b) *Continuity for the spaces $\mathcal{C}_b^m$.* This similarly follows from the inequality (c) of Theorem 3.9.

(c) *Sequential continuity for the spaces $\mathcal{K}^m$.* Let $f_{1,n} \to f_1$ in $\mathcal{K}^m(\Omega; E_1)$ and $f_{j,n} \to f_j$ in $\mathcal{C}^m(\Omega; E_j)$ for $2 \le j \le \ell$. Furthermore, let $\mu \in \mathcal{N}_F$, and let $N_j \subset \mathcal{N}_{E_j}$ and $c \in \mathbb{R}$ (again given by Theorem A.40) satisfy the hypotheses of Theorem 3.9. Then, by Definition 2.14 (e) of the semi-norms of $\mathcal{K}^m(\Omega; E)$,

$$\sup_{n\in\mathbb{N},\, \nu_1\in N_1} \ \sup_{|\alpha|\le m,\, x\in\Omega} q(x)\, \|\partial^\alpha f_{1,n}(x)\|_{E_1;\nu_1} = \\ = \sup_{n\in\mathbb{N},\, \nu_1\in N_1} \|f_{1,n}\|_{\mathcal{K}^m(\Omega;E_1);q,\nu_1} < \infty.$$

Therefore, by Theorem 1.23, there exists a compact set $K_\mu \subset \Omega$ such that:

$$\text{the } \|\partial^\alpha f_{1,n}\|_{E_1;\nu_1} \text{ are all zero outside of } K_\mu. \tag{3.12}$$

We can further choose K_μ so that f_1 is also zero on its complement. Now, decompose

$$T(f_{1,n}, \dots, f_{\ell,n}) - T(f_1, \dots, f_\ell) = T(f_{1,n} - f_1, f_{2,n}, \dots, f_{\ell,n}) + \\ + T(f_1, f_{2,n} - f_2, \dots, f_{\ell,n}) + \dots + T(f_1, \dots, f_{\ell-1}, f_{\ell,n} - f_\ell).$$

The first term of the right-hand side tends to 0 in $\mathcal{K}^m(\Omega; F)$ because the inequality (d) of Theorem 3.9 implies, because of (3.12),

$$\|T(f_{1,n} - f_1, f_{2,n}, \dots, f_{\ell,n})\|_{\mathcal{K}^m(\Omega;F);q,\mu} \le \\ \le c\ell^m \sup_{\nu_1\in N_1} \|f_{1,n} - f_1\|_{\mathcal{K}^m(\Omega;E_1);q,\nu_1} C^{\ell-1},$$

where $C = \sup_{j\ge 2,\, n\in\mathbb{N},\, \nu_j\in N_j} \|f_{j,n}\|_{\mathcal{C}^m(\Omega;E_j);K,\nu_j}$. The other terms on the left-hand side also tend to 0 because, similarly,

$$\|T(f_1, \dots, f_{j,n} - f_j, \dots, f_{\ell,n})\|_{\mathcal{K}^m(\Omega;F);q,\mu} \le \\ \le c\ell^m \sup_{\nu_j\in N_j} C_1 \|f_{j,n} - f_j\|_{\mathcal{C}^m(\Omega;E_j);K,\nu_j} C^{\ell-2},$$

where $C_1 = \sup_{\nu_1\in N_1} \|f_1\|_{\mathcal{K}^m(\Omega;E_1);q,\nu_1}$.

Therefore, $T(f_{1,n}, \dots, f_{\ell,n}) \to T(f_1, \dots, f_\ell)$ in $\mathcal{K}^m(\Omega; F)$.

(d) *Partial continuity for the space $\mathcal{K}^m$.* This follows from the inequality (d) in Theorem 3.9. □

3.5. Change of variables in a derivative

Let us calculate the partial derivatives after performing a change of variables. In space dimension one, this is the **chain rule theorem** (see part (c)).

THEOREM 3.12.– *Let f be a differentiable function from an open subset Ω of $\mathbb{R}^d$ into a separated semi-normed space E, and let T be a differentiable function from an open subset Λ of $\mathbb{R}^\ell$ into Ω. Then:*

(a) *$f \circ T$ is differentiable from Λ into E and, for every $i \in [\![1, \ell]\!]$,*

$$\partial_i(f \circ T) = \sum_{j=1}^{d} ((\partial_j f) \circ T)\, \partial_i T_j.$$

(b) *If T is also a bijection from Λ onto Ω and T^{-1} is differentiable, then*

$$(\partial_i f) \circ T = \sum_{j=1}^{d} \partial_j(f \circ T)\, ((\partial_i T_j^{-1}) \circ T)$$

and, for all $k \in [\![1, d]\!]$,

$$\sum_{j=1}^{d} \partial_j T_k\, ((\partial_i T_j^{-1}) \circ T) \text{ is } 1 \text{ if } i = k \text{ and zero otherwise.}$$

(c) *If $d = \ell = 1$,*

$$(f \circ T)' = (f' \circ T)\, T'. \blacksquare$$

Proof. (a) Let $\{\| \ \|_{E;\nu} : \nu \in \mathcal{N}_E\}$ be the family of semi-norms of E, $x \in \Omega$ and $z \in \mathbb{R}^d$ such that $x + z \in \Omega$. Decompose

$$(f \circ T)(x+z) - (f \circ T)(x) - \sum_{i=1}^{d}\sum_{j=1}^{d} z_i (\partial_j f \circ T)(x)\, \partial_i T_j(x) = A + \sum_{j=1}^{d} B_j,$$

where

$$A \stackrel{\text{def}}{=} f(T(x+z)) - f(T(x)) - \sum_{j=1}^{d} \partial_j f(T(x))\, (T_j(x+z) - T_j(x))$$

and

$$B_j \stackrel{\text{def}}{=} \partial_j f(T(x)) \Big(T_j(x+z) - T_j(x) - \sum_{i=1}^{d} \partial_i T_j(x) \Big).$$

Let $\nu \in \mathcal{N}_E$ and $\epsilon > 0$. Since T is differentiable at the point x, by Theorem 2.3, there exists $\zeta > 0$ and $c \geq 0$ such that if $|z| \leq \zeta$, then

$$|T(x+z) - T(x)| \leq c\,|z|. \tag{3.13}$$

By Definition 2.1 of differentiability, in this case where f is at the point $T(x)$, there exists $\eta > 0$ such that if $|T(x+z) - T(x)| \leq \eta$, then

$$\|A\|_{E;\nu} \leq \frac{\epsilon}{2c}|T(x+z) - T(x)|. \tag{3.14}$$

If $|z| \leq \inf\{\zeta, \eta/c\}$, the inequalities (3.13) and (3.14) are satisfied, so

$$\|A\|_{E;\nu} \leq \frac{\epsilon}{2}|z|.$$

Moreover, denote $\gamma_j = \|\partial_j f(T(x))\|_{E;\nu}$. Since the component T_j is differentiable by Theorem 2.4, if $\gamma_j > 0$, there exists $\eta_j > 0$ such that if $|z| \leq \eta_j$, then

$$\Big|T_j(x+z) - T_j(x) - \sum_{i=1}^{d} \partial_i T_j(x)\Big| \leq \frac{\epsilon}{2d\gamma_j}|z|,$$

and so

$$\|B_j\|_{E;\nu} \leq \frac{\epsilon}{2d}|z|.$$

If $\gamma_j = 0$, this inequality is satisfied for $\eta_j = \infty$.

If $|z| \leq \inf\{\zeta, \eta/c, \eta_1, \dots, \eta_d\}$, we therefore have $\|A + \sum_{j=1}^{d} B_j\|_{E;\nu} \leq \epsilon|z|$, which proves that $f \circ T$ is differentiable at the point x and that its derivatives can be expressed as claimed.

(b) If T is invertible and its inverse is differentiable, part (a) gives

$$\partial_i f = \partial_i\big((f \circ T) \circ T^{-1}\big) = \sum_{j=1}^{d} (\partial_j (f \circ T) \circ T^{-1})\, \partial_i T_j^{-1}.$$

Taking the composition with T gives

$$(\partial_i f) \circ T = \sum_{j=1}^{d} \partial_j (f \circ T)\, ((\partial_i T_j^{-1}) \circ T).$$

This is the first of the stated equalities. This equality, taking $f(y) = y_k$, then implies the second, since in this case $f \circ T = T_k$ and $\partial_i y_i = 1$ and $\partial_i y_k = 0$ if $i \neq k$.

(c) When $d = \ell = 1$, part (a) reduces to $\partial_1(f \circ T) = (\partial_1 f \circ T)\, \partial_1 T$, which is the stated equality because, in one dimension, $\partial_1 f = f'$ by Theorem 2.9. □

Remark. In part (b) of Theorem 3.12, we necessarily have $\ell = d$ (because here T is a homeomorphism from an open subset of $\mathbb{R}^\ell$ onto an open subset of $\mathbb{R}^d$), but we do not need to use this fact. □

Let us state an inequality describing the effect of a change of variable on the semi-norms of $\mathcal{C}^m(\Omega; E)$.

THEOREM 3.13.– *Let $f \in \mathcal{C}^m(\Omega; E)$, where Ω is an open subset of $\mathbb{R}^d$, E is a separated semi-normed space and $0 \leq m \leq \infty$, and let $T \in \mathcal{C}^m(\Lambda; \Omega)$, where Λ is an open subset of $\mathbb{R}^\ell$. Then:*

(a) $$f \circ T \in \mathcal{C}^m(\Lambda; E).$$

(b) *If $m < \infty$, denoting by $\{\| \ \|_{E;\nu} : \nu \in \mathcal{N}_E\}$ the family of semi-norms of E, then, for any compact set $Q \subset \Lambda$ and all $\nu \in \mathcal{N}_E$,*

$$\|f \circ T\|_{\mathcal{C}^m(\Lambda;E);Q,\nu} \leq c_m \, \|f\|_{\mathcal{C}^m(\Omega;E);K,\nu} \sup\Big\{1, d^m \sup_{1\leq i\leq d} \|T_i\|^m_{\mathcal{C}^m(\Omega);Q}\Big\},$$

where $K = T(Q)$ and $c_m = 2^{m(m-1)/2}$. ▮

Proof. (a) If f is continuous, then the composite function $f \circ T$ is continuous by Theorem A.35, which proves the property (a) for $m = 0$.

We will show this property for $m \geq 1$ by induction. Suppose that it holds for $m-1$ and that f and T are m times continuously differentiable. By Theorem 3.12, $f \circ T$ is differentiable and

$$\partial_i(f \circ T) = \sum_{j=1}^{d} ((\partial_j f) \circ T) \, \partial_i T_j. \tag{3.15}$$

Since $\partial_j f \in \mathcal{C}^{m-1}(\Omega; E)$, the induction hypothesis for $m - 1$ shows that $(\partial_j f) \circ T$ belongs to $\mathcal{C}^{m-1}(\Lambda; E)$. By Theorem 3.5 (a), each product $(\partial_j f) \circ T) \, \partial_i T_j$ also belongs to this space, and hence so does each $\partial_i(f \circ T)$. Therefore, $f \circ T \in \mathcal{C}^m(\Lambda; E)$ by Theorem 2.11, which proves the induction hypothesis for m.

Since the property (a) holds for $m = 0$, it holds for every $m \in \mathbb{N}^*$ by induction, and hence also for $m = \infty$.

(b) The stated inequality is satisfied for $m = 0$, because

$$\|f \circ T\|_{\mathcal{C}^0(\Lambda;E);Q,\nu} = \sup_{y\in Q} \|(f \circ T)(y)\|_{E;\nu} = \sup_{x\in K} \|f(x)\|_{E;\nu} = \|f\|_{\mathcal{C}^0(\Omega;E);K,\nu}.$$

Let us show that it also holds for $m \geq 1$, again by induction. Thus, suppose that it holds for $m - 1$. Observe that

$$\|f \circ T\|_{\mathcal{C}^m(\Lambda;E);Q,\nu} =$$
$$= \sup\{\|f \circ T\|_{\mathcal{C}^0(\Lambda;E);Q,\nu}, \sup_{1\leq i\leq d} \|\partial_i(f \circ T)\|_{\mathcal{C}^{m-1}(\Lambda;E);Q,\nu}\}. \tag{3.16}$$

By (3.15) and the bound on the norm of a product from Theorem 3.9 (b),

$$\|\partial_i(f \circ T)\|_{C^{m-1}(\Lambda;E);Q,\nu} \leq d\,\|((\partial_j f) \circ T)\,\partial_i T_j\|_{C^{m-1}(\Lambda;E);Q,\nu} \leq$$
$$\leq d2^{m-1} \sup_{1\leq j\leq d} \|(\partial_j f) \circ T\|_{C^{m-1}(\Lambda;E);Q,\nu} \|\partial_i T_j\|_{C^{m-1}(\Lambda);Q}.$$

With the inequality for $m-1$, this is bounded by

$$d2^{m-1} c_{m-1} \sup_{1\leq j\leq d} \|\partial_j f\|_{C^{m-1}(\Omega;E);K,\nu} \times$$
$$\times \sup\Big\{1, d^{m-1} \sup_{1\leq \ell\leq d} \|T_\ell\|^{m-1}_{C^{m-1}(\Omega);Q}\Big\} \|\partial_i T_j\|_{C^{m-1}(\Lambda);Q}.$$

Therefore,

$$\|\partial_i(f \circ T)\|_{C^{m-1}(\Lambda;E);Q,\nu} \leq c_m \|f\|_{C^m(\Omega;E);K,\nu} \sup\Big\{1, d^m \sup_{1\leq i\leq d} \|T_i\|^m_{C^m(\Omega);Q}\Big\}$$

(here, we are using the fact that $a \leq \sup\{1, a^m\}$, which holds for $a \geq 0$). Taking the supremum for $1 \leq i \leq d$ and applying (3.16) gives us the inequality for m. Since this inequality holds for $m = 0$, it holds for all $m \in \mathbb{N}^*$. □

3.6. Differentiation with respect to a separated variable

Let us differentiate an expression where the variables are separated. This will allow us to differentiate under the integral sign later.

THEOREM 3.14.– *Let f, g_1, ..., g_d be functions in $\mathbf{C}_b(\Omega \times \Lambda; E)$, where Ω is an open subset of $\mathbb{R}^d$, Λ is an open subset of $\mathbb{R}^\ell$ and E is a separated semi-normed space. Suppose that, for every $y \in \Lambda$, the function $x \mapsto f(x,y)$ is differentiable from Ω into E and its partial derivatives are given by the functions $x \mapsto g_i(x,y)$.*

Then, after separating the variables using Theorem 1.27, $\underline{f} \in \mathbf{C}^1_b(\Omega; \mathbf{C}_b(\Lambda; E))$ and, for every $i \in [\![1, d]\!]$,

$$\partial_i \underline{f} = \underline{g_i}. \blacksquare$$

Proof. Denote by $\{\| \ \|_{E;\nu} : \nu \in \mathcal{N}_E\}$ the family of semi-norms of E. Let $x \in \Omega$, $i \in [\![1, d]\!]$ and $r > 0$ such that $x + t\mathbf{e}_i \in \Omega$ for every $t \in \mathbb{R}$ satisfying $|t| \leq r$.

By the hypotheses, the ith partial derivative of the function $x \mapsto f(x,y)$ is the function $x \mapsto g_i(x,y)$, thus the characterization (2.7), p. 37, shows that this partial derivative is in fact the derivative of the function $t \mapsto f(x + t\mathbf{e}_i, y)$. The finite increment theorem (Theorem 2.5 (b), here in one dimension, t) implies that, for every $\nu \in \mathcal{N}_E$ and $y \in \Lambda$,

$$\|f(x + t\mathbf{e}_i, y) - f(x, y) - t\, g_i(x, y)\|_{E;\nu} \leq$$
$$\leq |t| \sup_{0 \leq s \leq 1} \|g_i(x + st\mathbf{e}_i, y) - g_i(x, y)\|_{E;\nu}.$$

But by separation of variables (Theorem 1.27),

$$\underline{f} \text{ and } \underline{g_i} \text{ belong to } \mathbf{C}_\mathrm{b}(\Omega; \mathbf{C}_\mathrm{b}(\Lambda; E)), \tag{3.17}$$

so $\underline{f}(x + t\mathbf{e}_i)$, $\underline{f}(x)$ and $\underline{g_i}(x)$ belong to $\mathbf{C}_\mathrm{b}(\Lambda; E)$, which is endowed with the semi-norms of $\mathcal{C}_\mathrm{b}(\Lambda; E)$ by Definition 1.4(b). Hence, by Definition 1.3 (b) of these semi-norms,

$$\|\underline{f}(x + t\mathbf{e}_i) - \underline{f}(x) - t\, \underline{g_i}(x)\|_{\mathbf{C}_\mathrm{b}(\Lambda;E);\nu} =$$
$$= \sup_{y \in \Lambda} \|f(x + t\mathbf{e}_i, y) - f(x, y) - t\, g_i(x, y)\|_{E;\nu} \leq$$
$$\leq |t| \sup_{0 \leq s \leq t} \sup_{y \in \Lambda} \|g_i(x + s\mathbf{e}_i, y) - g_i(x, y)\|_{E;\nu} =$$
$$= |t| \sup_{0 \leq s \leq t} \|\underline{g_i}(x + s\mathbf{e}_i) - \underline{g_i}(x)\|_{\mathbf{C}_\mathrm{b}(\Lambda;E);\nu}.$$

For every $\epsilon > 0$, there exists $\eta > 0$ such that the final expression is $\leq \epsilon\, |t|$ whenever $|t| \leq \eta$ because $\underline{g_i}$ is continuous. By the characterization of the partial derivatives from Definition 2.8, this proves that $\underline{f}$ has a partial derivative at the point x and that $\partial_i \underline{f} = \underline{g_i}(x)$.

This holds for all i and x, which implies that $\underline{f}$ is continuously differentiable from Ω into $\mathbf{C}_\mathrm{b}(\Lambda; E)$ by Theorem 2.13 (b). Together with (3.17), this finally gives $\underline{f} \in \mathbf{C}^1_\mathrm{b}(\Omega; \mathbf{C}_\mathrm{b}(\Lambda; E))$. □

3.7. Image under a differentiable mapping

Let us generalize the notion of differentiability[2] to mappings from one separated semi-normed space into another.

DEFINITION 3.15.– *Let S be a mapping from an open subset X of a separated semi-normed space E into a separated semi-normed space F with families of semi-norms $\{\| \, \|_{E;\nu} : \nu \in \mathcal{N}_E\}$ and $\{\| \, \|_{F;\mu} : \mu \in \mathcal{N}_F\}$, respectively.*

2 **History of the notion of differentiability.** Maurice FRÉCHET defined differentiable mappings from one normed space into another in 1906 [38].

We say that S is ***differentiable at the point*** $u \in X$ *if there exists a mapping* $\mathrm{d}S(u)$ *in* $\mathcal{L}(E;F)$ *and for every* $\mu \in \mathcal{N}_F$*, there exists a finite subset* N *of* $\mathcal{N}_E$ *such that: for every* $\epsilon > 0$*, there exists* $\eta > 0$ *such that* $w \in E$ *and* $\sup_{\nu \in N} \|w\|_{E;\nu} \leq \eta$ *imply* $u + w \in X$ *and*

$$\|S(u+w) - S(u) - \mathrm{d}S(u)(w)\|_{F;\mu} \leq \epsilon \sup_{\nu \in N} \|w\|_{E;\nu}.$$

If so, $\mathrm{d}S(u)$ *is said to be the* ***differential*** *of* S *at the point* u*.*

We say that S is ***differentiable*** *if it is differentiable at every point of X and we say that it is* ***continuously differentiable*** *if* $\mathrm{d}S$ *is also continuous from X into* $\mathcal{L}(E;F)$*.*

We say that S is m ***times differentiable****, where* $m \in \mathbb{N}$*, if it has the successive differentials* $\mathrm{d}S$*,* d^2S*, ...,* d^mS*, which are mappings from E into* $\mathcal{L}(E;F)$*,* $\mathcal{L}(E;\mathcal{L}(E;F))$*, ..., respectively. We say that S is m* ***times continuously differentiable*** *if its successive differentials are also continuous.*

We say that f is ***infinitely differentiable*** *if it is m times differentiable for every* $m \in \mathbb{N}$. ▮

Justification. The vector space $\mathcal{L}(E;F)$ of continuous linear mappings from E into F is endowed with the semi-norms, indexed by the bounded subsets B of E and $\mu \in \mathcal{N}_F$,

$$\|L\|_{\mathcal{L}(E;F);B,\mu} \stackrel{\text{def}}{=} \sup_{u \in B} \|Lu\|_{F;\mu}.$$

These semi-norms make this space a separated semi-normed space, which allows us to define the continuity of $\mathrm{d}S$ from X into $\mathcal{L}(E;F)$ and the successive differentials $\mathrm{d}^2S\ldots$ □

If E is a normed space, this topology is generated by the semi-norms, indexed by $\mu \in \mathcal{N}_F$,

$$|||L|||_{\mathcal{L}(E;F);\mu} = \sup_{\|u\|_E \leq 1} \|L(u)\|_{F;\mu}.$$

Indeed, for every bounded subset B of E, writing $t = \sup_{u \in B} \|u\|_E$,

$$\|L\|_{\mathcal{L}(E;F);B,\mu} \leq t\, |||L|||_{\mathcal{L}(E;F);\mu}$$

and, conversely, $|||L|||_{\mathcal{L}(E;F);\mu} = \|L\|_{\mathcal{L}(E;F);\mathcal{B},\mu}$, where $\mathcal{B} = \{u \in E : \|u\|_E \leq 1\}$.

Originality. This generalization [Vol. 1, Definition 19.1] of the notion of differentiability to any separated semi-normed space, E or F, and therefore to any separated locally convex topological vector space, is new.

It relies on the use of semi-norms because we can evaluate them numerically. Here, defining the topology with open sets is not suitable because we cannot evaluate these sets numerically. A metric d is also not suitable because $d(tu, 0)$ is not proportional to t.

These are the reasons why, if we are working solely in terms of topologies, metrics and norms as usual, differentiability can only be defined in a normed space, as well as in some special cases like *scalar differentiability*, which corresponds to differentiability for the weak topology. □

Let us show that for functions, that is, for mappings defined on a subset of $\mathbb{R}^d$, this definition of differentiability coincides with that of Chapter 1.

THEOREM 3.16.– *Let f be a function from an open subset Ω of $\mathbb{R}^d$ into a semi-normed space E. Then:*

(a) *The function f is differentiable at the point x in the sense of Definition 3.15 if and only if it is differentiable in the sense of Definition 2.1. If so, for all $z \in \mathbb{R}^d$,*

$$\mathrm{d}f(x)(z) = z \cdot \nabla f(x), \tag{3.18}$$

and

$$\nabla f(x) = (\mathrm{d}f(x)(\mathbf{e}_1), \ldots, \mathrm{d}f(x)(\mathbf{e}_d)). \tag{3.19}$$

(b) *The function $\mathrm{d}f$ is continuous from Ω into $\mathcal{L}(\mathbb{R}^d; E)$ if and only if ∇f is continuous from Ω into E^d.*

(c) *For any function from Ω into E, Definitions 2.1 and 3.15 provide the same notions of m times differentiability and m times continuous differentiability, for $m \in \mathbb{N}$.* ▮

Proof. (a) For the function f, Definition 3.15 of differentiability at the point x reduces to the existence of a function $\mathrm{d}f(x)$ in $\mathcal{L}(\mathbb{R}^d; E)$ such that, for every $\nu \in \mathcal{N}_E$ and $\epsilon > 0$, there exists $\eta > 0$ such that if $z \in \mathbb{R}^d$, $|z| \leq \eta$ and $x + z \in \Omega$, then

$$\|f(x+z) - f(x) - \mathrm{d}f(x)(z)\|_{E;\nu} \leq \epsilon\,|z|. \tag{3.20}$$

Defining $\nabla f(x) \in E^d$ to be $\nabla_i f(x) = \mathrm{d}f(x)(\mathbf{e}_i)$ for every $i \in [\![1, d]\!]$ gives

$$\mathrm{d}f(x)(z) = \sum_{i=1}^{d} \mathrm{d}f(x)(z_i \mathbf{e}_i) = \sum_{i=1}^{d} z_i \nabla_i f(x) = z \cdot \nabla f(x),$$

which is the characterization (2.3) of differentiability of Definition 2.1, namely

$$\|f(x+z) - f(x) - z \cdot \nabla f(x)\|_{E;\nu} \leq \epsilon\,|z|.$$

Conversely, if f is differentiable in that sense, it satisfies the characterization (3.20) of differentiability with $\mathrm{d}f(x)(z) = z \cdot \nabla f(x)$, since the mapping $z \mapsto z \cdot \nabla f(x)$ is continuous and linear from $\mathbb{R}^d$ into E^d by the inequality (2.2) because $\nabla f(x) \in E^d$.

(b) If ∇f is continuous from Ω into E^d, then $\mathrm{d}f$ is continuous from Ω into $\mathcal{L}(\mathbb{R}^d; E)$, since, for every $\nu \in \mathcal{N}_E$,

$$|||\mathrm{d}f(x) - \mathrm{d}f(y)|||_{\mathcal{L}(\mathbb{R}^d;E);\nu} = \sup_{|z|\le 1} \|(\mathrm{d}f(x) - \mathrm{d}f(y))(z)\|_{E;\nu} =$$
$$= \sup_{|z|\le 1} \|z \cdot (\nabla f(x) - \nabla f(y))\|_{E;\nu} \le \|\nabla f(x) - \nabla f(y)\|_{E^d;\nu}$$

because $\|z \cdot u\|_{E;\nu} \le |z|\, \|u\|_{E^d;\nu}$ for all $u \in E^d$ by the inequality (2.2), p. 31.

The converse follows from

$$\|\nabla f(x) - \nabla f(y)\|_{E^d;\nu} =$$
$$= \|\big((\mathrm{d}f(x) - \mathrm{d}f(y))(\mathbf{e}_1), \dots, (\mathrm{d}f(x) - \mathrm{d}f(y))(\mathbf{e}_d)\big)\|_{E^d;\nu} =$$
$$= \Big(\sum_{i=1}^d \|(\mathrm{d}f(x) - \mathrm{d}f(y))(\mathbf{e}_i)\|^2_{E;\nu}\Big)^{1/2} \le \sqrt{d}\, |||\mathrm{d}f(x) - \mathrm{d}f(y)|||_{\mathcal{L}(\mathbb{R}^d;E);\nu}$$

because $\|L(\mathbf{e}_i)\|_{E;\nu} \le \sup_{|z|\le 1} \|L(z)\|_{E;\nu} = |||L|||_{\mathcal{L}(\mathbb{R}^d;E);\nu}$ for every $L \in \mathcal{L}(\mathbb{R}^d; E)$.

(c) The properties of order m follow from (a) and (b) by Definitions 2.1 and 3.15. □

Derivative and gradient versus differential. The derivative $f'(x)$, being an element of E, is nicer to work with than the differential $\mathrm{d}f(x)$, which is an element of $\mathcal{L}(\mathbb{R}; E)$ in dimension $d = 1$. For higher dimensions, the gradient $\nabla f(x)$, which belongs to E^d, is also nicer than the differential, which belongs to $\mathcal{L}(\mathbb{R}^d; E)$. □

Exterior differential. The differential of a function f is a special case of the exterior differential of a differential form, which explains why we use the same notation $\mathrm{d}f$. Indeed, any function f is a differential form of degree 0 [CARTAN, Henri, 19, p. 189], in which case its exterior differential is its differential [19, p. 192], which CARTAN calls the derivative [19, p. 29] and denotes by f'. □

Let us calculate the partial derivatives of the image of a function under a differentiable mapping from one separated semi-normed space to another.

THEOREM 3.17.– *Let $f \in \mathcal{C}^1(\Omega; E)$, where Ω is an open subset of $\mathbb{R}^d$ and E is a separated semi-normed space. Furthermore, let S be a continuously differentiable mapping from E into a separated semi-normed space F.*

Then $S \circ f \in \mathcal{C}^1(\Omega; F)$ and, for every $i \in [\![1, d]\!]$ and $x \in \Omega$,

$$\partial_i (S \circ f)(x) = \mathrm{d}S(f(x))\,(\partial_i f(x)). \;\blacksquare$$

Meaning of the terms in Theorem 3.17.
— $\partial_i f$ is the partial derivative of f (Definition 2.8), so $\partial_i f(x) \in E$ and $\partial_i (S \circ f)(x) \in F$.
— $\mathrm{d}S$ is the differential of S (Definition 3.15), so $\mathrm{d}S(f(x)) \in \mathcal{L}(E; F)$. □

Proof of Theorem 3.17. The composite function $S \circ f$ is continuously differentiable (Theorem A.51 (b), since $\mathbb{R}^d$ is metrizable) and applying Theorems 3.16 and A.51 (a), and Theorem 3.16 a second time shows that its partial derivatives are given by

$$\partial_i (S \circ f)(x) = \mathrm{d}(S \circ f)(x)(\mathbf{e}_i) = \big(\mathrm{d}S(f(x)) \circ \mathrm{d}f\big)(x)(\mathbf{e}_i) =$$
$$= \big(\mathrm{d}S(f(x))\big)\big(\mathrm{d}f(x)(\mathbf{e}_i)\big) = \mathrm{d}S(f(x))(\partial_i f(x)). \quad \square$$

3.8. Differentiation and translation

The translated function $\tau_z f$ of a function f defined on a subset Ω of $\mathbb{R}^d$ by $z \in \mathbb{R}^d$ is defined on $\Omega + z \overset{\text{def}}{=} \{x + z : x \in \Omega\}$ by

$$\tau_z f(x) \overset{\text{def}}{=} f(x - z). \tag{3.21}$$

Let us state some differentiability properties of z.

THEOREM 3.18.– *Let $\mu \in \mathbf{C}^\ell_{\mathrm{b}}(\mathbb{R}^d)$, where $0 \leq \ell \leq \infty$. Then:*

(a) *The function $z \mapsto \tau_z \mu$ is uniformly continuous and bounded from $\mathbb{R}^d$ into $\mathbf{C}^\ell_{\mathrm{b}}(\mathbb{R}^d)$.*

(b) *If $\ell \geq 1$, the function $z \mapsto \tau_z \mu$ is differentiable from $\mathbb{R}^d$ into $\mathbf{C}^{\ell-1}_{\mathrm{b}}(\mathbb{R}^d)$ and*

$$\frac{\partial \tau_z \mu}{\partial z_i} = -\tau_z \partial_i \mu.$$

(c) *For $0 \leq m \leq \ell$, the function $z \mapsto \tau_z \mu$ belongs to $\mathbf{C}^m_{\mathrm{b}}(\mathbb{R}^d; \mathbf{C}^{\ell-m}_{\mathrm{b}}(\mathbb{R}^d))$.*

(d) *If $\ell = \infty$, the function $z \mapsto \tau_z \mu$ belongs to $\mathbf{C}^\infty_{\mathrm{b}}(\mathbb{R}^d; \mathbf{C}^\infty_{\mathrm{b}}(\mathbb{R}^d))$.* ∎

Proof. (a) Case $\ell < \infty$. Recall that, by Definition 2.14 (d), $\mathbf{C}^\ell_{\mathrm{b}}(\mathbb{R}^d)$ is endowed with the semi-norms of $\mathcal{C}^\ell_{\mathrm{b}}(\mathbb{R}^d)$. By Definition 2.14 (b) of these semi-norms,

$$\|\tau_z \mu - \tau_{z'} \mu\|_{\mathcal{C}^\ell_{\mathrm{b}}(\mathbb{R}^d)} = \sup_{|\beta| \leq \ell} \sup_{x \in \mathbb{R}^d} |\partial^\beta \mu(x - z) - \partial^\beta \mu(x - z')|,$$

which tends to 0 as $|z' - z| \to 0$ because the $\partial^\beta \mu$ belong to $\mathbf{C}_{\mathrm{b}}(\mathbb{R}^d)$, which means they are uniformly continuous. The function $z \mapsto \tau_z \mu$ is therefore uniformly continuous from $\mathbb{R}^d$ into $\mathbf{C}^\ell_{\mathrm{b}}(\mathbb{R}^d)$. It is bounded because, for all z,

$$\|\tau_z \mu\|_{\mathcal{C}^\ell_{\mathrm{b}}(\mathbb{R}^d)} = \|\mu\|_{\mathcal{C}^\ell_{\mathrm{b}}(\mathbb{R}^d)}.$$

Case $\mu \in \mathbf{C}_b^\infty(\mathbb{R}^d)$. The two equalities above show that the function $z \mapsto \tau_z\mu$ is uniformly continuous and bounded from $\mathbb{R}^d$ into $\mathbf{C}_b^\infty(\mathbb{R}^d)$ because, by Definition 2.15 (d), this space is endowed with the semi-norms $\|\mu\|_{\mathcal{C}_b^\infty(\mathbb{R}^d);\ell}$ indexed by $\ell \in \mathbb{N}^*$ and because we have $\|\mu\|_{\mathcal{C}_b^\infty(\mathbb{R}^d);\ell} = \|\mu\|_{\mathcal{C}_b^\ell(\mathbb{R}^d)}$ by Definitions 2.15 (b) and 2.14 (b).

(b) Here, $\ell \geq 1$. Let $i \in [\![1, d]\!]$. By the finite increment theorem (Theorem 2.5 (b)),

$$|\partial^\beta \mu(x-(z+t\mathbf{e}_i)) - \partial^\beta \mu(x-z) + t\partial_i\partial^\beta\mu(x-z)| \leq$$
$$\leq |t| \sup_{0\leq s\leq 1} |\partial_i\partial^\beta\mu(x-(z+s\mathbf{e}_i)) - \partial_i\partial^\beta\mu(x-z)|.$$

In other words,

$$|\partial^\beta\big(\tau_{z+t\mathbf{e}_i}\mu - \tau_z\mu + t\tau_z\partial_i\mu\big)(x)| \leq |t| \sup_{0\leq s\leq 1} |\partial_i\partial^\beta\big(\tau_{z+s\mathbf{e}_i}\mu - \tau_z\mu\big)(x)|.$$

Taking the supremum for $x \in \mathbb{R}^d$ and $|\beta| \leq \ell - 1$ gives

$$\|\tau_{z+t\mathbf{e}_i}\mu - \tau_z\mu + t\tau_z\partial_i\mu\|_{\mathcal{C}_b^{\ell-1}(\mathbb{R}^d)} \leq |t| \sup_{0\leq s\leq 1} \|\tau_{z+s\mathbf{e}_i}\mu - \tau_z\mu\|_{\mathcal{C}_b^\ell(\mathbb{R}^d)}.$$

As $t \to 0$, the right-hand side tends to 0 by part (a). The function $z \mapsto \tau_z\mu$ therefore has partial derivatives taking values in $\mathbf{C}_b^{\ell-1}(\mathbb{R}^d)$, which are namely the functions $z \mapsto -\tau_z\partial_i\mu$.

(c) Denote $F_\mu(z) \stackrel{\text{def}}{=} \tau_z\mu$. Then part (a) implies that $F_\mu \in \mathbf{C}_b(\mathbb{R}^d; \mathbf{C}_b^\ell(\mathbb{R}^d))$.

When $\ell \geq 1$, part (b) implies that $\partial_i F_\mu = -F_{\partial_i\mu}$. Thus, since $\partial_i\mu \in \mathbf{C}_b^{\ell-1}(\mathbb{R}^d)$, part (a) gives $\partial_i F_\mu \in \mathbf{C}_b(\mathbb{R}^d; \mathbf{C}_b^{\ell-1}(\mathbb{R}^d))$.

We will show by induction on m that $F_\mu \in \mathbf{C}_b^m(\mathbb{R}^d; \mathbf{C}_b^{\ell-m}(\mathbb{R}^d))$ and that, for every $\beta \in \mathbb{N}^{*d}$ such that $|\beta| \leq m$,

$$\partial^\beta F_\mu = (-1)^{|\beta|}\, F_{\partial^\beta\mu}. \tag{3.22}$$

We saw just above that this is true for $m = 0$ or 1. If it is true for $m - 1$, then, for every $i \in [\![1, d]\!]$ and $|\beta| \leq m - 1$,

$$\partial_i\partial^\beta F_\mu = \partial_i\big((-1)^{|\beta|}\, F_{\partial^\beta\mu}\big) = (-1)^{|\beta|+1}\, F_{\partial_i\partial^\beta\mu},$$

which belongs to $\mathbf{C}_b(\mathbb{R}^d; \mathbf{C}_b^{\ell-m}(\mathbb{R}^d))$. Thus, (3.22) and part (c) hold for $|\beta| = m$. Since they hold for $m = 1$, they hold for every $m \leq \ell$.

(d) *Case* $\ell = \infty$. In this case, part (c) shows that the function F_μ belongs to $\mathbf{C}_b^m(\mathbb{R}^d; \mathcal{C}_b^\infty(\mathbb{R}^d))$ for every $m \in \mathbb{N}^*$ and therefore to $\mathbf{C}_b^\infty(\mathbb{R}^d; \mathcal{C}_b^\infty(\mathbb{R}^d))$ by Definition 2.15 (d) of this space. □

3.9. Localizing functions

Let us construct a function $\mathcal{C}^\infty$ whose support is included in the unit ball[3].

THEOREM 3.19.– *There exists a function ρ such that:*

$$\rho \in \mathcal{K}^\infty(\mathbb{R}^d), \quad \rho(x) = 0 \text{ if } |x| \geq 1, \quad \rho(x) > 0 \text{ if } |x| < 1. \ \blacksquare$$

Proof. Let us show that one particular such function is given by

$$\rho(x) = \begin{cases} \exp\Big(\dfrac{-1}{1-|x|^2}\Big) & \text{if } |x| < 1, \\ 0 & \text{if } |x| \geq 1. \end{cases} \tag{3.23}$$

To do this, it suffices to check that the function g defined by

$$g(t) = \begin{cases} \exp\Big(\dfrac{-1}{1-t}\Big) & \text{if } t < 1, \\ 0 & \text{if } t \geq 1, \end{cases}$$

is infinitely differentiable. Indeed, since the function $x \mapsto |x|^2$ is infinitely differentiable, the composite function ρ will also be infinitely differentiable (Theorem 3.13).

The composite function g is infinitely differentiable for $t < 1$, again by Theorem 3.13, since the functions $t \mapsto (1-t)^{-1}$ and $s \mapsto e^s$ are differentiable (Theorems A.56 and A.58). The same is true for $t > 1$, so we simply need to check the case $t = 1$. To do this, observe that the mth derivative at any point $t < 1$ is

$$g^{(m)}(t) = \exp\Big(\frac{-1}{1-t}\Big) \sum_{i=m+1}^{2m} c_{i,m} \frac{1}{(1-t)^i},$$

where $c_{i,m} \in \mathbb{R}$. Indeed, this equality is satisfied for $m = 0$; furthermore, if it holds for $m-1$, then it also holds for m by the Leibniz rule and the chain rule (Theorems 3.6 and 3.12 (c)), since $\mathrm{d}e^x/\mathrm{d}x = e^x$ and $\mathrm{d}(1-t)^{-i}/\mathrm{d}t = i(1-t)^{-i-1}$ (Theorems A.58 and A.56).

Given that $e^{-x} \leq r^r x^{-r}$ for positive x and r (Theorem A.58), for $r = i + 2$, it follows that $\exp(-1/(1-t)) \leq (i+2)^{i+2}(1-t)^{i+2}$, so

$$|g^{(m)}(t)| \leq c_m |1-t|^2.$$

Therefore, if g is m times differentiable and $g^{(m)}(1) = 0$, then $g^{(m)}$ is differentiable at $t = 1$ and $g^{(m+1)}(1) = 0$ (by the characterization of differentiability in one dimension from Definition 2.1 (b)). By induction, this holds for all m, so g is infinitely differentiable as required. □

3 **History of constructing a $\mathcal{C}^\infty$ function whose support is included in the unit ball.** Jean LERAY proved Theorem 3.19 in 1934 [55, footnote 1, p. 206] by constructing the function ρ introduced here.

Let us define the notion of **compact inclusion** of a subset of $\mathbb{R}^d$ in another subset of $\mathbb{R}^d$.

DEFINITION 3.20.– *For two subsets ω and Ω of $\mathbb{R}^d$, we denote:*

$$\omega \Subset \Omega \quad \stackrel{\text{def}}{\Leftrightarrow} \quad \omega \text{ is bounded and } \overline{\omega} \subset \mathring{\Omega}. \ ▮$$

Terminology. The inclusion $\Subset$ in the above is said to be *compact* because it implies that $\overline{\omega}$ is a compact set by the Borel–Lebesgue theorem (Theorem A.23 (b)), since $\overline{\omega}$ is closed and bounded. □

Let us construct a *localizing* function that is a special case of **Urysohn's theorem**[4].

THEOREM 3.21.– *Let $\omega \Subset \Omega \subset \mathbb{R}^d$. Then there exists a function α such that:*

$$\alpha \in \mathcal{K}^\infty(\Omega), \quad 0 \leq \alpha \leq 1, \quad \alpha = 1 \text{ on } \omega. \ ▮$$

Proof. Given $s \in \mathbb{Z}^d$ and $n \in \mathbb{N}^*$, denote $\rho_{s,n}(x) = \rho((2^n x - s)/\sqrt{d})$, where ρ satisfies (3.23). Then

$$\rho_{s,n} \in \mathcal{K}^\infty(\mathbb{R}^d), \quad \rho_{s,n} > 0 \text{ on } \mathring{B}(2^{-n}s, 2^{-n}\sqrt{d}), \quad \rho_{s,n} = 0 \text{ otherwise.}$$

Let

$$G_n(x) = \sum_{s \in \mathbb{Z}^d} \rho_{s,n}(x).$$

This sum is well defined and

$$G_n \in \mathcal{C}^\infty(\mathbb{R}^d), \quad G_n > 0.$$

Indeed, for fixed x and n, only a finite number ($\leq (2\sqrt{d}+1)^d$) of the $\rho_{s,n}(x)$ are non-zero, and at least one of them is non-zero (because $\rho_{s,n} \geq e^{-4/3}$ on the closed cube $\Delta_{s,n}$ of side length 2^{-n} centered at $2^{-n}s$ and these cubes cover $\mathbb{R}^d$). Therefore,

$$\frac{1}{G_n} \in \mathcal{C}^\infty(\mathbb{R}^d),$$

since it is the composite function (Theorem 3.13) of G_n with the function $t \mapsto 1/t$, which belongs to $\mathcal{C}^\infty((0,\infty))$ (Theorem A.56).

4 **History of Urysohn's theorem (Theorem 3.21).** Pavel Samuilovitch URYSOHN showed in 1925 [80, pp. 290–291] that if U and V are two disjoint closed subsets of a metrizable space (or more generally a *normal* space), then there exists a continuous function that is equal to 0 on U, equal to 1 on V and between 0 and 1 elsewhere.

Let

$$g_{s,n} = \frac{\rho_{s,n}}{G_n}.$$

Then $g_{s,n} \in \mathcal{K}^\infty(\mathbb{R}^d)$, $g_{s,n} > 0$ on $\mathring{B}(2^{-n}s, 2^{-n}\sqrt{d})$ and $g_{s,n} = 0$ outside of this ball (the fact that it belongs to $\mathcal{C}^\infty(\mathbb{R}^d)$, and therefore to $\mathcal{K}^\infty(\mathbb{R}^d)$ in this case, follows from Theorem 3.5 (a) with $T(u_1, u_2) = u_1 u_2$).

For sufficiently large n, the stated properties are satisfied by

$$\alpha(x) = \sum_{s \in \mathbb{Z}^d : B(2^{-n}s, 2^{-n}\sqrt{d}) \cap \omega \neq \emptyset} g_{s,n}(x).$$

Indeed:
— $\alpha \in \mathcal{C}^\infty(\mathbb{R}^d)$ because, on any given ball, this is just the sum of finitely many $g_{s,n}$.
— $0 \leq \alpha(x) \leq 1$ because $g_{s,n}(x) \geq 0$ and $\sum_{s \in \mathbb{Z}^d} g_{s,n}(x) = 1$.
— $\alpha(x) = 1$ if $x \in \omega$ because then $\alpha(x) = \sum_{s \in \mathbb{Z}^d} g_{s,n}(x)$, since $g_{s,n}(x) = 0$ for every s such that $B(2^{-n}s, 2^{-n}\sqrt{d}) \cap \omega = \emptyset$.
— $\operatorname{supp} \alpha \subset \bigcup_{z \in \omega} B(z, 2^{-n}\sqrt{d})$, which is included in Ω for sufficiently large n. Indeed, the open sets $\mathcal{O}_n = \{z \in \mathbb{R}^d : B(z, 2^{-n}\sqrt{d}) \subset \Omega\}$ increase with n and cover Ω, so one of them must cover the compact subset $\overline{\omega}$; thus,

$$\operatorname{supp} \alpha \subset \bigcup_{z \in \omega} B(z, 2^{-n}\sqrt{d}) \subset \mathcal{O}_n \subset \Omega. \quad \square$$

Terminology. We say that α is a *localizing* function because we will use such functions to *localize* the support of functions before regularizing them (in Definition 7.13). □

Another proof of Urysohn's theorem (Theorem 3.21). One example of a function $\zeta \in \mathcal{K}(\Omega)$ such that $\zeta = 1$ on ω is given by

$$\zeta(x) = \begin{cases} 1 & \text{if } \operatorname{dist}_\omega(x) \leq r, \\ 1 - \operatorname{dist}_\omega(x)/r & \text{if } r \leq \operatorname{dist}_\omega(x) \leq 2r, \\ 0 & \text{otherwise,} \end{cases}$$

where $\operatorname{dist}_\omega(x) \stackrel{\text{def}}{=} \inf_{y \in \omega} |y - x|$ and $r = \inf_{x \in \mathbb{R}^d \setminus \Omega} \operatorname{dist}_\omega(x)/4$.

By weighting ζ with a regularizing function ρ_n whose support is included in the ball $B(0, r)$, by Definition 7.7, we obtain a function $\zeta \diamond \rho_n \in \mathcal{D}(\mathbb{R}^d)$ whose support is included in $\omega + B(0, 3r)$ and hence in Ω. Its restriction α satisfies Theorem 3.21.

The reason we did not use this method is because we have not yet constructed weightings (or convolutions, which are essentially the same) and because we will apply Urysohn's theorem to construct them via Theorems 4.20 and 4.21. □

Chapter 4

Integrating Uniformly Continuous Functions

The goal of this chapter is to state results about integration in the simplest possible form that is sufficient to construct primitives of functions (in Chapter 9) and distributions (in Volume 3).

We therefore restrict ourselves to integrals of uniformly continuous functions on open sets, using the simplified Cauchy method based on uniform pavings of $\mathbb{R}^d$. We only define measure for open sets, again using uniform pavings.

However, we need to integrate functions taking values in a Neumann space E, i.e. a sequentially complete space, which is new for us. It is the most general possible assumption that guarantees that the integral belongs to E, as discussed in § 4.3.

We extend some of the classical properties of functions, taking values in a Banach space, or a complete semi-normed space, within this simple framework. None of the proofs are surprising. General study of integration will be performed in Volume 4 for integrable distributions taking values in a Neumann space.

4.1. Measure of an open subset of $\mathbb{R}^d$

Let us define the Lebesgue measure[1] of an open subset of $\mathbb{R}^d$.

---

1 **History of the measure of an open set.** Definition 4.1 applies an elementary method that has been used since Ancient Greece. EUDOXUS showed around 370 BC that the volume of a cone is one-third of the volume of a cylinder and the volume of a pyramid is one-third of the volume of the prism with the same base and height. This had been claimed earlier without proof by DEMOCRITUS around 400 BC. The proof is given in Book XII of the *Elements* by EUCLID [33, p. 479], published around 250 BC and was attributed to EUDOXUS by ARCHIMEDES. Among other things, ARCHIMEDES himself [1] calculated the area of an ellipse and the volume of a ball around 250 BC. For more details, see Jean-Paul PIER [62, pp. 4–11]. Measure was defined for more general sets by Henri LEBESGUE in 1904 [53, p. 114].

DEFINITION 4.1.– *The **measure** of an open subset ω of $\mathbb{R}^d$ is the nonnegative real or infinite number*

$$|\omega| \stackrel{\text{def}}{=} \lim_{n\to\infty} 2^{-nd} k_n,$$

where $n \in \mathbb{N}^*$ and k_n is the number of $s \in \mathbb{Z}^d$ such that $\Delta_{s,n}$ is included in ω, where

$\Delta_{s,n}$ is the closed cube of $\mathbb{R}^d$ of side length 2^{-n} centered at $2^{-n}s$. ▮

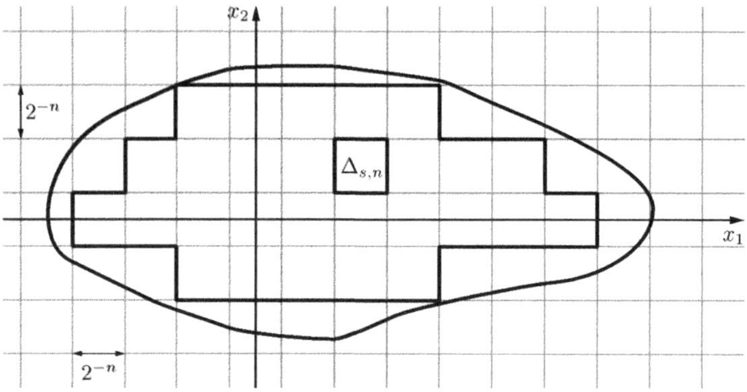

Figure 4.1. *Paving ω by the cubes $\Delta_{s,n}$*

Justification. When we increase the resolution from n to $m \geq n$, each cube $\Delta_{s,n} \subset \omega$ is divided into $(2^{-n}/2^{-m})^d$ cubes $\Delta_{r,m}$, so $k_m \geq 2^{(m-n)d} k_n$, and therefore

$$2^{-nd} k_n \text{ increases with } n. \tag{4.1}$$

As an increasing sequence, $2^{-nd}k_n$ has a limit $|\omega|$, which is finite if the sequence is bounded (Theorem A.4) and infinite otherwise. □

Note that, for every $n \in \mathbb{N}^*$, the property (4.1) implies

$$2^{-nd} k_n \leq |\omega|. \tag{4.2}$$

Choice of the cubes $\Delta_{s,n}$. The value of the measure would be the same if we chose to define $\Delta_{s,n}$ as the *open* cube of side length 2^{-n} centered at $2^{-n}s$ in Definition 4.1 (instead of the corresponding closed cube). However, the proofs of Theorems 4.5 and 4.6 would be much more complicated.

On the other hand, we would not obtain the same concept of measure if we replaced the cubes $\Delta_{s,n}$ included in ω by the cubes that cover it, i.e. by calculating $\|\omega\| = \lim_{n\to\infty} 2^{-nd} r_n$, where r_n is the number of $\Delta_{s,n}$ that intersect with ω.

For example, if $\omega = \bigcup_{n\in\mathbb{N}}(x_n - \epsilon 2^{-n}, x_n + \epsilon 2^{-n})$, where the sequence $(x_n)_{n\in\mathbb{N}}$ is dense in $(0, 1)$, then $\|\omega\| \geq 1$ (because every interval $\Delta_{s,n}$ included in $(0, 1)$ intersects with ω) but the measure that we defined above satisfies $|\omega| \leq \sum_{n\in\mathbb{N}} |(x_n - \epsilon 2^{-n}, x_n + \epsilon 2^{-n})| = \sum_{n\in\mathbb{N}} 2\epsilon 2^{-n} = 2\epsilon$, which is < 1 if $\epsilon < 1/2$. □

Let us show that measure is increasing and that it is finite for bounded open sets.

THEOREM 4.2.– *Let ω and Ω be open subsets of $\mathbb{R}^d$. Then:*

(a) $$\omega \subset \Omega \quad \Rightarrow \quad |\omega| \leq |\Omega|.$$

(b) $$\omega \text{ is bounded} \quad \Rightarrow \quad |\omega| < \infty.$$

(c) $$\omega \neq \emptyset \quad \Leftrightarrow \quad |\omega| > 0. \blacksquare$$

Proof. (a) If $\omega \subset \Omega$, then any cube $\Delta_{s,n}$ of Definition 4.1 included in ω is included in Ω, so $k_n(\omega) \leq k_n(\Omega)$, and hence $|\omega| \leq |\Omega|$.

(b) If ω is bounded, it is included in a cube Q of integer side length m centered at 0. This cube contains $(2^n m)^d$ cubes $\Delta_{s,n}$, so $2^{-nd}k_n(Q) = m^d$ and, by taking the limit, $|Q| = m^d$. Therefore, by part (a), $|\omega| \leq |Q| < \infty$.

(c) If ω is empty, it does not contain any cube $\Delta_{s,n}$, so $|\omega| = 0$.

Conversely, if ω is not empty, it contains a ball and hence one of the cubes $\Delta_{s,m}$ for sufficiently large m, so $k_m \geq 1$; therefore, by (4.2), $|\omega| \geq 2^{-md} > 0$. □

Remark. Measure is not strictly increasing with respect to open sets. For example, the measure is not changed by removing a single point (Theorem 5.6). □

Let us calculate the measure of a product of open sets.

THEOREM 4.3.– *Let ω_1 be an open subset of $\mathbb{R}^{d_1}$ and ω_2 an open subset of $\mathbb{R}^{d_2}$. Then:*

$$|\omega_1 \times \omega_2| = |\omega_1|\,|\omega_2|. \blacksquare$$

Proof. Since each cube from Definition 4.1 of the measure can be decomposed into a product $\Delta_{(s_1,s_2),n} = \Delta_{s_1,n} \times \Delta_{s_2,n}$, we have $k_n(\omega_1 \times \omega_2) = k_n(\omega_1)k_n(\omega_2)$. Therefore, $2^{-n(d_1+d_2)}k_n(\omega_1 \times \omega_2) = 2^{-nd_1}k_n(\omega_1)2^{-nd_2}k_n(\omega_2)$. Taking the limit gives the stated inequality. □

Let us calculate the measure of an interval or a parallelepiped.

THEOREM 4.4.– *Let $-\infty \leq a \leq b \leq \infty$ and $-\infty \leq a_i \leq b_i \leq \infty$ for $i = 1, 2, \ldots, d$.*

Then:

(a)
$$\big|(a,b)\big| = b - a.$$

(b)
$$\big|\{x \in \mathbb{R}^d : a_i < x_i < b_i,\ \forall i\}\big| = (b_1 - a_1) \times \cdots \times (b_d - a_d). \ ▮$$

Proof. (a) Here, $d = 1$ and the cubes $\Delta_{s,n}$ of Definition 4.1 are closed intervals of length 2^{-n}, so

$$k_n 2^{-n} < b - a \leq (k_n + 2)2^{-n}. \tag{4.3}$$

Taking the limit, $\big|(a,b)\big| \leq b - a \leq \big|(a,b)\big|$, so $\big|(a,b)\big| = b - a$.

(b) We need to calculate the measure of the product $(a_1, b_1) \times \cdots \times (a_d, b_d)$. By Theorem 4.3, this is the product of the measures of (a_i, b_i), namely the product of the $b_i - a_i$. □

Let us show that every open set of finite measure can be approximated in measure by a compact set.

THEOREM 4.5.– *Let ω be an open subset of $\mathbb{R}^d$ such that $|\omega| < \infty$ and $\epsilon > 0$. Then there exists a compact set $K \subset \omega$ such that*

$$|\omega \setminus K| \leq \epsilon.$$

For n sufficiently large, this is satisfied, in particular, by the union K_n of the closed cubes $\Delta_{s,n}$ of side length 2^{-n} centered at $2^{-n}s$ that are included in ω for $s \in \mathbb{Z}^d$. ▮

Proof. Let $n \in \mathbb{N}^*$, and let K_n be the union of the cubes $\Delta_{s,n}$ of Definition 4.1 that are included in ω. The number k_n of these $\Delta_{s,n}$ is finite (because $2^{-nd}k_n \leq |\omega|$, see (4.2)), so K_n is compact by the Borel–Lebesgue theorem (Theorem A.23 (b)) because it is closed and bounded in $\mathbb{R}^d$.

For $i \geq n$, each $\Delta_{s,n}$ may be divided into $2^{(i-n)d}$ sub-cubes $\Delta_{s,i}$, so K_n contains $2^{(i-n)d}k_n$ cubes $\Delta_{s,i}$. The number $\ell_{i,n}$ of $\Delta_{s,i}$ included in $\omega \setminus K_n$ is therefore bounded above by $k_i - 2^{(i-n)d}k_n$. Hence, by Definition 4.1 of the measure,

$$|\omega \setminus K_n| = \lim_{i \to \infty} \ell_{i,n} \leq \lim_{i \to \infty} 2^{-id}(k_i - 2^{(i-n)d}k_n) = |\omega| - 2^{-nd}k_n,$$

which is smaller than ϵ for sufficiently large n. □

Let us show that measure is continuous with respect to increasing open sets.

THEOREM 4.6.– *Let $(\omega_n)_{n\in\mathbb{N}}$ be an increasing sequence of open subsets of $\mathbb{R}^d$ and denote $\omega = \bigcup_{n\in\mathbb{N}} \omega_n$. Then, as $n \to \infty$,*

$$|\omega_n| \to |\omega|. \blacksquare$$

Proof. The measure $|\omega_n|$ increases with n (Theorem 4.2 (a)), so it has a limit ℓ that is finite if it is bounded from above (Theorem A.4) or infinite otherwise.

The set ω is open as a union of open sets (Theorem A.10), so its measure is defined. Since $\omega_n \subset \omega$, we have $|\omega_n| \leq |\omega|$ (Theorem 4.2), so

$$\ell \leq |\omega|.$$

Conversely, let $i \in \mathbb{N}$ and denote by $k_i(\omega)$ the number of (closed) cubes $\Delta_{s,i}$ that are included in ω, as in Definition 4.1 of the measure.

— If $k_i(\omega) < \infty$, the union K_i of these $\Delta_{s,i}$ is compact by the Borel–Lebesgue theorem (Theorem A.23 (b)) because it is closed and bounded in $\mathbb{R}^d$. Since the open sets ω_n cover K_i, they must have a finite subcovering. They increase with n, so one of them must cover K_i. For this particular n, all the $\Delta_{s,i}$ that make up K_i are included in ω_n, so $k_i(\omega_n) \geq k_i(\omega)$. Therefore (with the inequality (4.2)),

$$\ell \geq |\omega_n| \geq 2^{-id} k_i(\omega_n) \geq 2^{-id} k_i(\omega). \tag{4.4}$$

Letting $i \to \infty$, we obtain $\ell \geq |\omega|$ in the limit (Definition 4.1).

— If $k_i(\omega) = \infty$ (for one of the i), writing K_i for the union of an arbitrary finite number k of the $\Delta_{s,i}$ included in ω, we obtain (4.4) with k instead of $k_i(\omega)$. Therefore, $\ell = \infty$. Hence, in this case also, $\ell \geq |\omega|$. □

4.2. Integral of a uniformly continuous function

Let us define the vector space $\mathcal{B}(\Omega; E)$ of functions on which the Cauchy integral will be defined later.

DEFINITION 4.7.– *Let Ω be an open subset of $\mathbb{R}^d$ and E a separated semi normed space. We denote:*

$$\mathcal{B}(\Omega; E) \stackrel{\text{def}}{=} \{\textit{uniformly continuous functions from } \Omega \textit{ into } E \textit{ with bounded support}\}.$$

$\blacksquare$

Utility of $\mathcal{B}(\Omega; E)$. The integral of continuous functions is often introduced on $\mathcal{K}(\Omega; E)$, as in [BOURBAKI, 11, Chapter III, § 3, Definition 1, p. 75]. We prefer to define it on $\mathcal{B}(\Omega; E)$ so that we can apply it to functions like constants that are not zero in the neighborhood of the boundary. □

Notation $\mathcal{B}(\Omega; E)$**.** This notation is **non-standard** because the space itself is non-standard. □

Topology on $\mathcal{B}(\Omega; E)$**.** We do not endow $\mathcal{B}(\Omega; E)$ with semi-norms because we will not need them. □

Let us compare the space $\mathcal{B}(\Omega; E)$ with the spaces $\mathcal{K}(\Omega; E)$ (continuous functions with compact support), $\mathbf{C}_\mathrm{b}(\Omega; E)$ (uniformly continuous and bounded functions), and $\mathbf{C}_D(\Omega; E)$ (uniformly continuous functions with support in a compact set D).

THEOREM 4.8.– *Let Ω be an open subset of $\mathbb{R}^d$, D a compact subset of $\mathbb{R}^d$ and E a separated semi-normed space. Then:*

(a)
$$\mathcal{K}(\Omega; E) \subset \mathcal{B}(\Omega; E),$$
$$\mathbf{C}_D(\Omega; E) \subset \mathcal{B}(\Omega; E) \subset \mathbf{C}_\mathrm{b}(\Omega; E).$$

(b) *If Ω is bounded,*
$$\mathcal{B}(\Omega; E) = \mathbf{C}_\mathrm{b}(\Omega; E).$$

(c)
$$\mathcal{K}(\mathbb{R}^d; E) = \mathcal{B}(\mathbb{R}^d; E). \blacksquare$$

Proof. (a) *Inclusion* $\mathcal{K}(\Omega; E) \subset \mathcal{B}(\Omega; E)$. Every function in $\mathcal{K}(\Omega; E)$ is uniformly continuous (Theorem 1.19) and its support is compact (by Definition 1.17 of $\mathcal{K}(\Omega; E)$) and hence bounded (Theorem A.19 (a)), so it belongs to $\mathcal{B}(\Omega; E)$.

Inclusion $\mathbf{C}_D(\Omega; E) \subset \mathcal{B}(\Omega; E)$. This follows from Definitions 1.7 (b) and 4.7 of these spaces.

Inclusion $\mathcal{B}(\Omega; E) \subset \mathbf{C}_\mathrm{b}(\Omega; E)$. Every function f in $\mathcal{B}(\Omega; E)$ is uniformly continuous and the set $\mathcal{O} = \{x \in \Omega : f(x) \neq 0_E\}$ is bounded (by Definition 4.7 of $\mathcal{B}(\Omega; E)$) in $\mathbb{R}^d$ and hence precompact (Theorem A.23 (b)), so $f(\mathcal{O})$ is precompact (Theorem A.33) and hence bounded (Theorem A.19 (a)). Therefore, f is bounded and consequently belongs to $\mathbf{C}_\mathrm{b}(\Omega; E)$ (Definition 1.4 (b)).

(b) If Ω is bounded, any given function in $\mathbf{C}_\mathrm{b}(\Omega; E)$ belongs to $\mathcal{B}(\Omega; E)$ because its support is included in Ω and hence bounded. The reverse inclusion is given by (a).

(c) Every function in $\mathcal{B}(\mathbb{R}^d; E)$ belongs to $\mathcal{K}(\mathbb{R}^d; E)$ because its support is closed (by Definition 1.5 of the support, for a function defined on $\mathbb{R}^d$) and bounded on $\mathbb{R}^d$ (by Definition 4.7 of $\mathcal{B}(\Omega; E)$) and hence compact by the Borel–Lebesgue theorem (Theorem A.23 (b)). The reverse inclusion is given by (a). □

Let us define the **Cauchy integral**[2] of a uniformly continuous function with bounded support, taking values in a Neumann space.

DEFINITION 4.9.– *Let $f \in \mathcal{B}(\Omega; E)$, where Ω is an open subset of $\mathbb{R}^d$ and E is a Neumann space, and let ω be an open set included in Ω.*

(a) *The **approximate integral**[3] of f over ω is the element of E given, for $n \in \mathbb{N}^*$, by*

$$\mathbb{S}_\omega^n f \stackrel{\text{def}}{=} \sum_{s \in \mathbb{Z}^d : \Delta_{s,n} \subset \omega} 2^{-nd} f(2^{-n} s),$$

where $\Delta_{s,n}$ is the closed cube in $\mathbb{R}^d$ of side length 2^{-n} centered at $2^{-n}s$.

(b) *The **integral** of f over ω is the element of E defined by*[4]

$$\int_\omega f \stackrel{\text{def}}{=} \lim_{n \to \infty} \mathbb{S}_\omega^n f. \blacksquare$$

We write $\int_\omega f(x)\,\mathrm{d}x$ when we want to specify the variable with respect to which we are integrating (e.g. in the expression $\int_\omega f(x, y)\,\mathrm{d}x$).

---

2 **History of the integral of a continuous function. Real integral.** Augustin CAUCHY defined the integral of a real continuous function on a real interval in 1823 in *Leçons à l'École Royale Polytechnique sur le calcul infinitésimal* [21, pp. 122–126] by the method that we use in Definition 4.9. He showed [p. 125] that the approximate sums $\mathbb{S}_\omega^n f$ converge to a limit using the uniform continuity of the function (which was only shown later by Eduard HEINE [48] in 1872). He also introduced the integral of a function defined on $\mathbb{R}^2$ in 1825 [22].

Precursors. Johann BERNOULLI wrote the first treatise on differential and integral calculus between 1691–1692 [7], but the section on integral calculus was only published in 1742. In 1768, Leonhard EULER published his important treatise *Institutionum calculi integralis*, in which he used the sums $\mathbb{S}^n f$ [35, p. 178] with the assumption that they converge. He calculated integrals on $\mathbb{R}^2$ in 1770 [36]. For more details, see *Histoire de l'intégration* by Jean-Paul PIER [62, pp. 79–86].

Vector-valued integral. The integral of a function taking values in a Banach space E was defined for integrable functions in 1933 by Salomon BOCHNER [9] using a different method.
In 1966 [11, Chapter III, § 3, Definition 1, p. 75], Nicolas BOURBAKI defined the integral of a function in $\mathcal{K}(\Omega; E\text{-weak})$ where E is a *Hausdorff locally convex topological vector space*, a notion that is equivalent to *separated semi-normed space*. If E-weak is sequentially complete, this integral coincides with the Cauchy integral taking values in E-weak; if not, it takes values in the completion $\breve{E}$ of E, which coincides with the completion $\widehat{E\text{-weak}}$ of E-weak.

Originality. Extending the Cauchy integral to take values in a Neumann space E appears to be new, although it is entirely straightforward.

3 **Terminology.** The approximate integrals $\mathbb{S}_\omega^n f$ are sometimes called *Riemann sums* in reference to a later generalization by Bernhard RIEMANN in his habilitation dissertation of 1854, published posthumously in 1867 [65, p. 239]. For continuous functions, according to BOURBAKI [13, p. 65], "*we should instead call them Archimedes or Eudoxus sums*".

4 **History of notation for the integral.** The symbol $\int$, "*an elongated* s *for the first letter of the word* summa", was introduced by Gottfried von LEIBNIZ in 1675 in a letter to Henry OLDENBURG, reproduced by Florian CAJORI [16, Vol. 2, p. 243]. The notation $\int_a^b f$ was introduced by Joseph FOURIER [37, p. 252] in 1822. The term *integral* was introduced in 1690 by Jacob BERNOULLI [6] (see facsimile in [62, p. 78]).

Convention. The definition of the approximate integral does not make sense if no cube $\Delta_{s,n}$ included in ω exists, since the sum of zero elements is not defined. We can avoid this pitfall with the convention that

$$\mathbb{S}_\omega^n f \stackrel{\text{def}}{=} 0_E \text{ if no } \Delta_{s,n} \text{ is included in } \omega. \tag{4.5}$$

In particular, $\mathbb{S}_\emptyset^n f = 0_E$ and

$$\int_\emptyset f = 0_E. \ \square$$

The following result justifies the definition of the integral.

LEMMA 4.10.– *In Definition 4.9, $\mathbb{S}_\omega^n f$ is well defined and converges to a limit as $n \to \infty$.* ▌

Proof. The sum $\mathbb{S}_\omega^n f$ is indeed defined, if necessary by (4.5), because $f(2^{-n}s)$ vanishes except for, at most, a finite number of s (as f is zero outside of some bounded set).

To show that the sequence $(\mathbb{S}_\omega^n f)_{n\in\mathbb{N}^*}$ has a limit, it suffices to check that it is Cauchy because E is sequentially complete. First, observe that we can reduce to the case where ω is bounded, since the support of f is bounded and therefore included in some open cube of side length c centered at 0, and so we do not change the value of $\mathbb{S}_\omega^n f$ by replacing ω with $\omega \cap \lambda$, where λ is the open cube of side length $c+1$ centered at 0 (any such cube contains all the cubes $\Delta_{s,n}$ centered on the support of f, and the contribution of the others is zero).

Thus, suppose that ω is bounded, so $|\omega| < \infty$ (Theorem 4.2 (b)), and let us calculate $\mathbb{S}_\omega^m f - \mathbb{S}_\omega^n f$, where $m \geq n$. Each cube $\Delta_{s,n}$ may be divided into $2^{(m-n)d}$ cubes $\Delta_{r,m}$, so

$$\mathbb{S}_\omega^n f = \sum_{s\,:\,\Delta_{s,n}\subset\omega} 2^{-nd} f(2^{-n}s) = \sum_{s\,:\,\Delta_{s,n}\subset\omega} 2^{-nd} \sum_{r\,:\,\Delta_{r,m}\subset\Delta_{s,n}} 2^{(n-m)d} f(2^{-n}s).$$

There are also cubes $\Delta_{r,m} \subset \omega$ that were not in any of the $\Delta_{s,n} \subset \omega$. Denoting by $\mathcal{Y}_{m,n}$ the set of these r, it follows that

$$\begin{aligned}\mathbb{S}^m f &= \sum_{r\,:\,\Delta_{r,m}\subset\omega} 2^{-md} f(2^{-m}r) = \\ &= \sum_{s\,:\,\Delta_{s,n}\subset\omega} 2^{-nd} \sum_{r\,:\,\Delta_{r,m}\subset\Delta_{s,n}} 2^{(n-m)d} f(2^{-m}r) + \mathcal{Y}_{m,n},\end{aligned}$$

where

$$\mathcal{Y}_{m,n} = \sum_{r\in\mathcal{Y}_{m,n}} 2^{-md} f(2^{-m}r).$$

Therefore,

$$\mathbb{S}_\omega^m f - \mathbb{S}_\omega^n f = \mathcal{X}_{n,m} + \mathcal{Y}_{m,n}, \tag{4.6}$$

where

$$\mathcal{X}_{n,m} = \sum_{s\,:\,\Delta_{s,n}\subset\omega} 2^{-nd} \sum_{r\,:\,\Delta_{r,m}\subset\Delta_{s,n}} 2^{(n-m)d}\big(f(2^{-m}r) - f(2^{-n}s)\big).$$

The cube $\Delta_{s,n}$ is included in the ball centered at $2^{-n}s$ of radius $2^{-n-1}\sqrt{d}$, so, when $\Delta_{r,m} \subset \Delta_{s,n}$, we have $|2^{-m}r - 2^{-n}s| \le 2^{-n-1}\sqrt{d}$. Since there are k_n cubes $\Delta_{s,n}$ in ω and $2^{(m-n)d}$ cubes $\Delta_{r,m}$ in each $\Delta_{s,n}$,

$$\|\mathcal{X}_{n,m}\|_{E;\nu} \le 2^{-nd}k_n \sup_{|y-x|\le 2^{-n-1}\sqrt{d}} \|f(y) - f(x)\|_{E;\nu}, \tag{4.7}$$

which tends to 0 as $n \to \infty$ because $2^{-nd}k_n \le |\omega|$ (inequality (4.2)) and f is uniformly continuous by Definition 4.7 of $\mathcal{B}(\Omega; E)$.

Moreover, there are $2^{(m-n)d}k_n$ cubes $\Delta_{r,m}$ that are included in one of the k_n cubes $\Delta_{s,n} \subset \omega$, so there are $k_m - 2^{(m-n)d}k_n$ indexes r in $R_{m,n}$. Therefore,

$$\|\mathcal{Y}_{m,n}\|_{E;\nu} \le (2^{-md}k_m - 2^{-nd}k_n) \sup_{x\in\omega} \|f(x)\|_{E;\nu}, \tag{4.8}$$

which also tends to 0 as $n \to \infty$ because $2^{-nd}k_n \to |\omega|$ (by Definition 4.1 of the measure) and f is bounded (Theorem 4.8 (a)).

Hence, as $n \to \infty$,

$$\sup_{m\ge n} \|\mathbb{S}_\omega^m f - \mathbb{S}_\omega^n f\|_{E;\nu} \le \|\mathcal{X}_{n,m}\|_{E;\nu} + \|\mathcal{Y}_{n,m}\|_{E;\nu} \to 0.$$

This proves that $(\mathbb{S}_\omega^n f)_{n\in\mathbb{N}^*}$ is a Cauchy sequence. □

Examples of Neumann spaces. The following spaces are Neumann spaces, which allows us to use Cauchy integrals taking values in them (in Volume 4, we will define integrable distributions, valued in such spaces, more generally):

— Banach and Fréchet spaces, and in particular $\mathbb{R}$ and $\mathbb{R}^d$.

— The Lebesgue and Sobolev spaces $L^p(\Omega)$ and $W^{m,p}(\Omega)$ and their *local* variants $L^p_{\rm loc}(\Omega)$ and $W^{m,p}_{\rm loc}(\Omega)$ and vector-valued variants $L^p(\Omega; E)$, $W^{m,p}(\Omega; E)$, $L^p_{\rm loc}(\Omega; E)$ and $W^{m,p}_{\rm loc}(\Omega; E)$ for m any positive or negative integer, $1 \le p \le \infty$, and E a Neumann space [Vol. 4].

— The spaces of continuously differentiable functions $\mathcal{C}^m(\Omega)$, $\mathcal{C}_{\rm b}(\Omega)$, $\mathbf{C}_{\rm b}(\Omega)$, $\mathcal{C}^m_K(\Omega)$ and $\mathcal{K}^m(\Omega)$ for $m \in \mathbb{N}^*$ and $m = \infty$ (Theorem 2.24) and the distribution spaces $\mathcal{D}'(\Omega)$ [Vol. 3], and their variants taking values in a Neumann space.

— The *weak* space E-weak of any Hilbert, reflexive or semi-reflexive space E [Vol. 1, Theorems 17.7 and 17.12].

— The duals E', E'', E''', ... of any metrizable normed or semi-normed space E [Vol. 1, Theorem 13.8 (a) and Property (17.3), p. 272].

— The weak duals E'-weak, E''-weak, E'''-weak, ... of any Hilbert or reflexive space E [Vol. 1, Property (17.6), p. 272].

— The weak-$*$ duals E'-$*$weak, E''-$*$weak, E'''-$*$weak, ... of any Hilbert, Banach, Fréchet or reflexive space E [Vol. 1, Theorem 13.8 (b) and Properties (17.4) and (17.6)].

— Any of their closed or sequentially closed subspaces (Theorem A.26). □

4.3. Case where *E* is not a Neumann space

The assumption that E is a Neumann space is the most general hypothesis possible if we wish to guarantee that the integral belongs to E. Indeed, if we omit this hypothesis from Definition 4.9 of the Cauchy integral, we can no longer show that the approximate integrals $\mathbb{S}^n_\omega f$ converge for every f. If E is metrizable, we can even construct a function f for which they do not converge:

$$\left\{\begin{array}{l}\textit{If } \Omega \textit{ is a non-empty open subset of } \mathbb{R}^d \textit{ and } E \textit{ is a metrizable semi-normed space} \\ \textit{that is not sequentially complete, then there exists } f \in \mathcal{K}(\Omega; E) \textit{ for which the sums} \\ \mathbb{S}^n_\Omega f \textit{ stated in Definition 4.9 (a) do not converge in } E.\end{array}\right. \tag{4.9}$$

More precisely, let $\widehat{E}$ be a **sequential completion** of E, namely a Fréchet space in which E is sequentially dense and whose semi-norms extend the semi-norms of E, and let $\widehat{f}$ be the function f taking values in $\widehat{E}$. Then we can choose f in such a way that the following properties also hold:

$$\widehat{f} \in \mathcal{K}(\Omega; \widehat{E}), \quad \int_\Omega \widehat{f} \in \widehat{E} \setminus E, \quad \mathbb{S}^n_\Omega \widehat{f} \to \int_\Omega \widehat{f} \text{ in } \widehat{E}, \quad \mathbb{S}^n_\Omega f = \mathbb{S}^n_\Omega \widehat{f}. \tag{4.10}$$

Proof of (4.9) and (4.10). We will construct a function f whose sums $\mathbb{S}^n_\Omega f$ form a Cauchy sequence in E that converges in $\widehat{E}$ to a limit u that belongs to $\widehat{E} \setminus E$.

Since E is metrizable, it can [Vol. 1, Theorem 4.4] be endowed with an increasing sequence of semi-norms, say $(\| \; \|_m)_{m\in\mathbb{N}}$. Let $\widehat{E}$ be a sequential completion of E [Vol. 1, Theorem 4.25] with semi-norms $(\||\; \||_m)_{m\in\mathbb{N}}$ extending the semi-norms of E. Since E is not sequentially complete, it has a Cauchy sequence $(u_n)_{n\in\mathbb{N}}$ that does not converge. This sequence is also Cauchy in $\widehat{E}$, so it has a limit u that belongs to $\widehat{E} \setminus E$.

Let $z \in \Omega$ and $(B_n)_{n\in\mathbb{N}}$ a sequence of pairwise disjoint balls $B_n = \{x \in \mathbb{R}^d : |x - z_n| \le r_n\}$ included in Ω such that $r_n > 0$, $r_n \to 0$, and $z_n \to z$. Taking a subsequence of the u_n if necessary, we may assume

$$\|u_n - u_{n-1}\|_n \le \frac{1}{n}|r_n|^d.$$

Indeed, simply consider a subsequence $(u_{\sigma(n)})_{n\in\mathbb{N}}$ where, for each n, $\sigma(n)$ is chosen to satisfy

$$\||u_{\sigma(n)} - u\||_n \le \frac{1}{2}\inf\Big\{\frac{1}{n}|r_n|^d, \frac{1}{n+1}|r_{n+1}|^d\Big\}.$$

Finally, let $\phi \in \mathcal{C}(\mathbb{R}^d)$ such that $\operatorname{supp}\phi \subset \{x \in \mathbb{R}^d : |x| < 1\}$ and $\int_{\mathbb{R}^d} \phi = 1$. Now, define the function f by

$$f(x) = \begin{cases} |r_0|^{-d}\,\phi((x - z_0)/r_0)\,u_0 & \text{if } x \in B_0, \\ |r_n|^{-d}\,\phi((x - z_n)/r_n)\,(u_n - u_{n-1}) & \text{if } x \in B_n, n \ge 1, \\ 0_E & \text{if } x \notin \bigcup_n B_n. \end{cases} \tag{4.11}$$

This function is continuous at the point z because $\sup_{x\in B_n} \|f(x)\|_n \le \sup_{y\in\mathbb{R}^d} |\phi(y)|/n$ and is therefore continuous on the whole of Ω. Its support is compact and included in Ω, so it is uniformly continuous from Ω into E.

Denote by $\widehat{f}$ the same function taking values in $\widehat{E}$. It is uniformly continuous because $\||\;\||_m = \|\;\|_m$ on E, and its support is compact. In other words, $\widehat{f} \in \mathcal{K}(\Omega; \widehat{E})$. Moreover, $\int_{\mathring{B}_n} \widehat{f} = u_n - u_{n-1}$ and $\int_{\mathring{B}_0} \widehat{f} = u_0$, so, writing $\omega_n = \bigcup_{0\le i\le n} B_i$, by Theorem 4.21,

$$\int_{\omega_n} \widehat{f} = \sum_{1\le i\le n} \int_{\mathring{B}_i} \widehat{f} = u_n.$$

As $n \to \infty$, we have $\int_{\omega_n} \widehat{f} \to \int_\omega \widehat{f}$, where $\omega = \bigcup_{i\in\mathbb{N}} \mathring{B}_n$ (Theorem 4.19). Therefore, $\int_\omega \widehat{f} = u$. Since f is zero outside of ω, by Theorem 4.17, it follows that

$$\int_\Omega \widehat{f} = u.$$

By Definition 4.9 of the integral, $\mathbb{S}^n_\Omega \widehat{f} \to u$ in $\widehat{E}$, which proves (4.10). Since $\mathbb{S}^n_\Omega f = \mathbb{S}^n_\Omega \widehat{f}$, it does not converge in E because u does not belong to E, which proves (4.9). □

Integral in the sequential completion $\widehat{E}$ when E is not a Neumann space. If $f \in \mathcal{B}(\Omega; E)$, where E is a semi-normed space that is not sequentially complete, the simplest way to "integrate" f is to consider its image $\widehat{f}$ taking values in some sequential completion $\widehat{E}$, i.e. a space such that:

$$\begin{cases} \widehat{E} \textit{ is a Neumann space in which } E \textit{ is included and dense,} \\ \textit{whose semi-norms extend the semi-norms of } E, \textit{ and such that:} \\ \widehat{E} \textit{ is the only sequentially closed subspace containing } E. \end{cases} \qquad (4.12)$$

Then $\widehat{f} \in \mathcal{B}(\Omega; \widehat{E})$ and we can define $\int_\Omega \widehat{f}$.

Recall that any given separated semi-normed space E has a sequential completion [Vol. 1, Theorem 4.24, p. 81]; in fact, it has infinitely many, but between any two of them there is a bijection that is sequentially continuous, as well as its inverse. We could also consider a completion of E, i.e. a complete separated semi-normed space in which E is included and dense and whose semi-norms extend those of E.

Another possibility would be to define $\int_\Omega f$ as the limit in $\widehat{E}$ of the sums $\mathbb{S}^n_\Omega f$ from Definition 4.9 (a). This would give the same value as above but would significantly complicate the presentation; first, we would have to juggle between E and $\widehat{E}$, and, second, $\widehat{E}$ would only be determined up to isomorphism. But this complication would not provide any additional properties in return. □

4.4. Properties of the integral

Let us state some fundamental properties of the Cauchy integral.

THEOREM 4.11.– *Let $f \in \mathcal{B}(\Omega; E)$ and $g \in \mathcal{B}(\Omega; E)$, where Ω is an open subset of $\mathbb{R}^d$ and E is a Neumann space, ω is an open subset included in Ω and $t \in \mathbb{R}$. Then:*

(a)
$$\int_\omega f + g = \int_\omega f + \int_\omega g, \quad \int_\omega tf = t \int_\omega f.$$

(b) *For every semi-norm $\| \ \|_{E;\nu}$ of E, we have $\|f\|_{E;\nu} \in \mathcal{B}(\Omega)$ and*

$$\Big\| \int_\omega f \Big\|_{E;\nu} \le \int_\omega \|f\|_{E;\nu} \le \int_\Omega \|f\|_{E;\nu}. \blacksquare$$

Proof. (a) The stated equalities are obtained by taking the limit of the equalities $\mathbb{S}^n_\omega(f+g) = \mathbb{S}^n_\omega f + \mathbb{S}^n_\omega g$ and $\mathbb{S}^n_\omega(tf) = t\,\mathbb{S}^n_\omega f$ satisfied by the approximate integrals (Definition 4.9), which are clear.

(b) The function $\|f\|_{E;\nu}$ belongs to $\mathbf{C}_\mathrm{b}(\Omega)$ (Theorem 3.3, since $f \in \mathbf{C}_\mathrm{b}(\Omega; E)$, see Theorem 4.8 (a)) and its support is bounded (because it is included in the support of f); therefore, it belongs to $\mathcal{B}(\Omega)$ (Definition 4.7).

The stated inequalities are thus obtained by taking the limit of the inequalities $\|\mathbb{S}^n_\omega f\|_{E;\nu} \le \mathbb{S}^n_\omega(\|f\|_{E;\nu}) \le \mathbb{S}^n_\Omega(\|f\|_{E;\nu})$, which are clear. □

Let us state some elementary properties of the Cauchy integral.

THEOREM 4.12.– *Let $f \in \mathcal{B}(\Omega; E)$, where Ω is an open subset of $\mathbb{R}^d$ and E is a Neumann space, and let ω be an open set included in Ω. Then:*

(a) *The restriction to* ω *satisfies* $f|_\omega \in \mathcal{B}(\omega; E)$ *and*

$$\int_\omega f|_\omega = \int_\omega f.$$

(b) *If* ω *is bounded,*

$$|\omega| = \int_\omega 1.$$

(c) *If* $f \in \mathcal{B}(\Omega)$ *and* $e \in E$*, then* $fe \in \mathcal{B}(\Omega; E)$ *and*

$$\int_\omega fe = \Big(\int_\omega f\Big)e. \blacksquare$$

Proof. (a) Like f itself, the restriction $f|_\omega$ is uniformly continuous with bounded support, so it belongs to $\mathcal{B}(\omega; E)$. The stated equality is obtained by taking the limit of the equality $\mathbb{S}^n_\omega(f|_\omega) = \mathbb{S}^n_\omega f$ satisfied by the approximate integrals (Definition 4.9), which is clear.

(b) By Definition 4.9, $\mathbb{S}^n_\omega 1 = \sum_{s\in\mathbb{Z}^d:\Delta_{s,n}\subset\omega} 2^{-nd} = 2^{-nd}k_n$, where k_n is the number of cubes $\Delta_{s,n}$ included in ω. In the limit, $\int_\omega 1 = |\omega|$ by Definition 4.1 of the measure.

(c) The function fe is uniformly continuous, like f itself, and shares the same support (unless $e = 0_E$ in which case its support is empty), so it belongs to $\mathcal{B}(\Omega; E)$. The stated equality is obtained by taking the limit of the equality $\mathbb{S}^n_\omega(fe) = (\mathbb{S}^n_\omega f)e$, which is clear. □

Let us show that integration commutes with continuous linear mappings.

THEOREM 4.13.– *Let* $f \in \mathcal{B}(\Omega; E)$*, where* Ω *is an open subset of* $\mathbb{R}^d$ *and* E *is a Neumann space. Additionally, let* L *be a continuous, or sequentially continuous, linear mapping from* E *into a Neumann space* F*.*

Then $Lf \in \mathcal{B}(\Omega; F)$ *and, for every open set* ω *included in* Ω*,*

$$\int_\omega Lf = L\int_\omega f. \blacksquare$$

Proof. It suffices to assume that L is sequentially continuous, since continuity implies sequential continuity (Theorem A.29).

Then the function $Lf = L \circ f$ is uniformly continuous by Theorem A.36 as a uniformly continuous mapping (note that f is uniformly continuous by Definition 4.7 of $\mathcal{B}(\Omega; E)$) defined on a subset of a metrizable space (here, $\mathbb{R}^d$) composed with a sequentially continuous linear mapping. Furthermore, its support is bounded (because it is included in the support of f), and so the function Lf indeed belongs to $\mathcal{B}(\Omega; F)$.

The stated equality is obtained by taking the limit of the equality satisfied by the approximate integrals (Definition 4.9), namely $\mathbb{S}^n_\omega(Lf) = L(\mathbb{S}^n_\omega f)$, which is clear. □

Let us state some growth properties of the integrals of real-valued functions. Recall that $f \leq g$ means that $f(x) \leq g(x)$ for every x.

THEOREM 4.14.– *Let $f \in \mathcal{B}(\Omega)$ and $g \in \mathcal{B}(\Omega)$, where Ω is an open subset of $\mathbb{R}^d$, and let ω and λ be open sets included in Ω. Then:*

(a) $$f \leq g \quad \Rightarrow \quad \int_\omega f \leq \int_\omega g;$$

(b) $$\omega \subset \lambda,\ f \geq 0 \quad \Rightarrow \quad \int_\omega f \leq \int_\lambda f;$$

(c) $$f \neq g,\ f \leq g \quad \Rightarrow \quad \int_\Omega f < \int_\Omega g;$$

(d) $$f \neq 0,\ f \geq 0 \quad \Rightarrow \quad \int_\Omega f > 0. \blacksquare$$

Proof. (a) and (b). The stated inequalities may be obtained by taking the limit of the inequalities $\mathbb{S}^n_\omega f \leq \mathbb{S}^n_\omega g$ and $\mathbb{S}^n_\omega f \leq \mathbb{S}^n_\lambda f$ satisfied by the approximate integrals (Definition 4.9), which are clear.

(d) If $f \neq 0$, there exists $y \in \Omega$ such that $f(y) = a > 0$, so, since f is also continuous, there exists a (non-empty) open ball $\omega \subset \Omega$ on which $f \geq a/2$. Then, by (b), (a) and Theorems 4.11 (a), 4.12 (b) and 4.2 (c) respectively,

$$\int_\Omega f \geq \int_\omega f \geq \int_\omega \frac{a}{2} = \frac{a}{2} \int_\omega 1 = \frac{a}{2} |\omega| > 0.$$

(c) Here, linearity and (d) imply that $\int_\Omega g - \int_\Omega f = \int_\Omega g - f > 0$. □

4.5. Dependence of the integral on the domain of integration

Let us bound the semi-norms of the integral, in terms of the measure of the domain of integration.

THEOREM 4.15.– *Let $f \in \mathcal{B}(\Omega; E)$, where Ω is an open subset of $\mathbb{R}^d$ and E is a Neumann space, and let ω be an open subset included in Ω. Then, for every semi-norm $\| \, \|_{E;\nu}$ of E,*

$$\Big\| \int_\omega f \Big\|_{E;\nu} \leq |\omega| \sup_{x \in \omega} \|f(x)\|_{E;\nu}. \blacksquare$$

Proof. By Definition 4.9 (a), the approximate integral of f over the set ω is equal to $\mathbb{S}^n_\omega f = \sum_{s \in \mathbb{Z}^d : \Delta_{s,n} \subset \omega} 2^{-nd} f(2^{-n}s)$. Therefore,

$$\|\mathbb{S}^n_\omega f\|_{E;\nu} \leq 2^{-nd} k_n \sup_{x \in \omega} \|f(x)\|_{E;\nu},$$

where k_n is the number of cubes $\Delta_{s,n}$ included in ω. Taking the limit as $n \to \infty$ gives the stated inequality because, by Definition 4.1 of the measure, $2^{-nd} k_n \to |\omega|$. □

Let us state an analogous inequality for a difference of integrals.

THEOREM 4.16.– *Let $f \in \mathcal{B}(\Omega; E)$, where Ω is an open subset of $\mathbb{R}^d$ and E is a Neumann space. Additionally, let ω_1 and ω_2 be open subsets such that $\omega_1 \subset \omega_2 \subset \Omega$ and $|\omega_1| < \infty$ (or the stronger assumption $|\omega_2| < \infty$). Then, for every semi-norm $\| \, \|_{E;\nu}$ of E,*

$$\Big\| \int_{\omega_2} f - \int_{\omega_1} f \Big\|_{E;\nu} \leq (|\omega_2| - |\omega_1|) \sup_{x \in \omega_2 \setminus \omega_1} \|f(x)\|_{E;\nu}. \blacksquare$$

Remark. This inequality is not meaningful if $|\omega_1| = \infty$, because then $|\omega_2| = \infty$ and $|\omega_2| - |\omega_1|$ is not defined. □

Proof of Theorem 4.16. The approximate integrals (Definition 4.9) satisfy

$$\mathbb{S}^n_{\omega_2} f - \mathbb{S}^n_{\omega_1} f = \sum_{s \in \mathcal{S}_n} 2^{-nd} f(2^{-n}s),$$

where $\mathcal{S}_n = \{s \in \mathbb{Z}^d : \Delta_{s,n} \subset \omega_2, \ \Delta_{s,n} \not\subset \omega_1\}$. Therefore,

$$\|\mathbb{S}^n_{\omega_2} f - \mathbb{S}^n_{\omega_1} f\|_{E;\nu} \leq 2^{-nd}(k_n(\omega_2) - k_n(\omega_1)) \sup_{s \in \mathcal{S}_n} \|f(2^{-n}s)\|_{E;\nu}, \qquad (4.13)$$

where $k_n(\omega)$ is the number of cubes $\Delta_{s,n}$ included in ω. If a cube $\Delta_{s,n}$ is included in ω_2 but not in ω_1, then it contains some point $x \in \omega_2 \backslash \omega_1$; thus $|x - 2^{-n}s| \leq 2^{-n-1}\sqrt{d}$ because $\Delta_{s,n}$ is included in the ball centered at $2^{-n}s$ of radius $2^{-n-1}\sqrt{d}$, so

$$\sup_{s \in S_n} \|f(2^{-n}s)\|_{E;\nu} \leq \sup_{x \in \omega_2 \backslash \omega_1} \|f(x)\|_{E;\nu} + \sup_{|y-x| \leq 2^{-n-1}\sqrt{d}} \|f(y) - f(x)\|_{E;\nu}.$$

As $n \to \infty$, the second supremum tends to 0 because f is uniformly continuous, so the inequality (4.13) gives the stated inequality in the limit, since $2^{-nd}k_n(\omega) \to |\omega|$ by Definition 4.1 of the measure. □

Let us show that the integral does not change if we restrict attention to the domain where the function is non-zero.

THEOREM 4.17.– *Let $f \in \mathcal{B}(\Omega; E)$, where Ω is an open subset of $\mathbb{R}^d$ and E is a Neumann space. Additionally, let ω be an open set included in Ω. Suppose that*

$$f = 0_E \text{ on } \Omega \setminus \omega.$$

Then:

(a)
$$\int_\Omega f = \int_\omega f.$$

(b) *For every semi-norm $\| \ \|_{E;\nu}$ of E,*

$$\left\| \int_\Omega f \right\|_{E;\nu} \leq |\omega| \sup_{x \in \Omega} \|f(x)\|_{E;\nu}.$$

(c) *The properties (a) and (b) are satisfied if $\omega = \{x \in \Omega : f(x) \neq 0_E\}$ or if $\omega = \overset{\circ}{\operatorname{supp}} f$ (both of these sets are open and bounded).* ▮

Proof. (a) If Ω is bounded, then $|\Omega| < \infty$ by Theorem 4.2 (b), and by Theorem 4.16,

$$\left\| \int_\Omega f - \int_\omega f \right\|_{E;\nu} \leq (|\Omega| - |\omega|) \sup_{x \in \Omega \backslash \omega} \|f(x)\|_{E;\nu} = (|\Omega| - |\omega|) \times 0 = 0.$$

This holds for every semi-norm of E, so $\int_\Omega f = \int_\omega f$.

If Ω is not bounded, we can reduce to the case where it is by observing that

$$\int_\Omega f = \int_{\Omega \cap \lambda} f,$$

where λ is the open ball containing $\operatorname{supp} f + Q$ and Q is the cube of side length 1 centered at 0. Indeed, the approximate integral (Definition 4.9 (a)) then satisfies $\mathbb{S}^n_\Omega f = \mathbb{S}^n_{\Omega \cap \lambda} f$ because $f(2^{-n}s) = 0_E$ if $\Delta_{s,n} \not\subset \lambda$.

(b) This inequality follows from (a) and Theorem 4.15.

(c) The set $\lambda = \{x \in \Omega : f(x) \neq 0_E\}$ is open by Theorem 1.6 (b) because Ω is open and $f = 0_E$ on $\Omega \setminus \lambda$.

The set $\overset{\circ}{\operatorname{supp}} f$ is also open by Definition A.8 of the interior, and $f = 0_E$ on $\Omega \setminus \overset{\circ}{\operatorname{supp}} f$ by Theorem 1.6 (b), since Ω is open.

Therefore, (a) and (b) are satisfied if $\omega = \lambda$ or $\omega = \overset{\circ}{\operatorname{supp}} f$. These two sets are bounded, since they are included in the support of f (by Definition 1.5 of the support), which is bounded (by Definition 4.7 of $\mathcal{B}(\Omega; E)$). □

Let us show that, for a function with compact support, we can reduce to an integral over the whole of $\mathbb{R}^d$.

THEOREM 4.18.– *Let $f \in \mathcal{K}(\Omega; E)$, where Ω is an open subset of $\mathbb{R}^d$ and E is a Neumann space.*

Then the extension $\widetilde{f}$ of f by 0_E outside of Ω belongs to $\mathcal{K}(\mathbb{R}^d; E)$, and

$$\int_\Omega f = \int_{\mathbb{R}^d} \widetilde{f}. \blacksquare$$

Proof. This follows from Theorem 4.17 (a) because $\widetilde{f} = 0_E$ on $\mathbb{R}^d \setminus \Omega$. □

Let us show that integration is continuous with respect to increasing open sets.

THEOREM 4.19.– *Let $f \in \mathcal{B}(\Omega; E)$, where Ω is an open subset of $\mathbb{R}^d$ and E is a Neumann space. Additionally, let $(\omega_n)_{n\in\mathbb{N}}$ be an increasing sequence of open subsets of $\mathbb{R}^d$ that are included in Ω, and $\omega = \bigcup_{n\in\mathbb{N}} \omega_n$. Then, as $n \to \infty$,*

$$\int_{\omega_n} f \to \int_\omega f. \blacksquare$$

Proof. If $|\omega| < \infty$, the inequality from Theorem 4.16 gives, for every semi-norm $\| \|_{E;\nu}$ of E, setting $c_\nu = \sup_{x\in\omega} \|f(x)\|_{E;\nu}$,

$$\Big\| \int_{\omega_n} f - \int_\omega f \Big\|_{E;\nu} \leq c_\nu(|\omega| - |\omega_n|),$$

which tends to 0 by Theorem 4.6. Therefore, $\int_{\omega_n} f \to \int_\omega f$.

If $|\omega = \infty|$, we reduce to the above case by considering an open ball λ that contains the support of f, and observing that the contribution of the region outside of λ to the integral is zero (Theorem 4.17) and that $\omega \cap \lambda = \bigcup_{n \in \mathbb{N}} (\omega_n \cap \lambda)$ and $|\omega \cap \lambda| \leq |\lambda| < \infty$ (Theorem 4.2 (a) and (b)), so

$$\int_{\omega_n} f = \int_{\omega_n \cap \lambda} f \to \int_{\omega \cap \lambda} f = \int_\omega f. \;\square$$

4.6. Additivity with respect to the domain of integration

Let us show that every integral can be approximated by an integral of functions with compact support.

THEOREM 4.20.– *Let $f \in \mathcal{B}(\Omega; E)$, where Ω is an open subset of $\mathbb{R}^d$ and E is a Neumann space. Furthermore, let $(\omega_n)_{n \in \mathbb{N}}$ be a sequence of open subsets of $\mathbb{R}^d$ such that*

$$\omega_n \Subset \Omega, \quad \omega_n \textit{ increases with } n, \quad \bigcup_{n \in \mathbb{N}} \omega_n = \Omega,$$

and $(\alpha_n)_{n \in \mathbb{N}}$ a sequence of functions such that

$$\alpha_n \in \mathcal{K}(\Omega), \quad 0 \leq \alpha_n \leq 1, \quad \alpha_n = 1 \textit{ on } \omega_n.$$

(Such sequences exist.)

Then $\alpha_n f \in \mathcal{K}(\Omega; E)$ and, as $n \to \infty$,

$$\int_\Omega \alpha_n f \to \int_\Omega f. \;\blacksquare$$

Proof. Since Theorem 3.11 (c), in particular, implies that $\alpha_n f \in \mathcal{K}(\Omega; E)$, we only need to show convergence.

Let $\epsilon > 0$ and $\mathcal{O} = \{x \in \Omega : f(x) \neq 0_E\}$. The set $\mathcal{O}$ is open and bounded, so, by Theorem 4.5, there exists a compact set $K \subset \mathcal{O}$ such that $|\mathcal{O} \setminus K| \leq \epsilon$. The open sets $(\omega_n)_{n \in \mathbb{N}}$ cover K, so they have a finite subcovering, and they increase with n, so there exists $m \in \mathbb{N}$ such that, for every $n \geq m$, we have $K \subset \omega_n$. The open set $\mathcal{O} \setminus \overline{\omega_n}$ is then included in $\mathcal{O} \setminus K$, so, by Theorem 4.2 (a),

$$|\mathcal{O} \setminus \overline{\omega_n}| \leq |\mathcal{O} \setminus K| \leq \epsilon.$$

But $\alpha_n f = f$ outside of $\mathcal{O} \setminus \overline{\omega_n}$ (since $f = 0_E$ outside of $\mathcal{O}$ and $\alpha_n = 1$ on $\overline{\omega_n}$), so the inequality from Theorem 4.17 (b) states that, for every semi-norm of E,

$$\left\| \int_\Omega \alpha_n f - f \right\|_{E;\nu} \leq |\mathcal{O} \setminus \overline{\omega_n}| \sup_{x\in\Omega} \|(\alpha_n f - f)(x)\|_{E;\nu} \leq \epsilon \sup_{x\in\Omega} \|f(x)\|_{E;\nu}.$$

Since $\sup_{x\in\Omega} \|f(x)\|_{E;\nu} < \infty$ (because f is bounded, see Theorem 4.8 (a)), this shows that

$$\int_\Omega \alpha_n f - f \to 0_E.$$

Observe that the conditions on the ω_n are satisfied, for example, by

$$\omega_n = \Omega^n_{1/n} = \{x \in \Omega : |x| < n,\ B(x, 1/n) \subset \Omega\},$$

where $B(x, 1/n)$ is the closed ball centered at x of radius $1/n$, by Lemma 1.15. The α_n then follow from Urysohn's lemma (Theorem 3.21). □

Let us show that integration is additive with respect to disjoint open sets.

THEOREM 4.21.– *Let $f \in \mathcal{B}(\Omega; E)$, where Ω is an open subset of $\mathbb{R}^d$ and E is a Neumann space. Additionally, let ω_1 and ω_2 be two open sets included in Ω. Then:*

(a)
$$\int_{\omega_1\cup\omega_2} f + \int_{\omega_1\cap\omega_2} f = \int_{\omega_1} f + \int_{\omega_2} f.$$

(b) *If $\omega_1 \cap \omega_2 = \emptyset$,*
$$\int_{\omega_1\cup\omega_2} f = \int_{\omega_1} f + \int_{\omega_2} f.$$

(c) *If $f \in \mathcal{B}(\Omega)$ and $f \geq 0$,*
$$\int_{\omega_1\cup\omega_2} f \leq \int_{\omega_1} f + \int_{\omega_2} f. \blacksquare$$

Proof. (a) Let $(\alpha_{1n})_{n\in\mathbb{N}}$ and $(\alpha_{2n})_{n\in\mathbb{N}}$ be functions satisfying Theorem 4.20 for ω_1 and ω_2, respectively. Their extensions by 0 satisfy

$$\alpha_{1n} + \alpha_{2n} = \alpha_{1n}\alpha_{2n} + \beta_n, \quad \text{where } \beta_n = 1 - (1 - \alpha_{1n})(1 - \alpha_{2n}),$$

so

$$\int_{\omega_1\cup\omega_2} \alpha_{1n} f + \int_{\omega_1\cup\omega_2} \alpha_{2n} f = \int_{\omega_1\cup\omega_2} \alpha_{1n}\alpha_{2n} f + \int_{\omega_1\cup\omega_2} \beta_n f.$$

Since $\alpha_{1n} f$ vanishes outside of ω_1, we obtain $\int_{\omega_1 \cup \omega_2} \alpha_{1n} f = \int_{\omega_1} \alpha_{1n} f$ from Theorem 4.17 (a). Since $\alpha_{2n} f$ and $\alpha_{1n}\alpha_{2n} f$ similarly vanish outside of ω_2 and $\omega_1 \cap \omega_2$, respectively, it follows that

$$\int_{\omega_1} \alpha_{1n} f + \int_{\omega_2} \alpha_{2n} f = \int_{\omega_1 \cap \omega_2} \alpha_{1n}\alpha_{2n} f + \int_{\omega_1 \cup \omega_2} \beta_n f. \qquad (4.14)$$

But $(\alpha_{1n}\alpha_{2n})_{n\in\mathbb{N}}$ and $(\beta_n)_{n\in\mathbb{N}}$ satisfy Theorem 4.20 for the sets $\omega_1 \cap \omega_2$ and $\omega_1 \cup \omega_2$, respectively (because $\alpha_{1n}\alpha_{2n} = 1$ on $\omega_{1n} \cap \omega_{2n}$, which increases to $\omega_1 \cap \omega_2$, and $\beta_n = 1$ on $\omega_{1n} \cup \omega_{2n}$, which increases to $\omega_1 \cup \omega_2$); therefore, by Theorem 4.20, equality (4.14) implies (a) in the limit.

(b) This follows from (a) because $\int_{\omega_1 \cap \omega_2} f = \int_\emptyset f = 0_E$ in this case.

(c) This follows from (a) because, by Theorem 4.14 (a), $\int_{\omega_1 \cap \omega_2} f \geq \int_{\omega_1 \cap \omega_2} 0 = 0$ in this case. □

4.7. Continuity of the integral

Let us show that integrals are continuous in the space $\mathbf{C}_D(\Omega; E)$ of uniformly continuous functions whose support is included in a compact set D.

THEOREM 4.22.– *Let Ω and ω be two open subsets of $\mathbb{R}^d$ such that $\omega \subset \Omega$, E a Neumann space, and D a compact subset of $\mathbb{R}^d$. Then:*

(a) *The mapping $f \mapsto \displaystyle\int_\omega f$ is continuous from $\mathbf{C}_D(\Omega; E)$ into E.*

(b) *If Ω is bounded, $f \mapsto \displaystyle\int_\omega f$ is continuous from $\mathbf{C}_b(\Omega; E)$ into E.* ▮

Proof. (a) Let $f \in \mathbf{C}_D(\Omega; E)$, $\{\| \|_{E;\nu} : \nu \in \mathcal{N}_E\}$ the family of semi-norms of E, and λ a bounded open set containing D. By Definition 1.7 (b) of $\mathbf{C}_D(\Omega; E)$, f is zero outside of D and so in particular on $\omega \setminus \lambda$.

Since $\int_\omega f = \int_\omega f|_\omega$ (Theorem 4.12 (a)), the inequality of Theorem 4.17 (b) applied to $f|_\omega$ yields, together with Definition 1.3 (b) of the semi-norms of $\mathcal{C}_b(\Omega; E)$, for every $\nu \in \mathcal{N}_E$,

$$\Big\| \int_\omega f \Big\|_{E;\nu} = \Big\| \int_\omega f|_\omega \Big\|_{E;\nu} \leq |\omega \cap \lambda| \sup_{x\in\omega} \|f(x)\|_{E;\nu} \leq |\lambda| \|f\|_{\mathcal{C}_b(\Omega;E);\nu}.$$

Since $\mathbf{C}_D(\Omega; E)$ is endowed with the semi-norms of $\mathcal{C}_b(\Omega; E)$ by Definition 1.7 (b), this inequality implies that $\int_\omega$ is continuous from $\mathbf{C}_D(\Omega; E)$ into E by the characterization of continuous linear mappings from Theorem 1.25.

(b) If Ω is bounded, then $\overline{\Omega}$ is compact and the result follows from (a) applied to $D = \overline{\Omega}$ because $\mathbf{C}_{\overline{\Omega}}(\Omega; E) \mathrel{\overset{=}{\leftrightarrows}} \mathbf{C}_b(\Omega; E)$ in this case. Indeed, $\mathbf{C}_{\overline{\Omega}}(\Omega; E) \mathrel{\overset{=}{\leftrightarrows}} \mathbf{C}(\Omega; E)$ (by Definition 1.7 (b) of $\mathbf{C}_D(\Omega; E)$) and $\mathbf{C}(\Omega; E) \mathrel{\overset{=}{\leftrightarrows}} \mathbf{C}_b(\Omega; E)$ (Theorem 1.10 (d) because Ω is bounded). □

In order to bound the integral over a possibly unbounded domain in Theorem 4.24, let us give an upper bound of the integral of the function $x \mapsto (2 + |x|)^{-2-d}$ over a bounded open set that is independent of the set.

LEMMA 4.23.– *The function $x \mapsto 1/(2 + |x|)^{d+2}$ belongs to $\mathcal{B}(\Omega)$ if Ω is a bounded open subset of $\mathbb{R}^d$, and then*

$$\int_\Omega \frac{1}{(2 + |x|)^{d+2}} \, dx \leq 2^{d+1}. \blacksquare$$

Non-optimality. This inequality can be improved but is sufficient for our purposes. □

Proof of Lemma 4.23. The function $x \mapsto 1/(2 + |x|)^{d+2}$ is continuous on the whole of $\mathbb{R}^d$, so, by Heine's theorem (Theorem A.34), it is uniformly continuous on the compact set $\overline{\Omega}$ and its restriction therefore belongs to $\mathcal{B}(\Omega)$ (Definition 4.7).

Let $C_n = \{x \in \mathbb{R}^d : n - 1 < |x| < n + 1\}$. For N sufficiently large, Ω is included in the union $\bigcup_{0 \leq n \leq N} C_n$, so, by Theorems 4.21 (c) and 4.17 (b),

$$\int_\Omega \frac{1}{(2 + |x|)^{d+2}} \, dx \leq \sum_{n=0}^{N} \int_{C_n} \frac{1}{(2 + |x|)^{d+2}} \, dx \leq \sum_{n=0}^{N} \frac{|C_n|}{(n + 1)^{d+2}}.$$

Let λ_n be the open cube of side length $2(n + 1)$ centered at 0. Then Theorem 4.4 (b) gives $|C_n| \leq |\lambda_n| = (2(n + 1))^d$ and the result follows from the observation that

$$\sum_{n=1}^{N} \frac{1}{(n + 1)^2} \leq \sum_{n=1}^{N} \frac{1}{n(n + 1)} = \sum_{n=1}^{N} \frac{1}{n} - \frac{1}{n + 1} = 1 - \frac{1}{N + 1} \leq 1. \ \square$$

Let us give a weighted upper bound for the semi-norms of the integral that is independent of the measure of the domain.

THEOREM 4.24.– *Let $f \in \mathcal{B}(\Omega; E)$, where Ω is an open subset of $\mathbb{R}^d$ and E is a Neumann space. Additionally, let ω be an open set included in Ω and $\| \ \|_{E;\nu}$ a semi-norm of E. Then*

$$\Big\| \int_\omega f \Big\|_{E;\nu} \leq 2^{d+1} \sup_{x \in \omega} (2 + |x|)^{d+2} \|f(x)\|_{E;\nu}. \blacksquare$$

Proof. If ω is bounded, then, by Theorem 4.11 (a) and (b) and Theorem 4.14 (a),

$$\Big\| \int_\omega f \Big\|_{E;\nu} \leq \int_\omega \|f\|_{E;\nu} = \int_\omega \frac{(2+|x|)^{d+2}\|f(x)\|_{E;\nu}}{(2+|x|)^{d+2}}\, dx \leq$$
$$\leq \sup_{x\in\omega}(2+|x|)^{d+2}\|f(x)\|_{E;\nu} \int_\omega \frac{1}{(2+|x|)^{d+2}}\, dx$$

and the result follows from Lemma 4.23.

If ω is not bounded, then, by Theorem 4.12 (a) and 4.17 (a) and Theorem (c), observe that $\int_\omega f = \int_\omega f|_\omega = \int_\lambda f|_\omega$, where $\lambda = \{x \in \omega : f(x) \neq 0_E\}$. This open set λ is bounded by Definition 4.7 of $\mathcal{B}(\Omega; E)$ and therefore satisfies the stated inequality, which implies the inequality for ω. □

Let us show that the integral is continuous on $\mathcal{K}(\Omega; E)$.

THEOREM 4.25.– *Let Ω and ω be two open subsets of $\mathbb{R}^d$ such that $\omega \subset \Omega$, and let E be a Neumann space. Then:*

The mapping $f \mapsto \displaystyle\int_\omega f$ is continuous from $\mathcal{K}(\Omega; E)$ into E. ▮

Proof. Let $f \in \mathcal{K}(\Omega; E)$ and $\nu \in \mathcal{N}_E$ (set indexing the semi-norms of E). Then, by Theorem 4.8 (a), the fact that $f \in \mathcal{B}(\Omega; E)$ and the inequality from Theorem 4.24, together with Definition 1.17 of the semi-norms of $\mathcal{K}(\Omega; E)$, imply that

$$\Big\| \int_\omega f \Big\|_{E;\nu} \leq 2^{d+1} \sup_{x\in\omega} q(x)\|f(x)\|_{E;\nu} = 2^{d+1}\|f\|_{\mathcal{K}(\Omega;E);q,\nu},$$

where $q(x) = (2+|x|)^{d+2}$ and therefore $q \in \mathcal{C}^+(\Omega)$. This inequality implies the desired continuity by the characterization of continuous linear mappings from Theorem 1.25. □

4.8. Differentiating under the integral sign

Let us first express the **theorem on differentiation under the integral sign** in terms of a "function of functions".

THEOREM 4.26.– *Let $f \in \mathcal{C}^1(\Omega; \mathbf{C}_b(\Lambda; E))$, where Ω is an open subset of $\mathbb{R}^d$, Λ is a bounded open subset of $\mathbb{R}^\ell$ and E is a Neumann space.*

Then $\int_\Lambda f \in \mathcal{C}^1(\Omega; E)$ and, for every $i \in [\![1, d]\!]$,

$$\partial_i \int_\Lambda f = \int_\Lambda \partial_i f. \;▮$$

Proof. Since the integral $\int_\Lambda$ is linear and continuous from $\mathrm{C_b}(\Lambda; E)$ into E by Theorem 4.22 (b), the image $\int_\Lambda \underline{f}$ of $\underline{f}$ belongs to $\mathcal{C}^1(\Omega; E)$ by Theorem 3.2, and $\partial_i(\int_\Lambda \underline{f}) = \int_\Lambda \partial_i \underline{f}$ by Theorem 3.1. □

Let us now express the **theorem on differentiation under the integral sign**[5] in terms of a function of two variables, which is not that simple.

THEOREM 4.27.– *Let f, g_1, …, g_d be functions in $\mathrm{C_b}(\Omega \times \Lambda; E)$, where Ω is an open subset of $\mathbb{R}^d$, Λ is a bounded open subset of $\mathbb{R}^\ell$ and E is a Neumann space. Assume that, for every $y \in \Lambda$, the function $x \mapsto f(x,y)$ is differentiable from Ω into E and its partial derivatives are the functions $x \mapsto g_i(x,y)$, that is,*

$$\frac{\partial f}{\partial x_i}(x,y) = g_i(x,y).$$

Then the function $x \mapsto \int_\Lambda f(x,y)\,\mathrm{d}y$ belongs to $\mathrm{C}^1_\mathrm{b}(\Omega; E)$ and, for every $x \in \Omega$, the function $y \mapsto (\partial f/\partial x_i)(x,y)$ belongs to $\mathrm{C_b}(\Lambda; E)$. Furthermore,

$$\frac{\partial}{\partial x_i}\int_\Lambda f(x,y)\,\mathrm{d}y = \int_\Lambda \frac{\partial f}{\partial x_i}(x,y)\,\mathrm{d}y = \int_\Lambda g_i(x,y)\,\mathrm{d}y. \blacksquare$$

Proof. By Theorem 3.14, after separating the variables using Theorem 1.27, we obtain $\underline{f} \in \mathrm{C}^1_\mathrm{b}(\Omega; \mathrm{C_b}(\Lambda; E))$ and $\partial_i \underline{f} = \underline{g_i}$. Therefore, $\partial_i f(x) \in \mathrm{C_b}(\Lambda; E)$ and Theorem 4.26 on differentiating under the integral sign gives us $\int_\Lambda \underline{f} \in \mathrm{C}^1_\mathrm{b}(\Omega; E)$, as well as the stated equality in the form

$$\partial_i \int_\Lambda \underline{f} = \int_\Lambda \partial_i \underline{f} = \int_\Lambda \underline{g_i}. \ \square$$

One benefit of using vector values. Using a "function of functions" in Theorem 4.26, namely a function defined on Ω taking values in $\mathrm{C_b}(\Lambda; E)$, allows us to reduce differentiation under the integral sign to a special case of the commutativity of differentiation with linear mappings – in this case, the integral. However, even when $E = \mathbb{R}$, this requires vector-valued integration and differentiation, specifically in $\mathrm{C_b}(\Lambda)$.

Establishing the same result with just real-valued integration and differentiation, as in Theorem 4.27 for $E = \mathbb{R}$, is more laborious both to state and to prove (in fact, our own proof uses vector-valued differentiation and integration, since we reduced to Theorem 4.26 and appealed to separation of variables, which is a non-trivial result). □

5 **History of Theorem 4.27 on differentiating under the integral sign.** Augustin CAUCHY showed in 1823 [21, 33[e] leçon] that we can differentiate real integrals under the integral sign.

Chapter 5

Properties of the Measure of an Open Set

This chapter is dedicated to some of the classical and elementary properties of the measure of open subsets of $\mathbb{R}^d$ and determinants. The proofs are given for completeness, so that this book, supplemented by Volume 1, is self-contained; we could equally have omitted them.

5.1. Additivity of the measure

Let us show that the measure of open sets, introduced in Definition 4.1, is additive with respect to disjoint open sets.

THEOREM 5.1.– *Let ω_1 and ω_2 be two open subsets of $\mathbb{R}^d$. Then:*

(a)
$$|\omega_1 \cup \omega_2| + |\omega_1 \cap \omega_2| = |\omega_1| + |\omega_2|.$$

(b)
$$|\omega_1 \cup \omega_2| \leq |\omega_1| + |\omega_2|.$$

(c) *If $\omega_1 \cap \omega_2 = \emptyset$,*
$$|\omega_1 \cup \omega_2| = |\omega_1| + |\omega_2|. \blacksquare$$

Proof. (a) This follows from the additivity of the integral with respect to its domain (Theorem 4.21 (a) with $f = 1$), since $\int_\omega 1 = |\omega|$ (Theorem 4.12 (b)).

(b) This follows from (a), since $|\omega_1 \cap \omega_2| \geq 0$ (by Definition 4.1 of the measure).

(c) This follows from (a), since $|\emptyset| = 0$ (Theorem 4.2 (c)). $\square$

Another proof of (c). Alternatively, the property (c) immediately follows from Definition 4.1 of the measure (however, (a) and (b) use Theorem 4.21, which is non-trivial). □

Let us show that measure is invariant under translation.

THEOREM 5.2.– *Let ω be an open subset of $\mathbb{R}^d$ and $x \in \mathbb{R}^d$. Then*

$$|\omega + x| = |\omega|. \blacksquare$$

Proof. Let $n \in \mathbb{N}^*$. Since ω contains k_n (closed) cubes $\Delta_{s,n}$ from Definition 4.1 of the measure, the open set $\omega + x$ contains at least k_n (open) cubes $\mathring{\Delta}_{s,n} + x$. Each of them is of side length 2^{-n}, so, by Theorem 4.4 (b), $|\mathring{\Delta}_{s,n} + x| = 2^{-nd}$. These cubes are pairwise disjoint, so, by Theorems 4.2 (a) and 5.1 (c),

$$|\omega + x| \geq \left| \bigcup_{s:\mathring{\Delta}_{s,n}+x\subset\omega+x} \mathring{\Delta}_{s,n} + x \right| = \sum_{s:\mathring{\Delta}_{s,n}+x\subset\omega+x} |\mathring{\Delta}_{s,n} + x| \geq k_n 2^{-nd}.$$

As $n \to \infty$, in the limit, by Definition 4.1,

$$|\omega + x| \geq |\omega|.$$

Conversely, this inequality implies that

$$|\omega| = |(\omega + x) + (-x)| \geq |\omega + x|. \ \square$$

Let us calculate the measure of the image of an open set under a homothety.

THEOREM 5.3.– *Let ω be an open subset of $\mathbb{R}^d$, $t > 0$, and $t\omega = \{tx : x \in \omega\}$. Then*

$$|t\omega| = t^d|\omega|. \blacksquare$$

Proof. Let $n \in \mathbb{N}^*$. Since ω contains k_n (closed) cubes $\Delta_{s,n}$ from Definition 4.1 of the measure, the open set $t\omega$ contains at least k_n cubes $t\mathring{\Delta}_{s,n}$. Each of them is of side length $t2^{-n}$, so, by Theorem 4.4 (b), $|t\mathring{\Delta}_{s,n}| = t^d 2^{-nd}$. These cubes are pairwise disjoint, so, by Theorems 4.2 (a) and 5.1 (c),

$$|t\omega| \geq \left| \bigcup_{s:t\mathring{\Delta}_{s,n}\subset t\omega} t\mathring{\Delta}_{s,n} \right| = \sum_{s:t\mathring{\Delta}_{s,n}\subset t\omega} |t\mathring{\Delta}_{s,n}| \geq k_n t^d 2^{-nd}.$$

As $n \to \infty$, in the limit, by Definition 4.1,

$$|t\omega| \geq t^d|\omega|.$$

Conversely, applying this inequality to t^{-1} gives

$$|\omega| = |t^{-1}(t\omega)| \geq t^{-d}|t\omega|.$$

Therefore, $|t\omega| \leq t^d|\omega|$. □

Let us calculate the measure of a ball and a crown.

THEOREM 5.4.– *Let $a \in \mathbb{R}^d$, $0 \le \rho \le r \le \infty$, and*

$$\mathring{B}(a,r) = \{x \in \mathbb{R}^d : |x-a| < r\}, \quad C(a,\rho,r) = \{x \in \mathbb{R}^d : \rho < |x-a| < r\}.$$

Then, denoting $v_d = |\mathring{B}(0,1)|$:

(a) $$|\mathring{B}(a,r)| = v_d r^d.$$

(b) $$|C(a,\rho,r)| = v_d(r^d - \rho^d) \le d v_d r^{d-1}(r-\rho). \blacksquare$$

Proof. (a) Since $\mathring{B}(a,r) = a + r\mathring{B}(0,1)$, Theorems 5.2 and 5.3 imply that

$$|\mathring{B}(a,r)| = |r\mathring{B}(0,1)| = r^d\,|\mathring{B}(0,1)| = r^d v_d.$$

(b) Since $C(a,\rho,r)$ and $\mathring{B}(a,\rho)$ are disjoint and included in $\mathring{B}(a,r)$, Theorems 5.1 (c) and 4.2 (a) imply that $|C(a,\rho,r)|+|\mathring{B}(a,\rho)| \le |\mathring{B}(a,r)|$. Together with (a), this gives $|C(a,\rho,r)| \le v_d(r^d - \rho^d)$.

Conversely, since $\mathring{B}(a,r) \subset C(a,\rho,r) \cup \mathring{B}(a,\rho+\epsilon)$, Theorems 4.2 (a) and 5.1 (b) imply that $|\mathring{B}(a,r)| \le |C(a,\rho,r)| + |\mathring{B}(a,\rho+\epsilon)|$. Together with (a), this gives $|C(a,\rho,r)| \ge v_d(r^d-(\rho+\epsilon)^d)$. This holds for all $\epsilon > 0$, so $|C(a,\rho,r)| \ge v_d(r^d-\rho^d)$.

This proves the stated equality. The inequality $r^d - \rho^d \le dr^{d-1}(r-\rho)$ follows from Theorem A.56. □

5.2. Negligible sets

Let us define negligible sets.

DEFINITION 5.5.– *A subset σ of $\mathbb{R}^d$ is said to be* **negligible** *if, for every $\epsilon > 0$, there exists an open subset ω of $\mathbb{R}^d$ containing σ such that $|\omega| \le \epsilon$.* $\blacksquare$

Let us show that adjoining a negligible set to an open set does not change its measure.

THEOREM 5.6.– *Let ω be an open set and σ a negligible subset of $\mathbb{R}^d$ such that $\omega \cup \sigma$ is open. Then*

$$|\omega \cup \sigma| = |\omega|. \blacksquare$$

Proof. We have $\omega \subset \omega \cup \sigma$, so, by Theorem 4.2 (a), $|\omega| \leq |\omega \cup \sigma|$.

Conversely, for every $\epsilon > 0$, there exists an open set $\lambda \supset \sigma$ such that $|\lambda| \leq \epsilon$. Then $\omega \cup \sigma \subset \omega \cup \lambda$, so, by Theorems 4.2 (a) and 5.1 (b), $|\omega \cup \sigma| \leq |\omega| + |\lambda| \leq |\omega| + \epsilon$. This holds for all $\epsilon > 0$, so $|\omega \cup \sigma| \leq |\omega|$. □

Let us state a few conditions that guarantee that a set is negligible.

THEOREM 5.7.– *In* $\mathbb{R}^d$*:*

(a) *Every singleton is negligible.*

(b) *Every countable set is negligible.*

(c) *The image of any negligible set under a translation is negligible.*

(d) *Every set included in a negligible set is negligible.*

(e) *Every countable union of negligible sets is negligible.*

(f) *If* $\sigma \cap B$ *is negligible for every ball* B*, then* σ *is negligible.* ▌

Proof. (a) Every set $\{x\}$, where $x \in \mathbb{R}^d$, is negligible because it is included in $\mathring{B}(x, r)$ and, by Theorem 5.4 (a), $|\mathring{B}(x, r)| = v_d r^d$ which is $\leq \epsilon$ for sufficiently small r.

(d) This follows from Definition 5.5 of a negligible set.

(c) If σ is negligible, then, for every $\epsilon > 0$, there exists an open set $\omega \supset \sigma$ such that $|\omega| \leq \epsilon$. Given $x \in \mathbb{R}^d$, the translated set $\sigma + x$ is included in the open set $\omega + x$, and, by Theorem 5.2, $|\omega + x| = |\omega| \leq \epsilon$. Therefore, $\sigma + x$ is negligible.

(e) Let $\{\sigma_i : i \in I\}$ be a countable family of negligible sets. By Definition A.1 of a countable set, there exists a bijection f from I onto a subset N of $\mathbb{N}$. Let $\epsilon > 0$. For every i, since σ_i is negligible, there exists an open set ω_i such that $\sigma_i \subset \omega_i$ and $|\omega_i| \leq \epsilon 2^{-f(i)}$. The union $\omega = \bigcup_{i \in I} \omega_i$ is open by Theorem A.10.
— If I and therefore N are finite, then the sub-additivity of the measure (Theorem 5.1 (b)) and the bound on the sum of a geometric series (Theorem A.57) imply

$$|\omega| \leq \sum_{i \in I} |\omega_i| \leq \sum_{n \in N} \epsilon 2^{-n} \leq \epsilon. \tag{5.1}$$

Since the union $\bigcup_{i \in I} \sigma_i$ is included in ω, it is negligible.
— If I is infinite, then, for every $n \in \mathbb{N}$, let $\Omega_n = \bigcup_{i \in I : f(i) \leq n} \omega_i$. This set increases with n and $\bigcup_{n \in \mathbb{N}} \Omega_n = \omega$, so, by Theorem 4.6, $|\Omega_n| \to |\omega|$. The inequality (5.1) implies that $|\Omega_n| \leq \epsilon$, so, once again, $|\omega| \leq \epsilon$.

(f) If $\sigma \cap B$ is negligible for every ball B, then $\sigma = \bigcup_{n \in \mathbb{N}} \sigma \cap B(0, n)$ is negligible by (e).

(b) Every countable set is a countable union of singletons and is therefore negligible by (a) and (e). □

Let us show that the product of a negligible set and an arbitrary set is negligible.

THEOREM 5.8.– *Let $\sigma_1 \subset \mathbb{R}^{d_1}$ and $\sigma_2 \subset \mathbb{R}^{d_2}$ such that σ_1 or σ_2 is negligible. Then $\sigma_1 \times \sigma_2$ is negligible in $\mathbb{R}^{d_1+d_2}$.* ▮

Proof. By Theorem 5.7 (f), it is sufficient to check that $(\sigma_1 \times \sigma_2) \cap B$ is negligible for every ball $B = B(a, r)$ in $\mathbb{R}^{d_1+d_2}$.

If σ_1 is negligible, then, for every $\epsilon > 0$, by Definition 5.5, there exists an open set ω_1 such that $\sigma_1 \subset \omega_1$ and $|\omega_1| \leq \epsilon$. Then $(\sigma_1 \times \sigma_2) \cap B \subset \omega_1 \times B_2$, where B_2 is the ball $B(a_2, r)$ in $\mathbb{R}^{d_2}$ and (Theorem 4.3) $|\omega_1 \times B_2| = |\omega_1||B_2| \leq \epsilon|B_2|$ is arbitrarily small. Hence, $(\sigma_1 \times \sigma_2) \cap B$ is indeed negligible. We can proceed in the same way if σ_2 is negligible. □

Let us show that spheres are negligible.

THEOREM 5.9.– *Every sphere $S(a, r) = \{x \in \mathbb{R}^d : |x - a| = r\}$, where $a \in \mathbb{R}^d$ and $r > 0$, is negligible.* ▮

Proof. The sphere $S(a, r)$ is negligible because it is included in the crown $C(a, r - \eta, r + \eta)$, whose measure is less than $2dv_d(r + \eta)^{d-1}\eta$ by Theorem 5.4 (b) and therefore less than any given $\epsilon > 0$ for η sufficiently small. □

We say that (finitely many) vectors u_i of a vector space E are **linearly independent** if the linear combination $\sum_i c_i u_i$ only vanishes when all of the real numbers c_i are zero.

Let us show that proper vector subspaces, i.e. vector subspaces that are strictly smaller than the space itself, are negligible.

THEOREM 5.10.– *If $k < d$, the set $\{t_1 v^1 + \cdots + t_k v^k : t_i \in \mathbb{R}\}$ generated by the k vectors v^1, v^2, ..., v^k in $\mathbb{R}^d$ is negligible.* ▮

Proof. *Case where $v^1, \ldots, v^k$ are linearly independent.* Let us begin by showing that the set

$$P(v^1, \ldots, v^k) \stackrel{\text{def}}{=} \{t_1 v^1 + \cdots + t_k v^k : |t_i| < 1, \ \forall i\}$$

is negligible. Let $v^{k+1}, \ldots, v^d$ such that $v^1, \ldots, v^d$ are linearly independent. For $c > 0$, let

$$\omega(c) \stackrel{\text{def}}{=} \{t_1 v^1 + \cdots + t_{d-1} v^{d-1} + c t_d v^d : |t_i| < 1, \ \forall i\}.$$

This is a bounded open subset of $\mathbb{R}^d$. For every integer $m \geq 1$, let us show that its measure satisfies

$$\left|\omega\left(\frac{1}{m}\right)\right| \leq \frac{|\omega(1)|}{m}. \tag{5.2}$$

Indeed, $\omega(mc)$ contains the m disjoint open sets $\omega_i(c)$ obtained by translating $\omega(c)$ by $(2i - m)cv^d$ for $i = 0, 1, \ldots, m - 1$; these sets are pairwise disjoint and have the same measure as $\omega(c)$ (Theorem 5.2), so, by Theorems 4.2 (a) and 5.1 (c),

$$|\omega(mc)| \geq \left|\bigcup_{0 \leq i \leq m-1} \omega_i(c)\right| = \sum_{0 \leq i \leq m-1} |\omega_i(c)| = m\,|\omega(c)|.$$

For $c = 1/m$, this inequality implies (5.2).

Since $P(v^1, \ldots, v^k)$ is included in $\omega(1/m)$ for every m, the inequality (5.2) shows that it is negligible (Definition 5.5). This holds for every family of linearly independent v^i and hence, in particular, for $P(nv^1, \ldots, nv^k)$ for any $n \neq 0$. The union $\bigcup_{n \in \mathbb{N}} P(nv^1, \ldots, nv^k)$, namely $\{t_1 v^1 + \cdots + t_k v^k : t_i \in \mathbb{R}\}$, is therefore negligible (Theorem 5.7 (e)).

Case where $v^1, \ldots, v^k$ are not linearly independent. In this case, we can remove at least one of the v^i without changing the set that they generate. By repeating this operation until we obtain a linearly independent family, we reduce to the previous case. □

Let us show that the image of a negligible set under a regular bijection is negligible.

THEOREM 5.11.– *Let Ω and Λ be two open subsets of $\mathbb{R}^d$, T a bijection from Ω onto Λ such that $T \in C^1_b(\Omega; \Lambda)$ and $T^{-1} \in C(\Lambda; \Omega)$, and σ a negligible subset of Ω. Then $T(\sigma)$ is negligible.* ▌

Proof. Suppose for now that, for every open set $\omega \subset \Omega$,

$$|T(\omega)| \leq v_d (2\gamma\sqrt{d})^d |\omega|, \tag{5.3}$$

where $\gamma \overset{\text{def}}{=} \sup_{x\in\Omega} |\nabla T(x)|$ and v_d is the measure of the unit ball of $\mathbb{R}^d$.

By Definition 5.5 of a negligible set, for every $\epsilon > 0$, there exists an open set ω containing σ such that $|\omega| \leq \epsilon/(v_d(2\gamma\sqrt{d})^d)$. We can assume, in addition, that ω is included in Ω, replacing it by $\omega \cap \Omega$. Then $T(\omega)$ is open, as the preimage of an open set under a continuous mapping (Theorem A.31); indeed, $T(\omega) = S^{-1}(\omega)$, where $S = T^{-1}$ is continuous by the hypotheses. Furthermore, $T(\sigma) \subset T(\omega)$ and, together with (5.3), $|T(\omega)| \leq \epsilon$. Hence, $T(\sigma)$ is negligible.

We still need to check (5.3). Recall that $|\omega| = \lim_{n\to\infty} k_n 2^{-nd}$, where k_n is the number of $s \in \mathbb{Z}^d$ such that the closed cube $\Delta_{s,n}$ of side length 2^{-n} centered at $2^{-n}s$ is included in ω. Denote by $\mathcal{A}_n$ the set of these s, and

$$K_n \overset{\text{def}}{=} \bigcup_{s\in\mathcal{A}_n} \Delta_{s,n}, \quad \omega_n \overset{\text{def}}{=} \mathring{K}_n.$$

Then $\bigcup_{s\in\mathcal{A}_n} \omega_n = \omega$.

To obtain a covering of $T(\omega)$, we dilate the open cubes $\mathring{\Delta}_{s,n}$ so they cover their original boundaries. More precisely, let $\Delta'_{s,n}$ be the closed cube of side length 2^{1-n} (double that of $\Delta_{s,n}$) centered at $2^{-n}s$, and

$$\omega'_{s,n} \overset{\text{def}}{=} \mathring{\Delta}'_{s,n} \cap \omega, \quad \omega'_n \overset{\text{def}}{=} \bigcup_{s\in\mathcal{A}_n} \omega'_{s,n}.$$

Then $\omega_n \subset \omega'_n \subset \omega$, so $\bigcup_{n\in\mathbb{N}} \omega'_n = \omega$ and $\bigcup_{n\in\mathbb{N}} T(\omega'_n) = T(\omega)$. Furthermore, ω'_n increases with n, so $T(\omega'_n)$ does also; hence, by Theorem 4.6, as $n \to \infty$,

$$|T(\omega'_n)| \to |T(\omega)|. \tag{5.4}$$

The set $\Delta'_{s,n}$ and therefore $\omega'_{s,n}$ are included in the ball of radius $2^{1-n}\sqrt{d}$ centered at $2^{-n}s$. Hence, by the finite increment theorem (Theorem 2.5 (a)), $T(\Delta'_{s,n})$ is included in the open ball $\mathring{B}_{s,n}$ of radius $\gamma 2^{1-n}\sqrt{d}$ centered at $T(2^{-n}s)$. Therefore, by Theorem 5.4,

$$|T(\omega'_{s,n})| \leq |\mathring{B}_{s,n}| = v_d(\gamma 2^{1-n}\sqrt{d})^d.$$

Since $T(\omega'_n) = \bigcup_{s\in\mathcal{A}_n} T(\omega'_{s,n})$ and $\mathcal{A}_n$ consists of k_n points s, the sub-additivity of the measure (Theorem 5.1 (b)) implies

$$|T(\omega'_n)| \leq \sum_{s\in\mathcal{A}_n} |T(\omega'_{s,n})| \leq k_n v_d(\gamma 2^{1-n}\sqrt{d})^d.$$

Together with the convergence in (5.4) and $k_n 2^{-nd} \to |\omega|$, we obtain the inequality (5.3) in the limit. □

5.3. Determinant of d vectors

Let us state a few preliminary definitions.

DEFINITION 5.12.– (a) *A **permutation** of* $(1,\dots,d)$ *is any family that consists of the same elements ordered differently (or identically), say* $p = (p_1,\dots,p_d)$.

(b) *Two integers* i *and* j *such that* $1 \le i < j \le d$ *are said to be **inverted** by a permutation if* $p_i > p_j$.

(c) *The **signature** of a permutation, denoted by* $\varepsilon(p)$, *is defined as the number* 1 *if the permutation inverts an even number of integers and* -1 *otherwise. Therefore,*

$$\varepsilon(p) = \prod_{1\le i<j\le d} \operatorname{sign}(p_j - p_i). \ ■$$

Let us define the determinant of d vectors in $\mathbb{R}^d$.

DEFINITION 5.13.– *The **determinant** of* d *vectors* v^1, v^2, …, v^d *in* $\mathbb{R}^d$ *is the real number*

$$\det[v^1,\dots,v^d] \stackrel{\text{def}}{=} \sum_{p\in\mathcal{P}_d} \varepsilon(p)\, v^1_{p_1}\cdots v^d_{p_d},$$

where $\mathcal{P}_d$ *is the set of permutations of* $(1,\dots,d)$, $\varepsilon(p)$ *is the signature of* p, *and* v^j_i *is the* i*th component of* v^j, *namely* $v^j = (v^j_1,\dots,v^j_d)$. ■

Specifying d vectors of $\mathbb{R}^d$ is equivalent to specifying a $d\times d$ **matrix**, or in other words a family $[v] = [v^j_i]_{1\le i\le d,\, 1\le j\le d}$ of real numbers. The v^j are the **columns** of the matrix. We denote

$$\det[v] \stackrel{\text{def}}{=} \det[v^1,\dots,v^d].$$

Notation for matrices. We will use square brackets [ ] for $d\times d$ matrices, but often also for families of d vectors in $\mathbb{R}^d$ in order to emphasize that these families can be represented as a square array. This convention is somewhat uncommon. □

Let us state some of the properties of the determinant.

THEOREM 5.14.– *Let* v^1, v^2, …, v^d *be vectors in* $\mathbb{R}^d$.

(a) *The mapping* $(v^1,\dots,v^d) \mapsto \det[v^1,\dots,v^d]$ *is continuous and multilinear from* $(\mathbb{R}^d)^d$ *into* $\mathbb{R}$.

(b) *If* $\mathbf{e}_j$ *is the* j*th basis vector of* $\mathbb{R}^d$, *i.e.,* $(\mathbf{e}_j)_j = 1$ *and* $(\mathbf{e}_j)_i = 0$ *if* $i \neq j$, *then*

$$\det[\mathbf{e}_1, \ldots, \mathbf{e}_d] = 1.$$

(c) *If we exchange* v^k *and* v^ℓ, *where* $k \neq \ell$, *then* $\det[v^1, \ldots, v^d]$ *changes sign.*

(d) *If there exists* $k \neq \ell$ *such that* $v^k = v^\ell$, *then* $\det[v^1, \ldots, v^d] = 0$.

(e) *If* $v^1, \ldots, v^d$ *are not linearly independent, then* $\det[v^1, \ldots, v^d] = 0$.

(f) *If* $q = (q_1, \ldots, q_d)$ *is a permutation of* $(1, \ldots, d)$, *then*

$$\det[v^{q_1}, \ldots, v^{q_d}] = \varepsilon(q) \det[v^1, \ldots, v^d]. \blacksquare$$

Proof. (a) The mappings $v^j \mapsto \det[v^1, \ldots, v^d]$ are linear by Definition 5.13 of the determinant. Moreover,

$$|\det[v^1, \ldots, v^d]| \leq d! \left(\sup_{i=1\ldots d} |v_i^1| \right) \cdots \left(\sup_{i=1\ldots d} |v_d^1| \right) \leq d! \, |v^1| \cdots |v^d|,$$

since there are $d!$ permutations of $(1, \ldots, d)$ and $|v_i^j| \leq (\sum_{k=1}^d |v_k^j|^2)^{1/2} = |v^j|$. The mapping det is therefore continuous by the characterization of multilinear continuous mappings from Theorem A.40.

(b) We have $\det[\mathbf{e}_1, \ldots, \mathbf{e}_d] = \varepsilon(1, \ldots, d)\, (\mathbf{e}_1)_1 \cdots (\mathbf{e}_d)_d = 1$ because every other term in the determinant is zero. Indeed, $(\mathbf{e}_1)_{p_1} \cdots (\mathbf{e}_d)_{p_d} = 0$ for every permutation $(p_1, \ldots, p_d) \neq (1, \ldots, d)$ because $(\mathbf{e}_j)_{p_j} = 0$ if $p_j \neq j$.

(c) Exchanging v^k and v^ℓ is equivalent to exchanging p_k and p_ℓ in each signature $\varepsilon(p)$, which changes the sign. Hence, this changes the sign of $\det[v^1, \ldots, v^d]$.

(d) If there exists $k \neq \ell$ such that $v^k = v^\ell$, then $\det[v^1, \ldots, v^d] = -\det[v^1, \ldots, v^d]$ by (c), and hence $\det[v^1, \ldots, v^d] = 0$.

(e) If the vectors v^j are not linearly independent, then one of them is a linear combination of the others, say $v^1 = \sum_{j \geq 2} c_j v^j$; thus linearity and the property (d) imply

$$\det[v^1, \ldots, v^d] = \sum_{j \geq 2} c_j \det[v^j, v^2, \ldots, v^d] = 0.$$

(f) Every permutation q may be obtained by exchanging pairs of q_i some number r of times. Each exchange multiplies the determinant by -1 by (c). To conclude, simply observe that $\varepsilon(q) = (-1)^r$ to obtain

$$\det[v^{q_1}, \ldots, v^{q_d}] = (-1)^r \det[v^1, \ldots, v^d] = \varepsilon(q) \det[v^1, \ldots, v^d]. \square$$

Let us now consider the **product of two matrices** $d \times d$. Writing $[v] = [a][b]$, this product is defined by: for every $i \in [\![1, d]\!]$ and $j \in [\![1, d]\!]$,

$$v_i^j \stackrel{\text{def}}{=} \sum_{k=1}^{d} a_i^k b_k^j. \tag{5.5}$$

Like in Definition 5.13, $v = [v^1, \ldots, v^d]$, where $v^k = (v_1^k, \ldots, v_d^k)$ is the kth column of the matrix and $v_i = (v_i^1, \ldots, v_i^d)$ is the ith row.

Let us show that the determinant of a product is the product of the determinants.

THEOREM 5.15.– *Let $[a]$ and $[b]$ be two $d \times d$ matrices. Then*

$$\det([a][b]) = (\det[a])(\det[b]).$$

Proof. By Definition 5.13 of the determinant and equation (5.5) defining the product of matrices,

$$\det[v] = \sum_{p \in \mathcal{P}_d} \varepsilon(p)\, v_{p_1}^1 \cdots v_{p_d}^d = \sum_{p \in \mathcal{P}_d} \varepsilon(p) \Big(\sum_{k_1=1}^{d} a_{p_1}^{k_1} b_{k_1}^1 \Big) \cdots \Big(\sum_{k_d=1}^{d} a_{p_d}^{k_d} b_{k_d}^d \Big).$$

Inverting the sums gives

$$\det[v] = \sum_{k_1=1}^{d} \cdots \sum_{k_d=1}^{d} b_{k_1}^1 \cdots b_{k_d}^d \sum_{p \in \mathcal{P}_d} \varepsilon(p)\, a_{p_1}^{k_1} \cdots a_{p_d}^{k_d}. \tag{5.6}$$

For $k = (k_1, \ldots, k_d)$ in $([\![1, d]\!])^d$, we denote

$$D_k = \sum_{p \in \mathcal{P}_d} \varepsilon(p)\, a_{p_1}^{k_1} \cdots a_{p_d}^{k_d},$$

namely $D_k = \det[a^{k_1}, \ldots, a^{k_d}]$. If k is a permutation of $(1, \ldots, d)$, Theorem 5.14 (f) implies

$$D_k = \varepsilon(k) \det[a^1 \cdots a^d] = \varepsilon(k) \det[a].$$

If not, some two indices $k_1, \ldots, k_d$ are equal, so $D_k = 0$ (Theorem 5.14 (d)). We shall extend Definition 5.13 of the signature to this case by setting $\varepsilon(k) = 0$.

Then (5.6) may be stated as

$$\det[v] = \sum_{k_1=1}^{d} \cdots \sum_{k_d=1}^{d} \varepsilon(k)\, b_{k_1}^1 \cdots b_{k_d}^d \det[a].$$

To conclude, simply observe that, since $\varepsilon(k) = 0$ if $k \notin \mathcal{P}_d$,

$$\det[v] = \sum_{k \in \mathcal{P}_d} \varepsilon(k)\, b_{k_1}^1 \cdots b_{k_d}^d \det[a] = \det[b] \det[a]. \;\square$$

5.4. Measure of a parallelepiped

Let us calculate the measure of an open **parallelepiped**.

THEOREM 5.16.– *Let $v^1, v^2, \ldots, v^d$ be linearly independent vectors in $\mathbb{R}^d$ and*

$$P(v^1, \ldots, v^d) \stackrel{\text{def}}{=} \{t_1 v^1 + \cdots + t_d v^d : 0 < t_i < 1,\ 1 \le i \le d\}.$$

Then

$$|P(v^1, \ldots, v^d)| = |\det[v^1, \ldots, v^d]|. \blacksquare$$

Case of linearly dependent vectors. If $v^1, v^2, \ldots, v^d$ are not linearly independent, then $P(v^1, \ldots, v^d)$ is no longer open, so its measure is not defined for us. It is negligible (Theorem 5.10). For all vectors, we therefore have $|\mathring{P}(v^1, \ldots, v^d)| = |\det[v^1, \ldots, v^d]|$, a quantity that is zero if $v^1, v^2, \ldots, v^d$ are not linearly independent. □

Our proof will require the following two lemmas.

LEMMA 5.17.– *Let $v^1, \ldots, v^d$ be linearly independent vectors of $\mathbb{R}^d$ and $c \in \mathbb{R}$. Then*

$$|P(cv^1, v^2, \ldots, v^d)| = |c|\, |P(v^1, v^2, \ldots, v^d)|. \blacksquare$$

Proof. Denote $P(v) = P(v, v^2, \ldots, v^d)$.

Case where $c = n$ is an integer. The open set $P(nv)$ is "almost" equal to the open set

$$P' = \bigcup_{i=0}^{n-1} P(v) + iv.$$

More precisely, $P(nv) \setminus P'$ is the union of $n - 1$ hyperplane sections separating the $P(v) + iv$, and this union is negligible (Theorems 5.10 and 5.7 (d), (c), and (e)). Therefore, by Theorem 5.6,

$$|P(nv)| = |P'|.$$

Since the $P(v) + iv$ are pairwise disjoint and $|P(v) + iv| = |P(v)|$ (Theorem 5.2), by Theorem 5.1 (c), $|P'| = n|P(v)|$. Hence,

$$|P(nv)| = n\,|P(v)|.$$

Case where $c = m/n$ is rational. Given integers m and $n \neq 0$, we have

$$m|P(v)| = |P(mv)| = \left|P\left(n\frac{m}{n}v\right)\right| = n\left|P\left(\frac{m}{n}v\right)\right|.$$

Hence,

$$\left|P\Big(\frac{m}{n}v\Big)\right| = \frac{m}{n}\,|P(v)|.$$

Case where c is a positive real number. Given an integer $n \geq 1$, let m be the unique integer such that $m/n \leq c < (m+1)/n$. Now, $P(c)$ increases with c, so $|P(cv)|$ is also increasing (Theorem 4.2 (a)), and

$$\frac{m}{n}\,|P(v)| = \left|P\Big(\frac{m}{n}v\Big)\right| \leq |P(cv)| \leq \left|P\Big(\frac{m+1}{n}v\Big)\right| = \frac{m+1}{n}\,|P(v)|.$$

As $n \to \infty$, this gives

$$c\,|P(v)| \leq |P(cv)| \leq c\,|P(v)|.$$

Hence,

$$|P(cv)| = c\,|P(v)| = |c|\,|P(v)|.$$

Case where c is negative or zero. For $c < 0$, the open set $P(cv)$ coincides with the translated set $P(-cv) + 2cv$, so, by Theorem 5.2 and the positive case,

$$|P(cv)| = |P(-cv)| = |-c|\,|P(v)| = |c|\,|P(v)|.$$

For $c = 0$, $P(0v) = \emptyset$, so, by Theorem 4.2 (c), $|P(0v)| = 0 = |0|\,|P(v)|$. □

Lemma 5.18.– *Let $v^1, \ldots, v^d$ be linearly independent vectors of $\mathbb{R}^d$. Then*

$$|P(v^1 + v^2, v^2, \ldots, v^d)| = |P(v^1, v^2, \ldots, v^d)|. \blacksquare$$

Proof. Once again, denote $P(v) = P(v, v^2, \ldots, v^d)$. Consider the half-parallelepiped

$$D = \{t_1 v^1 + t_2 v^2 + \cdots + t_d v^d : 0 < t_i < 1,\ t_2 < t_1\}.$$

Then $P(v^1 + v^2) \cup D$ is "almost" equal to $P(v^1) \cup (D + v^2)$. More precisely, there exist hyperplane sections R and R' such that (see Figure 5.1)

$$P(v^1 + v^2) \cup D \cup R = P(v^1) \cup (D + v^2) \cup R'.$$

Indeed, both sides are equal to $P(v^1 + v^2) \cup P(v^1)$. Since R and R' are negligible (Theorems 5.10 and 5.7 (d) and (c)), Theorem 5.6 implies

$$|P(v^1 + v^2) \cup D| = |P(v^1) \cup (D + v^2)|.$$

Since we are considering a disjoint union of open sets, by Theorem 5.1 (c),

$$|P(v^1 + v^2)| + |D| = |P(v^1)| + |D + v^2|.$$

Finally, because $|D + v^2| = |D|$ (Theorem 5.2), we indeed have

$$|P(v^1 + v^2)| = |P(v^1)|. \ \square$$

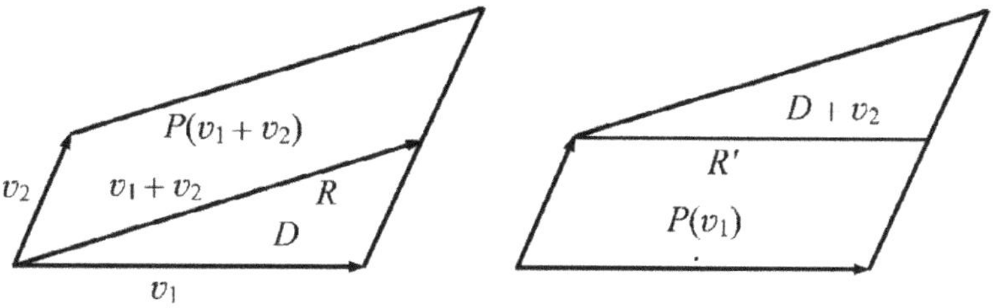

Figure 5.1. *Equivalent decompositions for "measuring" a parallelepiped*

Proof of Theorem 5.16. Denote

$$\phi(v^1,\ldots,v^d) \stackrel{\text{def}}{=} |P(v^1,\ldots,v^d)| - |\det[v^1,\ldots,v^d]|.$$

We have $\phi(cv^1,v^2,\ldots,v^d) = |c|\,\phi(v^1,v^2,\ldots,v^d)$ by Lemma 5.17 and Theorem 5.14 (a), and $\phi(v^1+v^2,v^2,\ldots,v^d) = \phi(v^1,v^2,\ldots,v^d)$ by Lemma 5.18 and Theorem 5.14 (a) and (d). Hence, for $c \neq 0$,

$$\begin{aligned}\phi(v^1+cv^2,v^2,\ldots,v^d) &= |c|\,\phi(c^{-1}v^1+v^2,v^2,\ldots,v^d) = \\ &= |c|\,\phi(c^{-1}v^1,v^2,\ldots,v^d) = \phi(v^1,v^2,\ldots,v^d).\end{aligned}$$

Similarly, $\phi(v^1+cv^i,v^2,\ldots,v^d) = \phi(v^1,v^2,\ldots,v^d)$ for every $i \neq 1$ because $\phi(v^1,\ldots,v^d)$ does not depend on the order of the v^i by the definition of P and by Theorem 5.14 (c). Hence,

$$\begin{aligned}\phi\Big(\sum_{i=1}^{d} c_i v^i, v^2,\ldots,v^d\Big) &= \phi\Big(\sum_{i=1}^{d-1} c_i v^i, v^2,\ldots,v^d\Big) = \\ &= \phi\Big(\sum_{i=1}^{d-2} c_i v^i, v^2,\ldots,v^d\Big) = \cdots = \\ &= \phi(c_1v^1,v^2,\ldots,v^d) = |c_1|\,\phi(v^1,v^2,\ldots,v^d). \qquad (5.7)\end{aligned}$$

Let us show by induction on k that

$$\phi\Big(\sum_{i=1}^{d} c_i^1 v^i,\ldots,\sum_{i=1}^{d} c_i^k v^i, v^{k+1},\ldots,v^d\Big) = |c_1^1\cdots c_k^k|\,\phi(v^1,v^2,\ldots,v^d). \qquad (5.8)$$

Suppose that this property holds for some fixed k. Together with the equality (5.7), since $\phi(v^1,\ldots,v^d)$ does not depend on the order of the v^j, this implies

$$\phi\Big(\sum_{i=1}^{d} c_i^1 v^i, \ldots, \sum_{i=1}^{d} c_i^{k+1} v^i, v^{k+2}, \ldots, v^d\Big) =$$

$$= |c_1^1 \cdots c_k^k|\, \phi\Big(v^1, \ldots, v^k, \sum_{i=1}^{d} c_i^{k+1} v^i, v^{k+2}, \ldots, v^d\Big) =$$

$$= |c_1^1 \cdots c_k^k||c_{k+1}^{k+1}|\, \phi(v^1, \ldots, v^d).$$

This is the property (5.8) for $k+1$. Since the property holds for $k = 1$ by (5.7), it holds for every $k \in [\![1, d]\!]$.

Since $v^j = \sum_i v_i^j \mathbf{e}_i$ for every j, the equality (5.8) for $k = d$ gives

$$\phi(v^1, \ldots, v^d) = |v_1^1 \cdots v_d^d|\, \phi(\mathbf{e}_1, \ldots, \mathbf{e}_d).$$

But the definition of ϕ gives

$$\phi(\mathbf{e}_1, \ldots, \mathbf{e}_d) = 0$$

because $|P(\mathbf{e}_1, \ldots, \mathbf{e}_d)| = 1$ (Theorem 4.4 (b), since $P(\mathbf{e}_1, \ldots, \mathbf{e}_d)$ is a cube of side length 1) and $\det[\mathbf{e}_1, \ldots, \mathbf{e}_d] = 1$ (Theorem 5.14 (b)). Hence,

$$\phi(v^1, \ldots, v^d) = 0.$$

In other words,

$$|P(v^1, \ldots, v^d)| = |\det[v^1, \ldots, v^d]|. \ \square$$

Chapter 6

Additional Properties of the Integral

This chapter states additional results about the Cauchy integral of a uniformly continuous function, including results about multiple integrals and changes of variables. Similarly to Chapter 5, we extend the classical properties to integrals taking values in a Neumann space, which is new but does not present any unexpected results.

This is just one step of the general study of integration performed in Volume 4 for integrable distributions. Some of the proofs are somewhat tedious because, as noted above, integrals taking values in a Neumann space have not yet been established.

6.1. Contribution of a negligible set to the integral

Let us show that adjoining a negligible set does not change the integral.

THEOREM 6.1.– *Let $f \in \mathcal{B}(\Omega; E)$, where Ω is an open subset of $\mathbb{R}^d$ and E is a Neumann space. Additionally, let ω be an open set and σ a negligible set such that $\omega \cup \sigma$ is open and included in Ω. Then*

$$\int_{\omega \cup \sigma} f = \int_{\omega} f. \blacksquare$$

Proof. Since $|\omega \cup \sigma| = |\omega|$ (Theorem 5.6), the bound on the semi-norms of a difference of integrals from Theorem 4.16 gives, if $|\omega| < \infty$, for every semi-norm $\| \ \|_{E;\nu}$ of E,

$$\Big\| \int_{\omega \cup \sigma} f - \int_{\omega} f \Big\|_{E;\nu} \leq (|\omega \cup \sigma| - |\omega|) \sup_{x \in \sigma} \|f(x)\|_{E;\nu} = 0.$$

This implies that $\int_{\omega \cup \sigma} f = \int_{\omega} f$.

If $|\omega| = \infty$, we reduce to the above case by considering an open ball λ containing the support of f and observing that the contribution of the points outside of λ to the integral is zero (Theorem 4.17), that $\sigma \cap \lambda$ is negligible (Theorem 5.7 (d)) and that $|\omega \cap \lambda| \leq |\lambda| < \infty$ (Theorem 4.2 (a) and (b)), so

$$\int_{\omega \cup \sigma} f = \int_{(\omega \cup \sigma) \cap \lambda} f = \int_{(\omega \cap \lambda) \cup (\sigma \cap \lambda)} f = \int_{\omega \cap \lambda} f = \int_{\omega} f. \square$$

6.2. Integration and differentiation in one dimension

Given real or infinite numbers s and t such that $s \leq t$, we denote, when these quantities are defined,

$$\int_s^t f \stackrel{\text{def}}{=} \int_{(s,t)} f, \quad \int_t^s f \stackrel{\text{def}}{=} -\int_{(s,t)} f = -\int_s^t f. \tag{6.1}$$

Let us show the additivity of the integral with respect to the interval of integration[1].

THEOREM 6.2.– *Let $f \in \mathcal{B}((a,b);E)$, where $-\infty \leq a \leq b \leq \infty$ and E is a Neumann space. Then, for every r, s and t in $[a,b]$,*

$$\int_r^t f = \int_r^s f + \int_s^t f. \blacksquare$$

Proof. If $r \leq s \leq t$, observe that the set $\{s\}$, being negligible (Theorem 5.7 (a)), does not contribute to the integral (Theorem 6.1), so the additivity with respect to the open sets (Theorem 4.21 (b)) gives

$$\int_{(r,t)} f = \int_{(r,s) \cup (s,t)} f = \int_{(r,s)} f + \int_{(s,t)} f.$$

This is the stated equality by the definition (6.1) of $\int_s^t f$.

If r, s and t are ordered differently, we reduce to the above case using the equality $\int_t^s f = -\int_s^t f$. For example, if $r \leq t \leq s$, the equality $\int_r^s f = \int_r^t f + \int_t^s f$ gives

$$\int_r^t f = \int_r^s f - \int_t^s f = \int_r^s f + \int_s^t f. \square$$

1 **History of the additivity with respect to the intervals.** The identity of Theorem 6.2 was shown by Augustin CAUCHY in 1823 [21].
It is one of the six conditions required by Henri LEBESGUE for his definition of the integral in one dimension, which he gave in 1904 [53, pp. 105 and 122]. The other conditions are additivity with respect to functions, continuity with respect to an increasing sequence of functions, stability under translation, positivity and normalization by $\int_0^1 1 = 1$. These properties are shown in [53, pp. 122–125].

Let us differentiate an integral with respect to its bounds. This is the first part of the **fundamental theorem of calculus**[2].

THEOREM 6.3.– *Let $f \in \mathcal{C}((a,b);E)$, where $-\infty \leq a < b \leq \infty$ and E is a Neumann space. Then:*

(a) *If $s \in (a,b)$, the functions $t \mapsto \int_s^t f$ and $t \mapsto \int_t^s f$ belong to $\mathcal{C}^1((a,b);E)$ and*

$$\frac{\mathrm{d}}{\mathrm{d}t}\int_s^t f = f(t), \quad \frac{\mathrm{d}}{\mathrm{d}t}\int_t^s f = -f(t).$$

(b) *If $s \in [a,b]$ and is finite and if $f \in \mathcal{C}([a,b];E)$ or $f \in \mathbf{C}((a,b);E)$, then the functions $t \mapsto \int_s^t f$ and $t \mapsto \int_t^s f$ belong to $\mathcal{C}^1([a,b];E)$.*

In other words, these functions are defined and continuous on the whole of $[a,b]$ and their derivatives, which are defined on (a,b), may be continuously extended to $[a,b]$. ▮

Proof. (a) Given $t \in (a,b)$, the closed interval $[s,t]$ (if $t \geq s$, or $[t,s]$ otherwise) is compact by the Borel–Lebesgue theorem (Theorem A.23 (b)), so f is uniformly continuous on this interval by Heine's theorem (Theorem A.34). The restriction of f to the open interval (s,t) (or (t,s)) is therefore uniformly continuous with bounded support, and hence the integral $\int_s^t$ is well defined (Definition 4.9).

Let $h \neq 0$ be such that $a < t+h < b$. By the additivity with respect to the interval of integration and the linearity (Theorems 6.2 and 4.11 (a)), and by the measure of an interval (Theorems 4.12 (b) and 4.4 (a)),

$$\int_s^{t+h} f - \int_s^t f - hf(t) = \int_t^{t+h} f - hf(t) = \int_t^{t+h} (f - f(t)).$$

2 **History of the fundamental theorem of calculus.** The fundamental theorem of calculus states that, in one dimension, the integral and the derivative are inverse mappings. The first part (Theorem 6.3) states that $\mathrm{d}\int_s^t f/\mathrm{d}t = f(t)$; the second part (Theorem 6.4) states that $\int_s^t f' = f(t) - f(s)$.

Differentiating an integral with respect to its upper bound. Augustin CAUCHY showed that the function $t \mapsto \int_s^t f$ is differentiable with derivative f in 1823 [21, p. 122] whenever f is a continuous real function by using the uniform continuity of f (which had not yet been proven!).

Precursors. The first partial, geometric form of this result was published by James GREGORY in 1668 [42, pp. 488–491]. A more general form, also geometric, was stated by Isaac BARROW in 1670 [5, Lecture X, § 11]. Isaac NEWTON, a student of BARROW, only finished developing the mathematical theory encompassing this result in 1670–1673 [60], even though he had discovered it before 1662. Gottfried Wilhelm LEIBNIZ established these results more systematically in the form of infinitesimal calculus and introduced the notation that we still use today.

Therefore, by Theorem 4.15, for every semi-norm of E,

$$\Big\|\int_s^{t+h} f - \int_s^t f - hf(t)\Big\|_{E;\nu} \leq |h| \sup_{|u|\leq |h|} \|f(t+u) - f(t)\|_{E;\nu}.$$

Let $\epsilon > 0$. Since f is continuous, there exists $\eta > 0$ such that $\|f(t+u)-f(t)\|_{E;\nu} \leq \epsilon$ if $|u| \leq \eta$. By the characterization (2.4) from Definition 2.1 (b) of differentiability in one dimension, this proves that the function $t \mapsto \int_s^t f$ is differentiable at the point t with derivative $f(t)$. By Theorem 2.2, this implies that it is continuous.

Since $\int_t^s f = -\int_s^t f$, it follows that $t \mapsto \int_t^s f$ is continuous and differentiable with derivative $-f$.

(b) Suppose now that $f \in \mathcal{C}([a,b];E)$ and $s \in [a,b]$ is finite. By (a), the function $t \mapsto \int_s^t f$ is differentiable on (a,b) with derivative f. To show that this function belongs to $\mathcal{C}^1([a,b];E)$, by Definition 2.26, we simply need to show that it is continuous on the whole of $[a,b]$ (differentiability only implies continuity on (a,b) by applying Theorem 2.3), since its derivative f is continuous on this interval by the hypotheses. This follows from Theorems 6.2 and 4.15, which give

$$\Big\|\int_s^t f - \int_s^{t'} f\Big\|_{E;\nu} = \Big\|\int_{t'}^t f\Big\|_{E;\nu} \leq |t'-t| \sup_{t'<u<t} \|f(u)\|_{E;\nu}.$$

If $f \in \mathbf{C}((a,b);E)$, it can be continuously extended to $[a,b]$ (Theorem A.38), which reduces to the previous case. □

Let us calculate the integral of a derivative. This is the second part of the **fundamental theorem of calculus**[3].

THEOREM 6.4.– *Let $f \in \mathcal{C}^1((a,b);E)$, where $-\infty \leq a < b \leq \infty$ and E is a Neumann space. Then:*

(a) *For every s and t in (a,b),*

$$\int_s^t f' = f(t) - f(s).$$

---

3 **History of the second part of the fundamental theorem of calculus.** Calculating the integral of a derivative, i.e. the equality $\int_s^t f' = f(t) - f(s)$, is closely related to calculating the derivative of an integral, whose history is discussed in footnote 2 (p. 121).
In Spain, this equality is known as *Barrow's rule*.

Vector values. James Andrew CLARKSON established the equality $\int_s^t f' = f(t) - f(s)$ in 1936 [23] for an absolutely continuous function f taking values in a uniformly convex Banach space.

(b) *If* $f \in \mathcal{C}^1([a,b];E)$, *the equality (a) holds for every finite* s *and* t *in* $[a,b]$. ▮

Proof. (a) The derivative of the function $t \mapsto f(t) - \int_s^t f'$ is zero by Theorem 6.3 (a). This function is therefore constant by Theorem 2.7. Since it is equal to $f(s)$ for $t = s$, it is equal to $f(s)$ everywhere.

(b) If $f \in \mathcal{C}^1([a,b];E)$, the equality (a) may be extended to all finite s and t in $[a,b]$, since the functions $s \mapsto \int_s^t f'$ and $t \mapsto \int_s^t f'$ are continuous on this interval (Theorem 6.3 (b)). □

6.3. Integration of a function of functions

Let us show that the order of integration does not matter for successive integrals[4].

THEOREM 6.5.– *Let*

$$f \in \mathbf{C}_{D_1}(\Omega_1; \mathbf{C}_{D_2}(\Omega_2; E)),$$

where Ω_1 *is an open subset of* $\mathbb{R}^{d_1}$, D_1 *is a compact subset of* $\mathbb{R}^{d_1}$, Ω_2 *is an open subset of* $\mathbb{R}^{d_2}$, D_2 *is a compact subset of* $\mathbb{R}^{d_2}$ *and* E *is a Neumann space.*

Then, for all open sets ω_1 *included in* Ω_1 *and* ω_2 *included in* Ω_2,

$$\int_{\omega_2}\int_{\omega_1} f = \int_{\omega_1}\int_{\omega_2} f. \ ▮$$

Meaning of the integrals in Theorem 6.5. *On the left-hand side:* $\int_{\omega_1} f$ is the integral of the function f, which takes values in $\mathbf{C}_{D_2}(\Omega_2;E)$. Thus, $\int_{\omega_1} f \in \mathbf{C}_{D_2}(\Omega_2;E)$ and $\int_{\omega_2}\int_{\omega_1} f \in E$ is its integral.

On the right-hand side: $\int_{\omega_2} f$ is the image of f under the mapping $\int_{\omega_2}$, which is linear and continuous from $\mathbf{C}_{D_2}(\Omega_2;E)$ into E. Thus, $\int_{\omega_2} f \in \mathbf{C}_{D_1}(\Omega_1;E)$ and $\int_{\omega_1}\int_{\omega_2} f \in E$ is its integral. □

Proof of Theorem 6.5. This is a special case of Theorem 4.13 on the commutativity of the integral, namely $\int_{\omega_1}$ in our case, with a continuous linear mapping, namely $\int_{\omega_2}$ in our case, which is continuous from $\mathbf{C}_{D_2}(\Omega_2;E)$ into E (Theorem 4.22). □

4 **History of Theorems 6.5 and 6.6 on changing the order of integration.** Augustin CAUCHY showed in 1823 [21, 33rd lesson] that, for real-valued functions, changing the order of successive integrals of real variables does not change the result. This is also known as the **theorem on integrating under the integral sign**.

Traditional formulation. The equality of Theorem 6.5 is traditionally stated for a function of two variables, say $\bar{f} \in \mathcal{B}(\Omega_1 \times \Omega_2; E)$, by using the partial integrals $F_1(x_1) = \int_{\omega_2} \bar{f}(x_1, x_2)\,\mathrm{d}x_2$ and $F_2(x_2) = \int_{\omega_1} \bar{f}(x_1, x_2)\,\mathrm{d}x_1$. Then $F_1 \in \mathcal{B}(\Omega_1; E)$ and $F_2 \in \mathcal{B}(\Omega_2; E)$, and their integrals coincide, namely $\int_{\omega_1} (\int_{\omega_2} \bar{f}(x_1, x_2)\,\mathrm{d}x_2)\,\mathrm{d}x_1 = \int_{\omega_2} (\int_{\omega_1} \bar{f}(x_1, x_2)\,\mathrm{d}x_1)\,\mathrm{d}x_2$.

This traditional formulation only uses successive integrals valued in E. In particular, for $E = \mathbb{R}$, it only uses real-valued integrals. □

Originality of Theorem 6.5. The use of a "function of functions", namely f defined on Ω_1 valued in the space of functions $\mathbf{C}_{D_2}(\Omega_2; E)$, is more original despite being easier to state and establish (see the comment *Benefit of vector values*, p. 125). It is easier even if $E = \mathbb{R}$, although it then makes use of a vector-valued integral, namely $\int_{\omega_1} f$ which is valued in $\mathbf{C}_{D_2}(\Omega_2)$ (observe that the functions $\int_{\omega_1} f$ and F_2 are equal, but are defined differently). □

Utility of the spaces $\mathbf{C}_D(\Omega; E)$. We cannot replace the hypothesis $f \in \mathbf{C}_{D_1}(\Omega_1; \mathbf{C}_{D_2}(\Omega_2; E))$ in Theorems 6.5 and 6.6 by $f \in \mathcal{B}(\Omega_1; \mathcal{B}(\Omega_2; E))$ because this space is not defined, since we did not endow $\mathcal{B}(\Omega_2; E)$ with semi-norms. Similarly, we cannot replace this hypothesis by $f \in \mathbf{C}_{\mathrm{b}}(\Omega_1; \mathbf{C}_{\mathrm{b}}(\Omega_2; E))$ because the integral $\int_{\omega_1} f$ is not defined if ω_1 is not bounded. □

Let us show that **permuting the variables** does not change the value of the integrals, regardless of whether one or all variables are integrated.

THEOREM 6.6.– *Let*

$$f \in \mathbf{C}_{D_1}(\Omega_1; \mathbf{C}_{D_2}(\Omega_2; E)),$$

where Ω_1 is an open subset of $\mathbb{R}^{d_1}$, D_1 is a compact subset of $\mathbb{R}^{d_1}$, Ω_2 is an open subset of $\mathbb{R}^{d_2}$, D_2 is a compact subset of $\mathbb{R}^{d_2}$ and E is a Neumann space.

Let $\check{f}$ be the function obtained by permuting the variables, namely

$$(\check{f}(x_2))(x_1) \stackrel{\text{def}}{=} (f(x_1))(x_2).$$

Then

$$\check{f} \in \mathbf{C}_{D_2}(\Omega_2; \mathbf{C}_{D_1}(\Omega_1; E))$$

and, for all open sets ω_1 included in Ω_1 and ω_2 included in Ω_2:

(a) $$\int_{\omega_1} \check{f} = \int_{\omega_1} f.$$

(b) $$\int_{\omega_2} \check{f} = \int_{\omega_2} f.$$

(c) $$\int_{\omega_1} \int_{\omega_2} \check{f} = \int_{\omega_2} \int_{\omega_1} \check{f} = \int_{\omega_2} \int_{\omega_1} f = \int_{\omega_1} \int_{\omega_2} f. \blacksquare$$

Meaning of the integrals in Theorem 6.6. In (a), both sides belong to $\mathbf{C}_{D_2}(\Omega_2; E)$: $\int_{\omega_1} \breve{f}$ is the image of $\breve{f}$ under the mapping $\int_{\omega_1}$, which is continuous and linear from $\mathbf{C}_{D_1}(\Omega_1; E)$ into E; and $\int_{\omega_1} f$ is the integral of f.

In (b), both sides belong to $\mathbf{C}_{D_1}(\Omega_1; E)$: $\int_{\omega_2} \breve{f}$ is the integral of $\breve{f}$; and $\int_{\omega_2} f$ is the image of f under the mapping $\int_{\omega_2}$, which is continuous and linear from $\mathbf{C}_{D_2}(\Omega_2; E)$ into E.

In (c), all terms belong to E. □

Proof of Theorem 6.6. The fact that $\breve{f}$ belongs to $\mathbf{C}_{D_2}(\Omega_2; \mathbf{C}_{D_1}(\Omega_1; E))$ was established in Theorem 1.28 (b).

(a) Here, $\int_{\omega_1} \breve{f}$ is the image of $\breve{f}$ under the mapping $\int_{\omega_1}$; in other words,

$$\Big(\int_{\omega_1} \breve{f}\Big)(x_2) = \int_{\omega_1} \breve{f}(x_2).$$

By Definition 4.9, the right-hand side is the limit of the approximate integrals

$$\begin{aligned} \mathbb{S}^n_{\omega_1}(\breve{f}(x_2)) &= \sum_{s_1 \in \mathbb{Z}^{d_1} : \Delta_{s_1,n} \subset \omega_1} 2^{-nd_1} (\breve{f}(x_2))(2^{-n} s_1) \\ &= \Big(\sum_{s_1 \in \mathbb{Z}^{d_1} : \Delta_{s_1,n} \subset \omega_1} 2^{-nd_1} f(2^{-n} s_1) \Big)(x_2) = (\mathbb{S}^n_{\omega_1} f)(x_2), \end{aligned}$$

where $\Delta_{s_1,n}$ is the closed cube of $\mathbb{R}^{d_1}$ of side length 2^{-n} centered at $2^{-n} s_1$. The limit of the final expression is $(\int_{\omega_1} f)(x_2)$, which is therefore equal to $(\int_{\omega_1} \breve{f})(x_2)$.

(b) This equality is just the equality (a) applied to $\breve{f}$ because $\breve{\breve{f}} = f$.

(c) The middle equality follows from (a), and the two others follow from Theorem 6.5. □

Benefit of vector values. Using vector values by considering a "function of functions" simplifies the proof of the properties of multiple integrals, even for real functions, namely when $E = \mathbb{R}$. This reduces the commutativity of the order of integration to a special case of the commutativity of the integral with linear mappings (see the proof of Theorem 6.5 and the comments about *Traditional formulation* and *Originality* that follow it). □

6.4. Integrating a function of multiple variables

Let us show that integrating with respect to multiple variables simultaneously is equivalent to integrating with respect to each variable successively, namely by **separating the variables**[5].

5 **History of the multiple integral calculation in Theorem 6.7**. Leonhard EULER calculated double integrals, namely on $\mathbb{R}^2$, in 1770 [36] by integrating each variable in turn. One example of his calculations

THEOREM 6.7.– *Let*

$$f \in \mathbf{C}_{D_1 \times D_2}(\Omega_1 \times \Omega_2; E),$$

where Ω_1 is an open subset of $\mathbb{R}^{d_1}$, Ω_2 is an open subset of $\mathbb{R}^{d_2}$, D_1 is a compact subset of $\mathbb{R}^{d_1}$, D_2 is a compact subset of $\mathbb{R}^{d_2}$ and E is a Neumann space.

Let $\underline{f}$ be the function obtained by separating the variables, namely

$$(\underline{f}(x_1))(x_2) \stackrel{\text{def}}{=} f(x_1, x_2).$$

Then

$$\underline{f} \in \mathbf{C}_{D_1}(\Omega_1; \mathbf{C}_{D_2}(\Omega_2; E))$$

and, for all open sets ω_1 included in Ω_1 and ω_2 included in Ω_2,

$$\int_{\omega_1 \times \omega_2} f = \int_{\omega_2} \int_{\omega_1} \underline{f} = \int_{\omega_1} \int_{\omega_2} \underline{f}. \blacksquare$$

Meaning of the integrals in Theorem 6.7. *In the second expression:* $\int_{\omega_1} \underline{f}$ is the integral of the function $\underline{f}$, which takes values in $\mathbf{C}_{D_2}(\Omega_2; E)$. Thus, $\int_{\omega_1} \underline{f} \in \mathbf{C}_{D_2}(\Omega_2; E)$ and $\int_{\omega_2} \int_{\omega_1} \underline{f} \in E$ is its integral.

In the third expression: $\int_{\omega_2} \underline{f}$ is the image of $\underline{f}$ under the mapping $\int_{\omega_2}$, which is linear and continuous from the space $\mathbf{C}_{D_2}(\Omega_2; E)$ into E. Thus, $\int_{\omega_2} \underline{f} \in \mathbf{C}_{D_1}(\Omega_1; E)$ and $\int_{\omega_1} \int_{\omega_2} \underline{f} \in E$ is its integral. □

Traditional formulation. The first equality of Theorem 6.7 is traditionally stated in the following form: $\int_{\omega_1 \times \omega_2} f(x_1, x_2) \,\mathrm{d}(x_1, x_2) = \int_{\omega_1} (\int_{\omega_2} f(x_1, x_2) \,\mathrm{d}x_2) \,\mathrm{d}x_1$. It only uses successive integrals valued in E, namely the integrals of $x_1 \mapsto f(x_1, x_2)$ for each fixed x_2, and $x_2 \mapsto \int_{\omega_1} f(x_1, x_2) \,\mathrm{d}x_1$. In particular, for $E = \mathbb{R}$, it only uses real-valued integrals. □

Originality of Theorem 6.7. The use of a "function of functions", namely $\underline{f}$ defined on Ω_1 valued in the space of functions $\mathbf{C}_{D_2}(\Omega_2; E)$, is more original despite being easier to state and establish. This may be seen by comparing the proof of Theorem 6.7 to that of Theorem 6.10 on integration between graphs, which is more complicated since it uses the traditional method (for the reason explained in the comment *A simple method...*, p. 130). □

The proof of Theorem 6.7 will use the following inequalities, which are shown later.

LEMMA 6.8.– *Let $f \in \mathcal{B}(\Omega; E)$, where Ω is an open subset of $\mathbb{R}^d$ and E is a Neumann space, ω is an open set included in Ω and $\mathbb{S}^n_\omega f$ is the approximate integral (Definition 4.9 (a)). Then, for every semi-norm $\| \;\|_{E;\nu}$ of E:*

is presented by Jean-Paul PIER [62, p. 50]. The integral was not yet defined rigorously (this was only done in 1823, see footnote 2, p. 89).
The equivalence of the double integral and two successive real integrals was established in 1907 by Guido FUBINI [39] in the significantly more delicate case of integrable functions.

(a) $$\|\mathbb{S}^n_\omega f\|_{E;\nu} \le |\omega| \sup_{x\in\omega} \|f(x)\|_{E;\nu}.$$

(b) $$\left\| \int_\omega f - \mathbb{S}^n_\omega f \right\|_{E;\nu} \le |\omega| \sup_{|y-x|\le 2^{-n-1}\sqrt{d}} \|f(y) - f(x)\|_{E;\nu} + \\ + (|\omega| - 2^{-nd}k_n) \sup_{x\in\omega} \|f(x)\|_{E;\nu},$$

where k_n is the number of closed cubes $\Delta_{s,n}$ of side length 2^{-n} centered at $2^{-n}s$ with $s \in \mathbb{Z}^d$ that are included in ω. ▮

Proof of Theorem 6.7. The fact that $\underline{f}$ belongs to $\mathbf{C}_{D_1}(\Omega_1; \mathbf{C}_{D_2}(\Omega_2; E))$ was established in Theorem 1.27 (c), and the equation $\int_{\omega_2}\int_{\omega_1} \underline{f} = \int_{\omega_1}\int_{\omega_2} \underline{f}$ is given by Theorem 6.5, so we simply need to establish the equality $\int_{\omega_1\times\omega_2} f = \int_{\omega_2}\int_{\omega_1} \underline{f}$, which we will do in two steps.

1. *Calculating the approximate integrals.* By Definition 4.9 of the approximate integral,

$$\mathbb{S}^n_{\omega_1}\underline{f} = \sum_{s_1\in\mathbb{Z}^{d_1}:\Delta_{s_1,n}\subset\omega_1} 2^{-nd_1}\underline{f}(2^{-n}s_1),$$

where $\Delta_{s_1,n}$ is the closed cube of $\mathbb{R}^{d_1}$ of side length 2^{-n} centered at $2^{-n}s_1$. This equality holds in $\mathbf{C}_{D_2}(\Omega_2; E)$, therefore, in E,

$$\begin{aligned}
&\mathbb{S}^n_{\omega_2}(\mathbb{S}^n_{\omega_1}\underline{f}) = \\
&= \sum_{s_2\in\mathbb{Z}^{d_2}:\Delta_{s_2,n}\subset\omega_2} 2^{-nd_2} \sum_{s_1\in\mathbb{Z}^{d_1}:\Delta_{s_1,n}\subset\omega_1} 2^{-nd_1}(\underline{f}(2^{-n}s_1))(2^{-n}s_2) = \\
&= \sum_{(s_1,s_2)\in\mathbb{Z}^{d_1+d_2}:\Delta_{(s_1,s_2),n}\subset\omega_1\times\omega_2} 2^{-n(d_1+d_2)} f(2^{-n}(s_1,s_2)) = \mathbb{S}^n_{\omega_1\times\omega_2} f.
\end{aligned}$$

Indeed, a cube $\Delta_{(s_1,s_2),n}$ is included in $\omega_1 \times \omega_2$ if and only if $\Delta_{s_1,n} \subset \omega_1$ and $\Delta_{s_2,n} \subset \omega_2$.

2. *Convergence.* By Definition 4.9 of the integral, $\mathbb{S}^n_{\omega_1\times\omega_2} f \to \int_{\omega_1\times\omega_2} f$, so it suffices to show that, as $n \to \infty$,

$$\mathbb{S}^n_{\omega_2}(\mathbb{S}^n_{\omega_1}\underline{f}) \to \int_{\omega_2}\int_{\omega_1} \underline{f}.$$

Decompose

$$\mathbb{S}^n_{\omega_2}(\mathbb{S}^n_{\omega_1}\underline{f}) - \int_{\omega_2}\left(\int_{\omega_1}\underline{f}\right) = \mathbb{S}^n_{\omega_2}\left(\mathbb{S}^n_{\omega_1}\underline{f} - \int_{\omega_1}\underline{f}\right) + \left(\mathbb{S}^n_{\omega_2} - \int_{\omega_2}\right)\left(\int_{\omega_1}\underline{f}\right).$$

Since $\int_{\omega_1}\underline{f} \in \mathbf{C}_{D_2}(\Omega_2; E)$, we have $(\mathbb{S}^n_{\omega_2} - \int_{\omega_2})(\int_{\omega_1}\underline{f}) \to 0$, so we just need to establish that

$$\mathbb{S}^n_{\omega_2}\left(\mathbb{S}^n_{\omega_1}\underline{f} - \int_{\omega_1}\underline{f}\right) \to 0. \qquad (6.2)$$

Since $\mathbb{S}^n_{\omega_1}\underline{f}$ and $\int_{\omega_1}\underline{f}$ belong to $\mathbf{C}_{D_2}(\Omega_2; E)$, which is endowed with the semi-norms of $\mathcal{C}_b(\Omega; E)$ by Definition 1.7 (b), the inequality from Lemma 6.8 (a) taking values in this space implies that, for every semi-norm $\| \ \|_{E;\nu}$ of E,

$$\left\|\mathbb{S}^n_{\omega_2}\left(\mathbb{S}^n_{\omega_1}\underline{f} - \int_{\omega_1}\underline{f}\right)\right\|_{E;\nu} \le |\omega_2|\left\|\mathbb{S}^n_{\omega_1}\underline{f} - \int_{\omega_1}\underline{f}\right\|_{\mathcal{C}_b(\Omega_2;E);\nu}.$$

Moreover, the inequality from Lemma 6.8 (b) gives

$$\left\|\mathbb{S}^n_{\omega_1}\underline{f} - \int_{\omega_1}\underline{f}\right\|_{\mathcal{C}_b(\Omega_2;E);\nu} \le |\omega_1| \sup_{|y_1-x_1|\le 2^{-n-1}\sqrt{d_1}} \|\underline{f}(y_1) - \underline{f}(x_1)\|_{\mathcal{C}_b(\Omega_2;E);\nu} +$$

$$+ \left(|\omega_1| - 2^{-nd_1}k_n(\omega_1)\right) \sup_{x_1\in\omega_1} \|\underline{f}(x_1)\|_{\mathcal{C}_b(\Omega_2;E);\nu}.$$

The right-hand side tends to 0 because $\underline{f}$ is uniformly continuous, bounded and takes values in $\mathbf{C}_{D_2}(\Omega_2; E)$ and $2^{-nd_1}k_n(\omega_1) \to |\omega_1|$ by Definition 4.1 of the measure. This implies (6.2) and hence

$$\int_{\omega_2}\int_{\omega_1}\underline{f} = \int_{\omega_1\times\omega_2} f. \ \square$$

Proof of Lemma 6.8. (a) By Definition 4.9 of the approximate integral,

$$\|\mathbb{S}^n_\omega f\|_{E;\nu} = \left\|\sum_{s\in\mathbb{Z}^d:\Delta_{s,n}\subset\omega} 2^{-nd}f(2^{-n}s)\right\|_{E;\nu} \le 2^{-nd}k_n \sup_{x\in\omega}\|f(x)\|_{E;\nu},$$

where k_n is the number of cubes $\Delta_{s,n}$ included in ω. This gives the stated inequality because $2^{-nd}k_n \le |\omega|$ (inequality (4.2), p. 84).

(b) The decomposition $\mathbb{S}^m_\omega f - \mathbb{S}^n_\omega f = \mathcal{X}_{m,n} + \mathcal{Y}_{m,n}$ of a difference of approximate integrals performed in (4.6), p. 91, and the bounds of $\|\mathcal{X}_{m,n}\|_{E;\nu}$ and $\|\mathcal{Y}_{m,n}\|_{E;\nu}$ given in (4.7) and (4.8) imply, for $m \ge n$,

$$\|\mathbb{S}^m_\omega f - \mathbb{S}^n_\omega f\|_{E;\nu} \le \|\mathcal{X}_{m,n}\|_{E;\nu} + \|\mathcal{Y}_{m,n}\|_{E;\nu} \le$$

$$\le 2^{-nd}k_n \sup_{|y-x|\le 2^{-n-1}\sqrt{d}} \|f(y) - f(x)\|_{E;\nu} +$$

$$+ (2^{-md}k_m - 2^{-nd}k_n) \sup_{x\in\omega}\|f(x)\|_{E;\nu}.$$

This gives the stated inequality by taking the limit as $m \to \infty$, since $2^{-nd}k_n \le |\omega|$ (inequality (4.2) once again) and $2^{-md}k_m \to |\omega|$ (Definition 4.1). $\square$

Successive integrals without using vector values. Theorem 6.10 establishes the result of Theorem 6.7 (and therefore also the result of Theorem 6.5) with a method that exclusively uses real-valued integration when $E = \mathbb{R}$ but which is more complicated. $\square$

Let us calculate the integral of a real **tensor product** $f_1 \otimes f_2$.

THEOREM 6.9.– *Let $f_1 \in \mathcal{B}(\Omega_1)$ and $f_2 \in \mathcal{B}(\Omega_2)$, where Ω_1 is an open subset of $\mathbb{R}^{d_1}$ and Ω_2 is an open subset of $\mathbb{R}^{d_2}$.*

We define $f_1 \otimes f_2 \in \mathcal{B}(\Omega_1 \times \Omega_2)$ by

$$(f_1 \otimes f_2)(x_1, x_2) \stackrel{\text{def}}{=} f_1(x_1) f_2(x_2).$$

For all open sets ω_1 included in Ω_1 and ω_2 included in Ω_2, it satisfies

$$\int_{\omega_1 \times \omega_2} f_1 \otimes f_2 = \Big(\int_{\omega_1} f_1\Big)\Big(\int_{\omega_2} f_2\Big). \blacksquare$$

Proof. By Definition 4.7 of $\mathcal{B}(\Omega_1 \times \Omega_2)$, we need to show that $f_1 \otimes f_2$ is uniformly continuous and its support is bounded. First, $f_1 \otimes f_2$ is uniformly continuous, like f_1 and f_2 are by the hypotheses, because, for all (x_1, x_2) and (y_1, y_2) in $\Omega_1 \times \Omega_2$,

$$\begin{aligned}&|(f_1 \otimes f_2)(x_1, x_2) - (f_1 \otimes f_2)(y_1, y_2)| = \\ &\quad = |(f_1(x_1) - f_1(y_1)) f_2(x_2)) + f_1(y_1)(f_2(x_2) - f_2(y_2))| \leq \\ &\quad \leq m_1 |f_1(x_1) - f_1(y_1)| + m_2 |f_2(x_2) - f_2(y_2)|,\end{aligned}$$

since f_1 and f_2 are bounded by Theorem 4.8 (a). Second, its support is bounded, like those of f_1 and f_2 are, since it is included in $\operatorname{supp} f_1 \times \operatorname{supp} f_2$. This proves that $f_1 \otimes f_2 \in \mathcal{B}(\Omega_1 \times \Omega_2)$.

Let us now calculate the integral of $f_1 \otimes f_2$. Every cube decomposes into a product $\Delta_{(s_1,s_2),n} = \Delta_{s_1,n} \times \Delta_{s_2,n}$, so Definition 4.9 of the approximate integral can be restated here as

$$\begin{aligned}\mathbb{S}^n_{\omega_1 \times \omega_2}(f_1 \otimes f_2) &= \sum_{(s_1,s_2) \in \mathbb{Z}^{d_1+d_2} : \Delta_{s_1,n} \times \Delta_{s_2,n} \subset \omega_1 \times \omega_2} 2^{-nd} f_1(2^{-n} s_1) f_2(2^{-n} s_2) = \\ &= \sum_{s_1 \in \mathbb{Z}^{d_1} : \Delta_{s_1,n} \subset \omega_1} 2^{-nd_1} f_1(2^{-n} s_1) \sum_{s_2 \in \mathbb{Z}^{d_2} : \Delta_{s_2,n} \subset \omega_2} 2^{-nd_2} f_2(2^{-n} s_2) = \\ &= \mathbb{S}^n_{\omega_1} f_1 \, \mathbb{S}^n_{\omega_2} f_2.\end{aligned}$$

In the limit, this gives the stated equality. □

6.5. Integration between graphs

Let us separate the variables in the integral of a function on a domain delimited by graphs.

THEOREM 6.10.– *Let*

$$\Omega = \{x \in \mathbb{R}^d : x = (x_*, x_d),\ x_* \in \Omega_*,\ \Upsilon_1(x_*) < x_d < \Upsilon_2(x_*)\},$$

where Ω_* is an open subset of $\mathbb{R}^{d-1}$, $\Upsilon_1 \in \mathbf{C}_\mathrm{b}(\Omega_*)$, $\Upsilon_2 \in \mathbf{C}_\mathrm{b}(\Omega_*)$ and $\Upsilon_1 \leq \Upsilon_2$. Additionally, let $f \in \mathcal{B}(\Omega; E)$, where E is a Neumann space.

Then the function $x_* \mapsto \int_{\Upsilon_1(x_*)}^{\Upsilon_2(x_*)} f(x_*, x_d)\,\mathrm{d}x_d$ belongs to $\mathcal{B}(\Omega_*; E)$ and

$$\int_\Omega f = \int_{\Omega_*} \Big(\int_{\Upsilon_1(x_*)}^{\Upsilon_2(x_*)} f(x_*, x_d)\,\mathrm{d}x_d \Big)\,\mathrm{d}x_*. \ \blacksquare$$

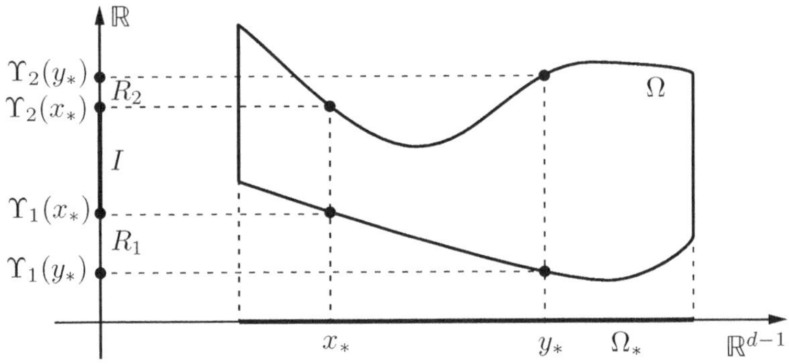

Figure 6.1. *Domain delimited by the graphs of Υ_1 and Υ_2*

A simple method for proving Theorem 6.10. If we could integrate discontinuous functions taking values in a Neumann space, we could prove Theorem 6.10 by extending f by 0_E outside of Ω and then separating the variables in the integral over $\mathbb{R}^d$ using Theorem 6.7. This would be much easier than the laborious proof given below.

Furthermore, we would be able to extend the result to arbitrary open subsets of $\mathbb{R}^d$ (instead of open sets delimited by two graphs) in the following form:

$$\int_\Omega f\,\mathrm{d}x = \int_{\mathbb{R}^{d-1}} \Big(\int_{\Omega_{x_*}} f(x_*, x_d)\,\mathrm{d}x_d \Big)\,\mathrm{d}x_*,$$

where $\Omega_{x_*} = \{x_d \in \mathbb{R} : (x_*, x_d) \in \Omega\}$.

We cannot use this method here because we considered that it would be more "economical" to construct distributions (in Volume 3) using just the Cauchy integral for uniformly continuous functions before constructing *integrable distributions* (which correspond to the extension of f by 0_E) in Volume 4 to fulfill the role usually played by *classes of almost everywhere equal integrable functions*. □

Proof of Theorem 6.10. Denote

$$F(x_*) = \int_{\Upsilon_1(x_*)}^{\Upsilon_2(x_*)} f(x_*, x_d)\,\mathrm{d}x_d. \tag{6.3}$$

Observe that the right-hand side is well defined because the function $x_d \mapsto f(x_*, x_d)$ belongs to $\mathcal{B}((\Upsilon_1(x_*), \Upsilon_2(x_*)); E)$.

Proof that F belongs to $\mathcal{B}(\Omega_*; E)$. By Definition 4.7 of $\mathcal{B}(\Omega_*; E)$, we need to show that F is uniformly continuous and its support is bounded. The latter property holds because Ω_* is bounded.

For every x_* and y_* in Ω_*, apply the additivity property (Theorem 6.2) to decompose

$$F(y_*) - F(x_*) = \int_I f(y_*, z) - f(x_*, z)\,\mathrm{d}z \pm \int_{R_1} f(u_*^1, z)\,\mathrm{d}z \pm \int_{R_2} f(u_*^2, z)\,\mathrm{d}z,$$

where $I = (\Upsilon_1(y_*), \Upsilon_2(y_*)) \cap (\Upsilon_1(x_*), \Upsilon_x(y_*))$ is the shared part of the line segments from y_* and x_* and R_1 and R_2 are the disjoint parts, which either point along the line segment from y_*, in which case $u_*^i = y_*$ and $\pm$ is $+$, or along the line segment from x_*, in which case $u_*^i = x_*$ and $\pm$ is $-$.

Observe that $\sup_{x_* \in \Omega_*} |I| \stackrel{\text{def}}{=} \ell$ is finite and $|R_i| = |\Upsilon_i(y_*) - \Upsilon_i(x_*)|$. For every semi-norm $\|\ \|_{E;\nu}$ of E, the bound on the semi-norms of the integral from Theorem 4.15 therefore implies

$$\begin{aligned}\|F(y_*) - F(x_*)\|_{E;\nu} \le \ell &\sup_{x\in\Omega,\, y\in\Omega,\, |x-y|=|x_*-y_*|} \|f(y) - f(x)\|_{E;\nu} + \\ &+ (|\Upsilon_1(y_*) - \Upsilon_1(x_*)| + |\Upsilon_2(y_*) - \Upsilon_2(x_*)|) \sup_{x\in\Omega} \|f(x)\|_{E;\nu}.\end{aligned}$$

We have $\sup_{x\in\Omega} \|f(x)\|_{E;\nu} < \infty$ because f is bounded by Theorem 4.4 (a), so F is uniformly continuous because f, Υ_1 and Υ_2 are uniformly continuous by the hypotheses. This proves that $F \in \mathcal{B}(\Omega_*; E)$.

Equality of the integrals. Let us show in three steps that the approximate integrals of f and F have the same limit.

1. *Calculating the difference of the approximate integrals.* By Definition 4.9 (a), the approximate integral of f is

$$\begin{aligned}\mathbb{S}_\Omega^n f &= \sum_{s\in\mathbb{Z}^d : \Delta_{s,n}\subset\Omega} 2^{-nd} f(2^{-n}s) = \\ &= \sum_{s_*\in\mathcal{S}_{*,n}} 2^{-n(d-1)} \sum_{s_d\in\mathcal{S}_{d,n}(s_*)} 2^{-n} f(2^{-n}s_*, 2^{-n}s_d),\end{aligned}$$

where $\Delta_{s,n}$ is the closed cube of $\mathbb{R}^d$ of side length 2^{-n} centered at $2^{-n}s$, which decomposes into $\Delta_{s,n} = \Delta_{s_*,n} \times \Delta_{s_d,n}$, and

$$\mathcal{S}_{*,n} = \{s_* \in \mathbb{Z}^{d-1} : \Delta_{s_*,n} \subset \Omega_*\},$$

$$\mathcal{S}_{d,n}(s_*) = \Big\{s_d \in \mathbb{Z} : \Delta_{s_d,n} \subset \Big(\sup_{z_* \in \Delta_{s_*,n}} \Upsilon_1(z_*), \inf_{z_* \in \Delta_{s_*,n}} \Upsilon_2(z_*)\Big)\Big\} =$$
$$= \Big\{s_d \in \mathbb{Z} : \sup_{z_* \in \Delta_{s_*,n}} \Upsilon_1(z_*) + 2^{-n-1} < 2^{-n}s_d < \inf_{z_* \in \Delta_{s_*,n}} \Upsilon_2(z_*) - 2^{-n-1}\Big\}.$$

On the other hand, the approximate integral of F is

$$\mathbb{S}^n_{\Omega_*} F = \sum_{s_* \in \mathcal{S}_{*,n}} 2^{-n(d-1)} F(2^{-n}s_*).$$

Denoting by $F_n(x_*)$ the approximate integral of the right-hand side of (6.3),

$$F_n(2^{-n}s_*) = \sum_{s_d \in \mathcal{S}'_{d,n}(s_*)} 2^{-n} f(2^{-n}s_*, 2^{-n}s_d),$$

where

$$\mathcal{S}'_{d,n}(s_*) = \{s_d \in \mathbb{Z} : \Delta_{s_d,n} \subset (\Upsilon_1(2^{-n}s_*), \Upsilon_2(2^{-n}s_*))\} =$$
$$= \{s_d \in \mathbb{Z} : \Upsilon_1(2^{-n}s_*) + 2^{-n-1} < 2^{-n}s_d < \Upsilon_2(2^{-n}s_*) - 2^{-n-1}\}.$$

Therefore,

$$\mathbb{S}^n_{\Omega_*} F - \mathbb{S}^n_{\Omega} f = \sum_{s_* \in \mathcal{S}_{*,n}} 2^{-n(d-1)} \Big(F(2^{-n}s_*) - F_n(2^{-n}s_*) +$$
$$+ \sum_{s_d \in \mathcal{S}'_{d,n}(s_*) \setminus \mathcal{S}_{d,n}(s_*)} 2^{-n} f(2^{-n}s_*, 2^{-n}s_d)\Big).$$

2. *Estimation.* For every semi-norm $\| \ \|_{E;\nu}$ of E, the inequality from Lemma 6.8 (b) gives

$$\|F(2^{-n}s_*) - F_n(2^{-n}s_*)\|_{E;\nu} \leq$$
$$\leq h \sup_{|y-x| \leq 2^{-n-1}} \|f(y) - f(x)\|_{E;\nu} + c_n \sup_{x \in \Omega} \|f(x)\|_{E;\nu},$$

with $h = \sup_{x_* \in \Omega_*} |\Upsilon_2(x_*) - \Upsilon_1(x_*)|$ and $c_n = |\Upsilon_2(2^{-n}s_*) - \Upsilon_1(2^{-n}s_*)| - 2^{-n}k_n$, where k_n is the number of segments $\Delta_{s_d,n} = [2^{-n}s_d - 2^{-n-1}, 2^{-n}s_d + 2^{-n-1}]$ included in $(\Upsilon_1(2^{-n}s_*), \Upsilon_2(2^{-n}s_*))$. By the inequalities (4.3), p. 86, it satisfies $c_n \leq 2^{1-n}$.

Denoting by $k_{*,n}$, $k_{d,n}(s_*)$ and $k'_{d,n}(s_*)$, respectively, the number of elements of $\mathcal{S}_{*,n}$, $\mathcal{S}_{d,n}(s_*)$ and $\mathcal{S}'_{d,n}(s_*)$, it therefore follows that

$$\begin{aligned} &\|\mathbb{S}^n_{\Omega_*} F - \mathbb{S}^n_\Omega f\|_{E;\nu} \le \\ &\le 2^{-n(d-1)} k_{*,n} \big(h\delta_n + 2^{1-n} c + 2^{-n} c \sup_{s_* \in \mathcal{S}_{*,n}} k'_{d,n}(s_*) - k_{d,n}(s_*)\big), \end{aligned} \tag{6.4}$$

where $\delta_n = \sup_{|y-x| \le 2^{-n-1}} \|f(y) - f(x)\|_{E;\nu}$ and $c = \sup_{x \in \Omega} \|f(x)\|_{E;\nu}$. Observe that $2^{-n(d-1)} k_{*,n} \le |\Omega_*|$ (inequality (4.2), p. 84). Furthermore, $\delta_n \to 0$ as $n \to \infty$ because f is uniformly continuous.

By the definitions of $\mathcal{S}_{d,n}(s_*)$ and $\mathcal{S}'_{d,n}(s_*)$,

$$\begin{aligned} &2^{-n} k'_{d,n}(s_*) - 2^{-n} k_{d,n}(s_*) \le \\ &\quad \le \Big|\Upsilon_2(2^{-n} s_*) - \inf_{z_* \in \Delta_{s_*,n}} \Upsilon_2(z_*)\Big| + \Big|\Upsilon_1(2^{-n} s_*) - \sup_{z_* \in \Delta_{s_*,n}} \Upsilon_1(z_*)\Big| \le \\ &\quad \le \sup_{|z_* - x_*| < 2^{-n-1}\sqrt{d-1}} |\Upsilon_2(z_*) - \Upsilon_2(x_*)| + |\Upsilon_1(z_*) - \Upsilon_1(x_*)|. \end{aligned}$$

As $n \to \infty$, the right-hand side tends to 0 because Υ_1 and Υ_2 are uniformly continuous. Therefore, (6.4) implies

$$\|\mathbb{S}^n_{\Omega_*} F - \mathbb{S}^n_\Omega f\|_{E;\nu} \to 0.$$

3. *Conclusion.* In the limit, by Definition 4.9 (b) of the Cauchy integral, this gives $\| \int_{\Omega_*} F - \int_\Omega f\|_{E;\nu} = 0$. This holds for every semi-norm of E, so

$$\int_{\Omega_*} F = \int_\Omega f. \ \square$$

6.6. Integration by parts and weak vanishing condition for a function

Let us state an elementary version of **Ostrogradsky's formula**[6].

THEOREM 6.11.– *Let $f \in \mathcal{K}^1(\Omega; E)$, where Ω is an open subset of $\mathbb{R}^d$ and E is a Neumann space. Then, for every $i \in [\![1, d]\!]$,*

$$\int_\Omega \partial_i f = 0_E. \ \blacksquare$$

6 **History of Ostrogradsky's formula.** Mikhail Vasilyevitch OSTROGRADSKY showed in 1831 [61] that $\int_\Omega \partial_1 f_1 + \partial_2 f_2 + \partial_3 f_3 \, dx = \int_\Omega n_1 f_1 + n_2 f_2 + n_3 f_3 \, ds$, which gives us the formula from Theorem 6.11 for $i = 1$ by choosing $f_2 = f_3 = 0$ and $f_1 = f$.

Proof. Since the support of f is compact, its extension $\widetilde{f}$ by 0_E belongs to $\mathcal{K}^1(\mathbb{R}^d; E)$. Let $\Lambda = \{x \in \mathbb{R}^d : a_i < x_i < b_i, \forall i\}$ be a parallelepiped containing its support. Since $\partial_i \widetilde{f}(x_1, x_2 \dots, x_d) = \mathrm{d}\widetilde{f}(x_1, x_2 \dots, x_d)/\mathrm{d}x_i$ (Definition 2.8), the expression of the integral of a derivative (Theorem 6.4 (a)) gives

$$\int_{a_i}^{b_i} \partial_i \widetilde{f}(x_1, \dots, x_d)\, \mathrm{d}x_i = \widetilde{f}(x_1, \dots, x_{i-1}, b_i, x_{i+1}, \dots, x_d) - \\ - \widetilde{f}(x_1, \dots, x_{i-1}, a_i, x_{i+1}, \dots, x_d) = 0_E.$$

Integrating with respect to other x_j, we obtain, since integrals may be merged and interchanged (Theorem 6.7),

$$\int_\Lambda \partial_i \widetilde{f} = \int_{a_d}^{b_d} \cdots \int_{a_1}^{b_1} \partial_i \widetilde{f} = \int_{a_d}^{b_d} \cdots \int_{a_{i+1}}^{b_{i+1}} \int_{a_{i-1}}^{b_{i-1}} \cdots \int_{a_1}^{b_1} \int_{a_i}^{b_i} \partial_i \widetilde{f} = 0_E.$$

Since the domain where the function is zero does not contribute to the integral (Theorem 4.18),

$$\int_\Omega \partial_i f = \int_{\mathbb{R}^d} \partial_i \widetilde{f} = \int_\Lambda \partial_i \widetilde{f} = 0_E. \ \square$$

Let us state a formula of **integration by parts**.

THEOREM 6.12.– *Let $f \in \mathcal{C}^1(\Omega; E)$ and $\varphi \in \mathcal{K}^1(\Omega)$, where Ω is an open subset of $\mathbb{R}^d$ and E is a Neumann space. Then, for every $i \in [\![1, d]\!]$,*

$$\int_\Omega \partial_i f\, \varphi = -\int_\Omega f\, \partial_i \varphi. \ \blacksquare$$

Proof. The product $f\varphi$ belongs to $\mathcal{K}^1(\Omega; E)$ (Theorem 3.11 (c)), and the Leibniz rule and Ostrogradsky's formula (Theorems 3.6 and 6.11) imply

$$\int_\Omega \partial_i f\, \varphi + f\, \partial_i \varphi = \int_\Omega \partial_i (f\varphi) = 0_E. \ \square$$

Let us state a **weak condition** that guarantees that a continuous function vanishes, namely the **Du Bois-Reymond lemma**, also known as the **fundamental lemma of the calculus of variations**[7].

---

7 **History of the Du Bois-Reymond lemma.** Friedrich Ludwig STEGMANN stated Theorem 6.13 for a real function on an interval in 1854 [77] but justified it incorrectly. Correct proofs were given by Eduard HEINE in 1870 [47] and Paul DU BOIS-REYMOND in 1879 [30].

Theorem 6.13.– *Let $f \in \mathcal{C}(\Omega; E)$, where Ω is an open subset of $\mathbb{R}^d$ and E is a Neumann space such that, for every $\varphi \in \mathcal{K}^\infty(\Omega)$,*

$$\int_\Omega f\varphi = 0_E.$$

Then

$$f = 0. \blacksquare$$

Proof. Suppose that f is not zero. Then there exists $y \in \Omega$ and a semi-norm of E such that $\|f(y)\|_{E;\nu} = \epsilon > 0$. Since Ω is open and f is continuous, there exists $r > 0$ such that $|x - y| \le r$ implies $x \in \Omega$ and

$$\|f(x) - f(y)\|_{E;\nu} \le \epsilon/2.$$

Let $\varphi \in \mathcal{K}^\infty(\Omega)$ with support in the ball $B = \{x \in \mathbb{R}^d : |y - x| \le r\}$ such that $\varphi > 0$ on $\mathring{B}$. These conditions are, for example, fulfilled by $\varphi(x) = \rho((x - y)/r)$, where the function ρ satisfies Theorem 3.19. Such a function φ satisfies (Theorems 4.17 (a) and 4.14 (d))

$$\int_\Omega \varphi = \int_{\mathring{B}} \varphi = b > 0.$$

By the properties of the integral (Theorems 4.17 (a), 4.11 (b), and 4.14 (a)),

$$\left\| \int_\Omega (f - f(y))\, \varphi \right\|_{E;\nu} = \left\| \int_{\mathring{B}} (f - f(y))\, \varphi \right\|_{E;\nu} \le \int_{\mathring{B}} \|f - f(y)\|_{E;\nu}\, \varphi \le \frac{\epsilon}{2} b,$$

$$\left\| \int_\Omega f(y)\, \varphi \right\|_{E;\nu} = \left\| f(y) \int_\Omega \varphi \right\|_{E;\nu} = \epsilon b,$$

Therefore,

$$\left\| \int_\Omega f\varphi \right\|_{E;\nu} \ge \epsilon/2.$$

This contradicts the hypothesis $\int_\Omega f\varphi = 0_E$ because E is separated. Hence, $f = 0$. $\square$

6.7. Change of variables in an integral

Let us perform a **change of variables** in an integral.

Theorem 6.14.– *Let $f \in \mathcal{B}(\Omega; E)$, where Ω is an open subset of $\mathbb{R}^d$ and E is a Neumann space, and T a bijection from an open subset Λ of $\mathbb{R}^d$ onto Ω such that*

$$T \in \mathbf{C}^1_{\mathrm{b}}(\Lambda; \Omega) \text{ and } T^{-1} \in \mathbf{C}^1_{\mathrm{b}}(\Omega; \Lambda).$$

Then, by denoting $\det[\nabla T]$ *as the determinant of the matrix* $[\partial_i T_j]$*:*

(a) $(f \circ T)\,|\det[\nabla T]| \in \mathcal{B}(\Lambda; E)$ *and, for every open set* λ *included in* Λ, *the set* $T(\lambda)$ *is open and*

$$\int_{T(\lambda)} f = \int_\lambda (f \circ T)\,|\det[\nabla T]|.$$

(b) $f\,|\det[\nabla T^{-1}]| \in \mathcal{B}(\Omega; E)$ *and, for every open set* λ *included in* Λ,

$$\int_\lambda f \circ T = \int_{T(\lambda)} f\,|\det[\nabla T^{-1}]|.$$

(c) *The equality (a) is still satisfied if, instead of* $T^{-1} \in \mathbf{C}^1_b(\Omega; \Lambda)$, *we assume:*

$$T^{-1} \in \mathcal{C}^1(\Omega; \Lambda), \quad \textit{and } \lambda \textit{ or } \operatorname{supp} f \circ T \textit{ is bounded.} \ ▮$$

Reminder. By Definition 2.29, $\mathcal{C}^1_b(\Omega; \Lambda)$ is the set of functions from Ω into Λ that are continuously differentiable and bounded, and $\mathbf{C}^1_b(\Omega; \Lambda)$ is the set of such functions that are uniformly continuous and whose derivatives are also uniformly continuous. □

Terminology. The function $\det[\nabla T]$ is called the **Jacobian determinant** or the **Jacobian** of T. □

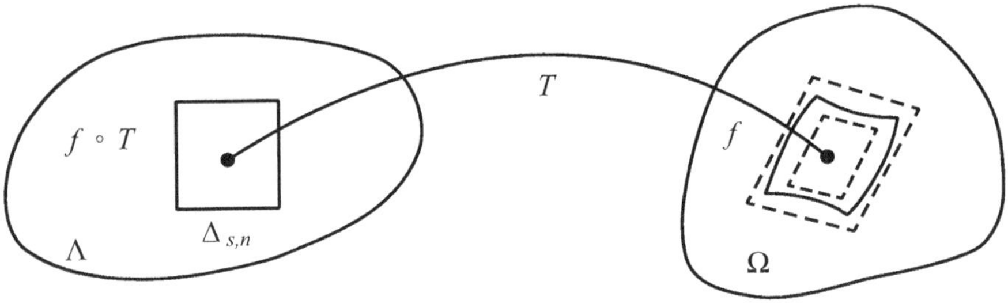

Figure 6.2. *Change of variables.*
The contour of $T(\Delta_{s,n})$ *is shown in bold, the contours of the parallelepipeds* $P^{1-\eta}_{s,n}$ *and* $P^{1+\eta}_{s,n}$ *that enclose it are shown with dashes*

Proof of Theorem 6.14. (a) **Integrability.** By Definition 4.7 of $\mathcal{B}(\Lambda; E)$, we need to show that $(f \circ T)\,|\det[\nabla T]|$ is uniformly continuous with bounded support.

The composite mapping $f \circ T$ is uniformly continuous (Theorem A.35) since f and T are assumed to be uniformly continuous. The mapping $|\det[\nabla T]|$ is also uniformly continuous as the composition of the following mappings:
— ∇T, which is uniformly continuous and bounded by the hypotheses;
— the determinant, which is multilinear and continuous (Theorem 5.14 (a)) and therefore uniformly continuous on bounded sets (Theorem A.41);

— the absolute value mapping, which is uniformly continuous (because $\big||t| - |s|\big| \le |t - s|$).

Their product $(f \circ T)\,|\det[\nabla T]|$ is therefore also uniformly continuous by Theorem A.35 applied to the composition with the product from $E \times \mathbb{R}$ into E, since the latter product is continuous and bilinear (Theorem A.30) and hence uniformly continuous on bounded sets (Theorem A.41).

Moreover, the support of $(f \circ T)\,|\det[\nabla T]|$ is bounded because it is included in the closure $\overline{\Lambda'}$ of the set $\Lambda' = \{x \in \Lambda : f \circ T(x) \neq 0_E\}$, which is bounded. Indeed, as the image of $\{y \in \Omega : f(y) \neq 0_E\}$, which is bounded (by the hypotheses) and therefore precompact on $\mathbb{R}^d$ (Theorem A.23 (b)), under the uniformly continuous mapping T^{-1}, the set Λ' is precompact (Theorem A.33) and therefore bounded (Theorem A.19 (a)).

This proves that $(f \circ T)\,|\det[\nabla T]| \in \mathcal{B}(\Lambda; E)$ by Definition 4.7 of this space.

Meaning of the integrals. The integrals in the statement are meaningfully defined because the domains of integration are open: λ is open by the hypotheses and $T(\lambda)$ is open as the preimage $S^{-1}\lambda$ of an open set under a continuous mapping (Theorem A.31), namely $S = T^{-1}$.

Equality of the integrals. We will proceed in six steps.

1. *Removing a neighborhood from the boundary.* Denote by $\mathcal{A}_n$ the set of $s \in \mathbb{Z}^d$ such that the closed cube $\Delta_{s,n}$ of side length 2^{-n} centered at $2^{-n}s$ is included in λ, and define

$$K_n \stackrel{\text{def}}{=} \bigcup_{s \in \mathcal{A}_n} \Delta_{s,n}, \quad \lambda_n \stackrel{\text{def}}{=} \mathring{K}_n.$$

Then $\bigcup_{n \in \mathbb{N}} \lambda_n = \lambda$, so $\bigcup_{n \in \mathbb{N}} T(\lambda_n) = T(\lambda)$. Furthermore, $T(\lambda_n)$ is open and increases with n, so (Theorem 4.6)

$$\int_{T(\lambda)} f = \lim_{n \to \infty} \int_{T(\lambda_n)} f. \tag{6.5}$$

2. *Removing a lattice.* Let

$$\lambda'_n \stackrel{\text{def}}{=} \bigcup_{s \in \mathcal{A}_n} \mathring{\Delta}_{s,n}.$$

Then $\lambda_n = \lambda'_n \cup \sigma_n$, where σ_n is included in $\bigcup_{s \in \mathcal{A}_n} \partial\Delta_{s,n}$. Each boundary $\partial\Delta_{s,n}$ is negligible as the union of the 2^d faces of $\Delta_{s,n}$, which are negligible by Theorem 5.10. Their union for $s \in \mathcal{A}_n$ is negligible and therefore so is the subset σ_n of their union (Theorem 5.7 (e) and (d)). The image $T(\sigma_n)$ is negligible by Theorem 5.11 and $T(\lambda_n) = T(\lambda'_n) \cup T(\sigma_n)$, so, by Theorem 6.1,

$$\int_{T(\lambda_n)} f = \int_{T(\lambda'_n)} f. \tag{6.6}$$

Moreover, $T(\lambda'_n)$ is the union of the $T(\mathring{\Delta}_{s,n})$ for $s \in \mathcal{A}_n$. Since these sets are pairwise disjoint, the additivity of the integral with respect to the domain (Theorem 4.21 (a)) gives

$$\int_{T(\lambda'_n)} f = \sum_{s \in \mathcal{A}_n} \int_{T(\mathring{\Delta}_{s,n})} f. \tag{6.7}$$

3. *Approximating* f. Since the cube $\Delta_{s,n}$ is included in the ball of radius $2^{-n}\sqrt{d}$ centered at $2^{-n}s$, the bound on the semi-norms of the integral from Theorem 4.15 implies that, for every semi-norm of E,

$$\Big\| \int_{T(\mathring{\Delta}_{s,n})} (f - f(T(2^{-n}s))) \Big\|_{E;\nu} \le \gamma_n \, |T(\mathring{\Delta}_{s,n})|,$$

where

$$\gamma_n \overset{\text{def}}{=} \sup_{|x'-x| \le 2^{-n}\sqrt{d}} \|f(x') - f(x)\|_{E;\nu}. \tag{6.8}$$

Furthermore, $\int_{T(\mathring{\Delta}_{s,n})} f(T(2^{-n}s)) = |T(\mathring{\Delta}_{s,n})| \, f(T(2^{-n}s))$ by Theorem 4.12 (c) and (b), so

$$\Big\| \sum_{s \in \mathcal{A}_n} \int_{T(\mathring{\Delta}_{s,n})} f - \sum_{s \in \mathcal{A}_n} |T(\mathring{\Delta}_{s,n})| \, f(T(2^{-n}s)) \Big\|_{E;\nu} \le \gamma_n \sum_{s \in \mathcal{A}_n} |T(\mathring{\Delta}_{s,n})|. \tag{6.9}$$

Finally, since the $T(\mathring{\Delta}_{s,n})$ are pairwise disjoint and included in $T(\lambda)$, the additivity and the increasingness of the measure (Theorems 5.1 (a) and 4.2 (a)) imply

$$\sum_{s \in \mathcal{A}_n} |T(\mathring{\Delta}_{s,n})| \le \gamma_n |T(\lambda)|. \tag{6.10}$$

4. *Approximating* $|T(\mathring{\Delta}_{s,n})|$. The image $T(\mathring{\Delta}_{s,n})$ of the cube $\mathring{\Delta}_{s,n}$ approximately coincides with the parallelepiped $P^1_{s,n}$, where, for $r > 0$,

$$P^r_{s,n} \overset{\text{def}}{=} \Big\{ T(2^{-n}s) + \sum_{i=1}^{d} t_i 2^{-n-1} \partial_i T(2^{-n}s) : |t_i| < r \Big\}. \tag{6.11}$$

More precisely, as we will verify in Lemma 6.15, for every $\eta > 0$, there exists $n_\eta \in \mathbb{N}$ such that (see Figure 6.2 on page 136), for every $n \ge n_\eta$ and every $s \in \mathcal{A}_n$,

$$P^{1-\eta}_{s,n} \subset T(\mathring{\Delta}_{s,n}) \subset P^{1+\eta}_{s,n}. \tag{6.12}$$

With the notation of Theorem 5.16 on calculating the measure of a parallelepiped,

$$P^r_{s,n} = a_{s,n} + P(r2^{-n}\partial_1 T(2^{-n}s), \ldots, r2^{-n}\partial_d T(2^{-n}s)),$$

where $a_{s,n}$ is a translation vector (equal to $T(2^{-n}s) - r2^{-n-1}\nabla T(2^{-n}s)$ because P is not centered at 0, whereas $P^r_{s,n}$ is centered at $T(2^{-n}s)$, but this fact is not useful). Therefore,

$$|P^r_{s,n}| = |\det[r2^{-n}\partial_1 T(2^{-n}s), \ldots, r2^{-n}\partial_d T(2^{-n}s)]| =$$
$$= r^d 2^{-nd}|\det[\nabla T(2^{-n}s)|.$$

(The second equality follows from the multilinearity of the determinant, see Theorem 5.14 (a).) Denoting $\tau = |T(\mathring{\Delta}_{s,n})|$ and $\delta = 2^{-nd}|\det[\nabla T(2^{-n}s)|$, the inclusion (6.12) therefore implies that $(1-\eta)^d\delta \leq \tau \leq (1+\eta)^d\delta$, so $|\tau - \delta| \leq ((1+\eta)^d - 1)\delta \leq c_\eta \tau$, where

$$c_\eta \stackrel{\text{def}}{=} \frac{((1+\eta)^d - 1)}{(1-\eta)^d}. \tag{6.13}$$

In other words,

$$\big|\,|T(\mathring{\Delta}_{s,n})| - 2^{-nd}|\det[\nabla T(2^{-n}s)|\,\big| \leq c_\eta\,|T(\mathring{\Delta}_{s,n})|.$$

Therefore, summing over s,

$$\Big\|\sum_{s\in\mathcal{A}_n}\big(|T(\mathring{\Delta}_{s,n})| - 2^{-nd}|\det[\nabla T(2^{-n}s)|\big)\, f(T(2^{-n}s))\Big\|_{E;\nu} \leq$$
$$\leq c_\eta \sum_{s\in\mathcal{A}_n}|T(\mathring{\Delta}_{s,n})|. \tag{6.14}$$

5. *Integral of* $(f\circ T)|\det[\nabla T]|$. By Definition 4.9 of the approximate integral,

$$\sum_{s\in\mathcal{A}_n} 2^{-nd}|\det[\nabla T(2^{-n}s)|\, f(T(2^{-n}s)) = \mathbb{S}^n_\lambda((f\circ T)|\det[\nabla T]|) \tag{6.15}$$

and

$$\int_\lambda (f\circ T)|\det[\nabla T]| = \lim_{n\to\infty} \mathbb{S}^n_\lambda((f\circ T)|\det[\nabla T]|). \tag{6.16}$$

6. *Conclusion*. The inequalities (6.9), (6.14) and (6.10) give

$$\Big\|\sum_{s\in\mathcal{A}_n}\int_{T(\mathring{\Delta}_{s,n})} f - \sum_{s\in\mathcal{A}_n} 2^{-nd}|\det[\nabla T(2^{-n}s)|\, f(T(2^{-n}s))\Big\|_{E;\nu}$$
$$\leq (\gamma_n + c_\eta)|T(\lambda)|.$$

With the equalities (6.6), (6.7) and (6.15), this reads

$$\Big\|\Big(\int_{T(\lambda_n)} f\Big) - \mathbb{S}^n_\lambda((f\circ T)|\det[\nabla T]|)\Big\|_{E;\nu} \leq (\gamma_n + c_\eta)|T(\lambda)|,$$

where γ_n and c_η are defined by (6.8) and (6.13). For every $\epsilon > 0$, there exists $n_\epsilon > 0$ such that, for every $n \geq n_\epsilon$, the right-hand side is smaller than ϵ because:
— first, f is uniformly continuous, so there exists n_γ such that $\gamma_n \leq \epsilon/2$ for every $n \geq n_\gamma$;
— second, we can choose $\eta > 0$ such that $c_\eta \leq \epsilon/(2|T(\lambda)|)$;
— finally, we can choose $n_\epsilon = \sup\{n_\gamma, n_\eta\}$.
This holds for every semi-norm of E, so

$$\lim_{n\to\infty}\Big(\Big(\int_{T(\lambda_n)} f\Big) - \mathbb{S}_\lambda^n((f \circ T)|\det[\nabla T]|)\Big) = 0_E.$$

With the convergences (6.5) and (6.16), we obtain the stated inequality

$$\int_{T(\lambda)} f = \int_\lambda (f \circ T)|\det[\nabla T]|.$$

(b) The property (b) is the property (a) applied to the function $f \circ T$ with the change of variables T^{-1}, since $(f \circ T) \circ T^{-1} = f$. Here, (a) holds because $f \circ T$ is uniformly continuous with bounded support, as we saw at the start of this proof, that is, $f \circ T \in \mathcal{B}(\Lambda; E)$.

(c) Here, instead of $T^{-1} \in \mathbf{C}_\mathrm{b}^1(\Omega; \Lambda)$, we assume that $T^{-1} \in \mathcal{C}_\mathrm{b}^1(\Omega; \Lambda)$ (meaning that the function is no longer necessarily uniformly continuous) and that λ or the support of $f \circ T$ is bounded.

Integrability. The function $(f \circ T)\,|\det[\nabla T]|$ still belongs to $\mathcal{B}(\Lambda; E)$ because the proof of its uniform continuity given at the start of this proof remains valid and the support of this function is bounded, since it is included in the support of $f \circ T$, which is bounded by the hypotheses.

Equality of the integrals. Let $(\omega_n)_{n\in\mathbb{N}}$ be an increasing sequence of open subsets of $\mathbb{R}^d$ with union $T(\lambda)$ such that $\omega_n \Subset T(\lambda)$, i.e. such that $\overline{\omega_n}$ is compact and included in $T(\lambda)$. The continuous functions T^{-1} and ∇T^{-1} are uniformly continuous and bounded on $\overline{\omega_n}$ by Heine's theorem (Theorem A.34), so $T^{-1} \in \mathbf{C}_\mathrm{b}^1(\omega_n; E)$.

Each $\lambda_n = T^{-1}(\omega_n)$ is open (as the preimage of an open set under a continuous mapping, see Theorem A.31), so (a) implies

$$\int_{\lambda_n} f = \int_{T(\lambda_n)} (f \circ T)|\det[\nabla T]|.$$

The λ_n increase with n, like the ω_n, and their union is λ, so, as $n \to \infty$, these integrals converge to the integrals over λ and $T(\lambda)$, respectively, by Theorem 4.19. In the limit, we obtain the stated equality. □

We still need to prove the inclusions (6.12), i.e. establish lower and upper bounds on the image $T(\mathring{\Delta}_{s,n})$. This is done by the following lemma.

LEMMA 6.15.– *With the hypotheses of Theorem 6.14, for every $\eta > 0$, there exists $n_\eta \in \mathbb{N}$ such that, for every $n \geq n_\eta$ and every $s \in \mathbb{Z}^d$ such that $\Delta_{s,n} \subset \Lambda$,*

$$P^{1-\eta}_{s,n} \subset T(\mathring{\Delta}_{s,n}) \subset P^{1+\eta}_{s,n}, \tag{6.17}$$

where $\Delta_{s,n}$ is the closed cube of side length 2^{-n} centered at $2^{-n}s$ and $P^r_{s,n}$ is the open parallelepiped defined by (6.11). ❚

Proof. Define $a \stackrel{\text{def}}{=} 2^{-n}s$ and denote by L the *linearized* mapping of T in the neighborhood of a. In other words, for every $x \in \mathbb{R}^d$,

$$L(x) \stackrel{\text{def}}{=} T(a) + (x-a) \cdot \nabla T(a). \tag{6.18}$$

Additionally, let

$$\Delta^r_{s,n} \text{ be the closed cube of side length } r2^{-n} \text{ centered at } a.$$

Then $P^r_{s,n} = L(\mathring{\Delta}^r_{s,n})$, which means that the first inclusion in (6.17) is equivalent to $L(\mathring{\Delta}^{1-\eta}_{s,n}) \subset T(\mathring{\Delta}^1_{s,n})$ and therefore to

$$T^{-1}(L(\mathring{\Delta}^{1-\eta}_{s,n})) \subset \mathring{\Delta}^1_{s,n}.$$

Since $\mathring{\Delta}^1_{s,n}$ contains every ball of radius $\eta 2^{-n-1}$ centered at a point x of $\mathring{\Delta}^{1-\eta}_{s,n}$, it suffices to have, for any such x,

$$|T^{-1}(L(x)) - x| \leq \eta 2^{-n-1}. \tag{6.19}$$

But if $x \in \mathring{\Delta}^1_{s,n}$, and therefore in particular if $x \in \mathring{\Delta}^{1-\eta}_{s,n}$, then $|x-a| \leq 2^{-n-1}\sqrt{d}$. By (6.18), the finite increment theorem (Theorem 2.5 (b)) therefore implies

$$|L(x) - T(x)| = |T(a) - T(x) - (a-x) \cdot \nabla T(a)| \leq \delta_n |a-x|,$$

where

$$\delta_n = \sup_{|x-a| \leq 2^{-n-1}\sqrt{d}} |\nabla T(x) - \nabla T(a)|.$$

Hence, applying again the finite increment theorem (Theorem 2.5 (a)), now to T^{-1},

$$|T^{-1}(L(x)) - x| = |T^{-1}(L(x)) - T^{-1}(T(x))| \leq$$
$$\leq \theta |L(x) - T(x)| \leq \theta \delta_n 2^{-n-1}\sqrt{d},$$

where

$$\theta \stackrel{\text{def}}{=} \sup_{y\in\Omega} |\nabla T^{-1}(y)|.$$

Therefore, (6.19) and consequently the first inclusion of (6.17) are satisfied whenever n is sufficiently large that $\delta_n \leq \eta/(\theta\sqrt{d})$.

Similarly, the second inclusion of (6.17) is equivalent to $T(\mathring{\Delta}^1_{s,n}) \subset L(\mathring{\Delta}^{1+\eta}_{s,n})$. But, as we will verify later, L is invertible and

$$\nabla L^{-1}(y) = \nabla T^{-1}(T(a)). \tag{6.20}$$

This inclusion is therefore equivalent to $L^{-1}(T(\mathring{\Delta}^1_{s,n})) \subset \mathring{\Delta}^{1+\eta}_{s,n}$, which is satisfied if $|L^{-1}(T(x)) - x| \leq \eta 2^{-n-1}$ for every $x \in \mathring{\Delta}^1_{s,n}$. As above, we have the bound

$$|L^{-1}(T(x)) - x| = |L^{-1}(T(x)) - L^{-1}(L(x))| \leq \theta'\delta_n 2^{-n-1}\sqrt{d},$$

where $\theta' = \sup_{y\in\Omega} |\nabla L^{-1}(y)| = |\nabla T^{-1}(T(a))|$. Therefore, $\theta' \leq \theta$, and so the second inclusion of (6.17) is also satisfied whenever $\delta_n \leq \eta/(\theta\sqrt{d})$.

We still need to check (6.20). It suffices to show that $L^{-1} = M$, where

$$M(y) \stackrel{\text{def}}{=} a + (y - b) \cdot \nabla T^{-1}(b)$$

and $b \stackrel{\text{def}}{=} T(a)$. But, by Definition (6.18), $L_j(x) - b_j = \sum_{i=1}^d (x_i - a_i)\, \partial_i T_j(a)$, so

$$M_k(L(x)) = a_k + \sum_{j=1}^d (L_j(x) - b_j)\, \partial_j T_k^{-1}(b) =$$

$$= a_k + \sum_{i=1}^d \sum_{j=1}^d (x_i - a_i)\, \partial_i T_j(a)\, \partial_j T_k^{-1}(b) = x_k$$

because $\sum_{j=1}^d \partial_i T_j(a)\, \partial_j T_k^{-1}(b) = 1$ if $i = k$ and this expression is zero otherwise by Theorem 3.12 (b). This proves that $M = L^{-1}$ and thus establishes (6.20). □

Non-bijective change of variables in an integral. Even though the formula in Theorem 6.14 (a) does not explicitly involve T^{-1}, it cannot be extended to the case where T is not invertible.

For example, if $\Lambda = (-2, 2)$, $T(x) = x^3 - x$, $\Omega = (-6, 6)$ and $f \equiv 1$, then $\int_\Omega f = \int_{-6}^6 1 = 12$, whereas $\int_\Lambda (f \circ T)\, |\det[\nabla T]| = \int_{-2}^2 |3x^2 - 1|\, dx = 12 + 8\sqrt{3}/9$. □

6.8. Some particular changes of variables in an integral

Let us observe that translations preserve the integral[8].

---

8 **History of the invariance of the integral under translation.** The invariance of the integral under translation is similar to the eighth axiom of the *Elements* of EUCLID [33], the equality of congruent sets.

THEOREM 6.16.– *Let $f \in \mathcal{B}(\Omega; E)$, where Ω is an open subset of $\mathbb{R}^d$ and E is a Neumann space. Then, for every $z \in \mathbb{R}^d$,*

$$\int_\Omega f(x)\,\mathrm{d}x = \int_{\Omega - z} f(x+z)\,\mathrm{d}x,$$

where $\Omega - z = \{x - z : x \in \Omega\}$. ▮

Proof. This follows from the change of variables theorem (Theorem 6.14 (a)) by setting $T(x) = x + z$ and $\lambda = \Omega - z$, and therefore $\det[\nabla T] = \det[\mathbf{e}_1, \ldots, \mathbf{e}_d] = 1$ (Theorem 5.14 (b)). □

Let us perform a linear change of variables in an integral.

THEOREM 6.17.– *Let $f \in \mathcal{B}(\Omega; E)$, where Ω is an open subset of $\mathbb{R}^d$ and E is a Neumann space, and suppose that*

T is a linear bijection from $\mathbb{R}^d$ onto itself.

Then $f \circ T \in \mathcal{B}(T^{-1}(\Omega); E)$ and

$$\int_\Omega f = \kappa \int_{T^{-1}(\Omega)} f \circ T,$$

where $\kappa = |\det[T(\mathbf{e}_1), \ldots, T(\mathbf{e}_d)]|$. ▮

Proof. This follows from the change of variables theorem (Theorem 6.14 (a) with $\lambda = T^{-1}(\Omega)$) because T, being linear, is differentiable and

$$\nabla T(x) = (T(\mathbf{e}_1), \ldots, T(\mathbf{e}_d)).$$

Indeed, Definition 2.1 of differentiability is satisfied here because, for every $z \in \mathbb{R}^d$,

$$T(x+z) - T(x) - \sum_{i=1}^{d} z_i T(\mathbf{e}_i) = 0,$$

since $T(z) = T(\sum_{i=1}^d z_i \mathbf{e}_i) = \sum_{i=1}^d z_i T(\mathbf{e}_i)$. □

In particular, symmetries do not change the integral.

THEOREM 6.18.– *Let $f \in \mathcal{B}(\Omega; E)$, where Ω is an open subset of $\mathbb{R}^d$ and E is a Neumann space. Then*

$$\int_{\Omega} f(x)\,\mathrm{d}x = \int_{-\Omega} f(-x)\,\mathrm{d}x. \blacksquare$$

Proof. This is a special case of Theorem 6.17 with $T(x) = -x$, and therefore $\kappa = |\det[-\mathbf{e}_1, \ldots, -\mathbf{e}_d]| = |\det[\mathbf{e}_1, \ldots, \mathbf{e}_d]| = 1$ (Theorem 5.14 (c)). □

Homotheties have the following effect on the integral.

THEOREM 6.19.– *Let $f \in \mathcal{B}(\Omega; E)$, where Ω is an open subset of $\mathbb{R}^d$ and E is a Neumann space. Then, for every $t > 0$,*

$$\int_{\Omega} f(x)\,\mathrm{d}x = t^d \int_{\Omega/t} f(ty)\,\mathrm{d}y,$$

where $\Omega/t = \{x/t : x \in \Omega\}$. ▮

Proof. This is a special case of Theorem 6.17 with $T(x) = tx$, and therefore $\kappa = |\det[t\mathbf{e}_1, \ldots, t\mathbf{e}_d]| = t^d |\det[\mathbf{e}_1, \ldots, \mathbf{e}_d]| = t^d$ (Theorem 5.14 (a) and (b)). □

Let us show that the Jacobian determinant of a $\mathcal{C}^1$ bijection whose inverse is also $\mathcal{C}^1$ does not vanish.

THEOREM 6.20.– *Let T be a bijection from an open subset Ω of $\mathbb{R}^d$ onto an open subset Λ of $\mathbb{R}^d$ such that*

$$T \in \mathcal{C}^1(\Lambda; \Omega) \text{ and } T^{-1} \in \mathcal{C}^1(\Omega; \Lambda).$$

Then, denoting by $\det[\nabla T]$ *the determinant of the matrix* $[\partial_i T_j]$ *and by* $[\mathrm{I}]$ *the identity matrix, i.e.* $\mathrm{I}_i^i = 1$ *and* $\mathrm{I}_i^j = 0$ *if* $i \neq j$:

(a)
$$[\nabla T^{-1}] \circ T \times [\nabla T] = [\mathrm{I}] \text{ on } \Lambda,$$
$$[\nabla T] \circ T^{-1} \times [\nabla T^{-1}] = [\mathrm{I}] \text{ on } \Omega.$$

(b)
$$\det[\nabla T^{-1}] \circ T \times \det[\nabla T] = 1 \text{ on } \Lambda,$$
$$\det[\nabla T] \circ T^{-1} \times \det[\nabla T^{-1}] = 1 \text{ on } \Omega.$$

(c) *Furthermore, if Λ or Ω is connected, then both are connected,* $\det[\nabla T]$ *does not change sign on Λ, and* $\det[\nabla T^{-1}]$ *does not change sign on Ω.* ▮

Proof. (a) By Theorem 3.12 (b), $\sum_{j=1}^{d}((\partial_j T_k^{-1}) \circ T)\, \partial_i T_j = (\mathbf{e}_i)_k$, where $(\mathbf{e}_i)_i = 1$ and $(\mathbf{e}_i)_k = 0$ if $k \neq i$. In matrix notation, this is the first equality. The second equality follows from the first by taking the composition with T^{-1} or by applying the first equality to T^{-1}.

(b) These equalities follow from (a) because the determinant of a product of matrices is the product of their determinants (Theorem 5.15) and $\det[\mathrm{I}] = 1$ (Theorem 5.14 (b)).

(c) If Λ is connected, its image Ω under the continuous mapping T is connected on $\mathbb{R}^d$ (Theorem A.33). The converse holds because T^{-1} is also continuous.

The image $U = \det[\nabla T](\Lambda)$ of Λ under the continuous mapping $\det[\nabla T]$ is connected on $\mathbb{R}$ (Theorem A.33 once again). If U contains a point $a < 0$ and another point $b > 0$, then it must contain 0; if not, the disjoint open sets $(-\infty, 0)$ and $(0, +\infty)$ would cover U, contradicting its connectedness (Definition A.15). In other words, there would exist $y \in \Lambda$ such that $\det[\nabla T](y) = 0$, which is impossible by the first equality of (b). Therefore, $\det[\nabla T]$ is either always positive or always negative.

Similarly, $\det[\nabla T^{-1}]$ is either always positive or always negative, since T^{-1} satisfies the same hypotheses as T. □

Chapter 7

Weighting and Regularization of Functions

In this chapter, we define the *weighted function* of $f \in \mathcal{C}(\Omega; E)$ given a weight $\mu \in \mathcal{C}_D(\mathbb{R}^d)$ with compact support as $(f \diamond \mu)(x) = \int_{\mathring{D}} f(x+y)\mu(y)\,dy$. This notion will play a central role later. It requires E to be a Neumann space.

Weighting plays a similar role for Ω as convolution does for $\mathbb{R}^d$. But since f cannot necessarily be extended, $f \diamond \mu$ is only defined on $\Omega_D = \{x : x + D \subset \Omega\}$. When f and μ are differentiable, the choice of signs explained on page 149 leads to $\partial_i(f \diamond \mu) = \partial_i f \diamond \mu = -f \diamond \partial_i \mu$ (Theorem 7.4).

Weighting is a regularizing operation. If f is $\mathcal{C}^m$ and μ is $\mathcal{C}^l$, then $f \diamond \mu$ is $\mathcal{C}^{m+\ell}$ (Theorem 7.5). If ρ_n is a *regularizing function*, namely a $\mathcal{C}^\infty$ function with support in the ball $B(0, 1/n)$ and whose integral is 1, then $f \diamond \rho_n$ is $\mathcal{C}^\infty$ and converges to f except outside of an arbitrarily small neighborhood of the boundary, where it is not defined (Theorem 7.9).

To obtain convergence on the whole of Ω, we define *global regularization* in § 7.5. We finish by introducing the *covering by crown-shaped sets* of Ω and partition of unity in § 7.6, as well as by constructing a sequence that is dense in $\mathcal{K}^\infty(\Omega)$ in § 7.7.

7.1. Weighting

Let us define the notion of weighting[1] for continuous functions taking values in a Neumann space.

DEFINITION 7.1.– *Let*

$$f \in \mathcal{C}(\Omega; E) \quad \textit{and} \quad \mu \in \mathcal{C}_D(\mathbb{R}^d),$$

1 **History of the weighting of functions.** Definition 7.1 was given by Jacques SIMON in 1993 [72, p. 203] for a Banach space E. It generalizes the definition of the convolution to an open subset Ω of $\mathbb{R}^d$, while replacing $x - y$ by $x + y$.

Convolution. Vito VOLTERRA introduced convolution in one dimension, namely $d = 1$, [81]. Jean LERAY later used it in three dimensions in 1934 [55, p. 206].

where Ω is an open subset of $\mathbb{R}^d$, E is a Neumann space and D is a compact subset of $\mathbb{R}^d$.

*The **weighted function** of f by (the weight) μ is the function $f \diamond \mu$ defined on the set*

$$\Omega_D \stackrel{\text{def}}{=} \{x \in \mathbb{R}^d : x + D \subset \Omega\}$$

by

$$(f \diamond \mu)(x) \stackrel{\text{def}}{=} \int_{\mathring{D}} f(x+y)\,\mu(y)\,\mathrm{d}y. \blacksquare$$

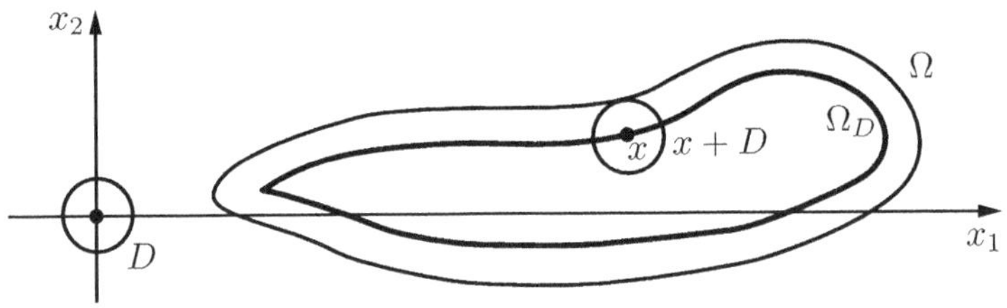

Figure 7.1. *Domain Ω_D of the weighted function $f \diamond \mu$*

Justification. The Cauchy integral in Definition 7.1 makes sense (according to Definition 4.9) because the function $y \mapsto f(x+y)\,\mu(y)$ is uniformly continuous with bounded support.

Indeed, it is continuous as a product of continuous functions (Theorem 3.5 (a)) on $\Omega - x$ and is therefore uniformly continuous on the compact set D by Heine's theorem (Theorem A.34). Furthermore, its support is bounded, since it is included in D, which is bounded (Theorem A.19). □

Terminology. The *weighted function* is a **new concept** whose definition should be recalled by readers who wish to use it. We say that $f \diamond \mu$ is the weighted function of f by μ because $(f \diamond \mu)(x)$ is the integral with weight μ of f over $x + D$. □

Need for E to be a Neumann space. It is necessary that E be a Neumann space to define weighting. Indeed, if we omit this hypothesis from Definition 7.1, the Cauchy integral $\int_{\mathring{D}} f(x+y)\,\mu(y)\,dy$ is not defined in a satisfactory manner. In particular, see § 4.3, *Case where E is not a Neumann space*, if f was given by (4.1) and μ was equal to 1 on its support, then, for $x = 0$, the above "integral", that is $(f \diamond \mu)(0)$, would not belong to E. □

Utility of assuming that Ω is open. It is not necessary to require Ω to be open when defining the weighting of functions, but it will be essential in Volume 3 for extending this notion to the weighting of distributions, which are only defined on open sets. □

Remarks about the domain Ω_D. The domain Ω_D is *"relatively" smaller* than Ω in the sense that it is contained by it after a suitable translation. If D is not a singleton, Ω_D is even strictly included in Ω after a translation. If $0 \in D$, the translation is not necessary because we then have $\Omega_D \subset \Omega$.

If Ω_D is non-empty, then D is necessarily *"relatively" smaller* than Ω, i.e. contained by it after a suitable translation. □

Choice of D. In this volume, D is always a ball centered at 0, in which case Ω_D is Ω with a neighborhood of the boundary removed, as shown in Figure 7.1.

However, in subsequent volumes, to define the weighted function in the neighborhood of the boundary, we will use a conic sector K, in which case Ω_K extends up to part of the boundary if the boundary is regular, as shown in Figure 7.2. □

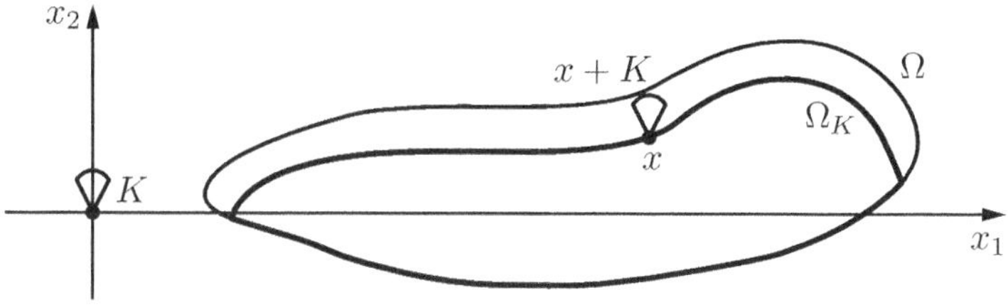

Figure 7.2. *Domain Ω_K extending up to part of the boundary. The contour of Ω_K is shown in bold*

Utility of μ being extendable. It must be possible to extend μ by 0 outside of D, or, equivalently, for it to be zero on the boundary ∂D, in order to obtain the expressions $(f \diamond \mu)(x) = \int_\Omega f(y)\mu(y - x)\,\mathrm{d}x$ and $\partial_i(f \diamond \mu) = -f \diamond \partial_i \mu$ in Theorems 7.2 (c) and 7.4 (b), as well as to extend the notion of weighting to a distribution f in the form $f \diamond \mu = \langle f, \tau_x \mu \rangle_\Omega$ [Vol. 3]. □

Weighting versus convolution. When $\Omega = \mathbb{R}^d$, weighting is equal to **convolution**[2] up to symmetry, namely

$$(f \star \mu)(x) \stackrel{\text{def}}{=} \int_{\mathbb{R}^d} f(x - y)\, \mu(y)\, \mathrm{d}y. \tag{7.1}$$

The originality and the benefit of weightings arise from the fact that they can be used with functions that are not defined on the whole of $\mathbb{R}^d$. The two operations of weighting and convolution satisfy the relationship

$$f \star \mu = f \diamond \check{\mu},$$

where $\check{\mu}(x) = \mu(-x)$. □

Choice of signs in the definition of weighting. When $\Omega = \mathbb{R}^d$ and $E = \mathbb{R}$, convolution is pleasant to work with because, given suitable hypotheses,

$$f \star \mu = \mu \star f \text{ and } \partial_i(f \star \mu) = \partial_i f \star \mu = f \star \partial_i \mu.$$

With the same hypotheses, weighting gives (Theorem 7.4 and Vol. 3) the less agreeable equalities

$$f \diamond \mu = (\mu \diamond f)\check{} = \check{\mu} \diamond \check{f} \text{ and } \partial_i(f \diamond \mu) = \partial_i f \diamond \mu = -f \diamond \partial_i \mu,$$

where $\check{\mu}(x) = \mu(-x)$. The symmetry $\check{}$ and the minus sign occur because the term $x - y$ in the definition of convolution given in Equation (7.1) is replaced by $x + y$ in Definition 7.1 of weighting.

---

2 **History of convolution.** Vito VOLTERRA introduced the convolution of real functions on $\mathbb{R}$ in 1913 [81]. He called it *composition of the first kind*. The convolution of functions on $\mathbb{R}^d$ was used in 1934 by Jean LERAY [55, Formula (1.18), p. 206].

Why make this choice? The reason is because, for a general domain Ω, extending convolution in the form

$$(f \tilde{\star} \mu)(x) \stackrel{\text{def}}{=} \int_D f(x-y)\,\mu(y)\,\mathrm{d}y = \int_{-D} f(x+y)\,\mu(-y)\,\mathrm{d}y \tag{7.2}$$

is awkward, since it involves the integral of f with the weight $\check{\mu}$ over $x - D$. In other words, $(f \tilde{\star} \mu)(x)$ is the average of f weighted by the symmetrized function $\check{\mu}$ of μ, whose support is included in $-D$.

This is the same thing if D and μ are symmetric but is confusing and quickly becomes inconvenient when D is not symmetric. This case is essential in some applications, such as immersions and extension properties in Sobolev spaces, which are obtained by considering integrals of f weighted over a cone D that is **interior** to Ω. If we used (7.2), such an integral would be expressed as the convolution of f with a weight whose support is in the **exterior cone** $-D$.

Furthermore, whenever Ω is not $\mathbb{R}^d$, the roles of f and μ can no longer coincide, because f may not necessarily be extendable outside of Ω, which is the key benefit of weighting, whereas μ must be extendable by 0 outside of D. Additionally, f is vector-valued, but μ is not. In practice, in the applications that we will consider, the roles of f and μ will be entirely distinct. □

7.2. Properties of weighting

Let us state some properties of weighting.

THEOREM 7.2.– *Let*

$$f \in C(\Omega; E) \quad \text{and} \quad \mu \in C_D(\mathbb{R}^d),$$

where Ω is an open subset of $\mathbb{R}^d$, E is a Neumann space and D is a compact subset of $\mathbb{R}^d$. Then:

(a)
$$\Omega_D \text{ is open.}$$

(b)
$$f \diamond \mu \in C(\Omega_D; E).$$

(c) *For every $x \in \Omega_D$,*

$$(f \diamond \mu)(x) = \int_{\Omega - x} f(x+y)\,\mu(y)\,\mathrm{d}y = \int_{\Omega} f(y)\,\mu(y-x)\,\mathrm{d}y.$$

(d)
$$\operatorname{supp}(f \diamond \mu) \subset \operatorname{supp} f - \operatorname{supp} \mu.$$

(e) *If $f \in \mathbf{C}(\Omega; E)$,*

$$f \diamond \mu \in \mathbf{C}(\Omega_D; E).$$

(f) *If $f \in C(\Omega)$, $f \geq 0$ and $\mu \geq 0$,*

$$f \diamond \mu \geq 0.$$

(g) *If* $\varphi \in \mathcal{K}(\Omega_D)$, *denoting* $\check{\mu}(x) = \mu(-x)$,

$$\int_{\Omega_D} (f \diamond \mu)\, \varphi = \int_{\Omega} f\, (\varphi \diamond \check{\mu}). \; \blacksquare$$

Proof. Denote by $\{\| \; \|_{E;\nu} : \nu \in \mathcal{N}_E\}$ the semi-norms of E.

(a) *Property of* Ω_D. If $x \in \Omega_D$, then $x + D$ is a compact set included in Ω, so the strong inclusion theorem (Theorem A.22) shows that there exists a ball $B(0, r)$ such that $x + D + B \subset \Omega$, i.e. $B(x, r)$ is included in Ω_D, which is therefore open.

(b) *Continuity of* $f \diamond \mu$. For any given $x \in \Omega_D$, let $r > 0$ be such that the ball $B = \{x' \in \mathbb{R}^d : |x' - x| \leq r\}$ is included in Ω_D. The bound on the semi-norms of the integral from Theorem 4.15 gives, for every $x' \in B$,

$$\|(f \diamond \mu)(x') - (f \diamond \mu)(x)\|_{E;\nu} = \left\| \int_{\mathring{D}} (f(x' + y) - f(x + y))\, \mu(y)\, \mathrm{d}y \right\|_{E;\nu} \leq$$

$$\leq |\mathring{D}| \sup_{y \in D} \|f(x' + y) - f(x + y)\|_{E;\nu} \sup_{y \in D} |\mu(y)|. \qquad (7.3)$$

The set $B + D$ is compact and included in Ω, so f is uniformly continuous on it by Heine's theorem (Theorem A.34) and therefore the final expression in (7.3) tends to 0 as $x' \to x$. This proves that $f \diamond \mu \in \mathcal{C}(\Omega_D; E)$.

(c) *Expression of* $(f \diamond \mu)(x)$. The first expression follows from Definition 7.1 because μ is zero outside of the interior of its support D (Theorem 1.6 (b)) and therefore is certainly zero outside of $\mathring{D}$, and so (Theorem 4.17 (a)) the contribution of $(\Omega - x) \setminus \mathring{D}$ to the integral is zero.

The second expression then follows from the first because translations, here $y \mapsto y - x$, preserve integrals (Theorem 6.16).

(d) *Support.* Let $x \in \Omega_D \setminus (S_f - S_\mu)$, where $S_f = \operatorname{supp} f$ and $S_\mu = \operatorname{supp} \mu$. Then

$$(f \diamond \mu)(x) = \int_{\Omega - x} f(x + y)\, \mu(y)\, \mathrm{d}y = 0_E. \qquad (7.4)$$

Indeed, if $y \notin S_\mu$, then $\mu(y) = 0$, and if $y \in S_\mu$, then $x + y \notin S_f$ and so $f(x + y) = 0_E$. Therefore, the function integrated above is always zero.

Now, let us show that

$$\Omega_D \setminus (S_f - S_\mu) \text{ is open.} \qquad (7.5)$$

Indeed, if $x \in \Omega_D \setminus (S_f - S_\mu)$, then $x + S_\mu \subset x + D \subset \Omega$ and $x + S_\mu \not\subset S_f$, so

$$x + S_\mu \subset \Omega \setminus S_f.$$

By the properties of the support, S_μ is compact (Theorem 1.18, because it is included in the compact set D), and therefore $x + S_\mu$ is compact; and $\Omega \setminus S_f$ is open (Theorem 1.6 (a), because Ω is open). Therefore, by the strong inclusion theorem (Theorem A.22), there exists a ball $B(0, \epsilon)$, where $\epsilon > 0$, such that

$$x + S_\mu + B(0, \epsilon) \subset \Omega \setminus S_f.$$

In other words, $B(x, \epsilon) \subset \Omega \setminus (S_f - S_\mu)$. Additionally we can choose ϵ in such a way that $B(x, \epsilon)$ is also included in Ω_D because this set is open by (a). Thus,

$$B(x, \epsilon) \subset \Omega_D \setminus (S_f - S_\mu).$$

This proves (7.5).

By (7.4) and (7.5), $\Omega_D \setminus (S_f - S_\mu)$ is an open set on which $f \diamond \mu$ vanishes. It is therefore included in $\Omega_D \setminus \operatorname{supp} f \diamond \mu$ (Theorem 1.6 (a) once again, because Ω_D is open). Therefore,

$$\operatorname{supp} f \diamond \mu \subset S_f - S_\mu.$$

(e) *Uniform continuity.* If f is uniformly continuous, then so is $f \diamond \mu$ by the inequality (7.3).

(f) *Positivity.* If $E = \mathbb{R}$ and f and μ are positive functions, then Theorem 4.14 (a) gives $\int_{\mathring{D}} f(x + y)\mu(y)\,\mathrm{d}y \geq 0$; in other words, $(f \diamond \mu)(x) \geq 0$, for every $x \in \Omega_D$.

(g) *Integration of the product with φ.* Let $\varphi \in \mathcal{K}(\Omega_D)$ with support K. The function F defined by

$$F(x, y) \overset{\text{def}}{=} f(y)\mu(y - x)\varphi(x)$$

is continuous on $\Omega \times \Omega_D$. Its support is included in $(K + D) \times K$, which is a compact set included in $\Omega \times \Omega_D$, so F is uniformly continuous (Theorem 1.10 (c)). Theorem 6.7 on the separation of variables in an integral therefore gives

$$\int_{\Omega_D \times \Omega} F = \int_{\Omega_D} \Big(\int_\Omega f(y)\mu(y - x)\,\mathrm{d}y \Big) \varphi(x)\,\mathrm{d}x =$$
$$= \int_\Omega f(y) \Big(\int_{\Omega_D} \mu(y - x)\varphi(x)\,\mathrm{d}x \Big)\,\mathrm{d}y.$$

In other words, by the second expression of (c) applied to $(f \diamond \mu)(x)$ and $(\varphi \diamond \check{\mu})(y)$ respectively,

$$\int_{\Omega_D} (f \diamond \mu)(x)\,\varphi(x)\,\mathrm{d}x = \int_\Omega f(y)\,(\varphi \diamond \check{\mu})(y)\,\mathrm{d}y. \ \square$$

Let us state some properties of the domain of the weighted function.

THEOREM 7.3.– *Let Ω be an open subset of $\mathbb{R}^d$.*

(a) *If D and D' are two compact subsets of $\mathbb{R}^d$ such that $D' \subset D$, then*

$$\Omega_D \subset \Omega_{D'}.$$

(b) *If K is a compact subset of $\mathbb{R}^d$ included in Ω, then there exists $r > 0$ such that the ball $B = B(0,r) = \{x \in \mathbb{R}^d : |x| \leq r\}$ satisfies $K + B \subset \Omega$, which implies*

$$K \subset \Omega_B \text{ and } B \subset \Omega_K. \blacksquare$$

Proof. (a) By Definition 7.1 of Ω_D, if $x \in \Omega_D$, then $x + D \subset \Omega$. If $D' \subset D$, then $x + D' \subset x + D \subset \Omega$, so $x \in \Omega_{D'}$.

(b) The strong inclusion theorem (Theorem A.22) gives $r > 0$ such that $K + B \subset \Omega$. Definition 7.1 of Ω_B then shows that $K \subset \Omega_B$ and $B \subset \Omega_K$. $\square$

7.3. Weighting of differentiable functions

Let us show that the weighted function is differentiable whenever the function or the weight is differentiable.

THEOREM 7.4.– *Let Ω be an open subset of $\mathbb{R}^d$, E a Neumann space and D a compact subset of $\mathbb{R}^d$.*

(a) *If $f \in \mathcal{C}^1(\Omega; E)$ and $\mu \in \mathcal{C}_D(\mathbb{R}^d)$, then $f \diamond \mu \in \mathcal{C}^1(\Omega_D; E)$ and*

$$\partial_i(f \diamond \mu) = \partial_i f \diamond \mu.$$

(b) *If $f \in \mathcal{C}(\Omega; E)$ and $\mu \in \mathcal{C}^1_D(\mathbb{R}^d)$, then $f \diamond \mu \in \mathcal{C}^1(\Omega_D; E)$ and*

$$\partial_i(f \diamond \mu) = -f \diamond \partial_i \mu. \blacksquare$$

Proof. Denote by $\{\| \ \|_{E;\nu} : \nu \in \mathcal{N}_E\}$ the family of semi-norms of E.

(a) *Case where $f \in \mathcal{C}^1(\Omega; E)$ and $\mu \in \mathcal{C}_D(\mathbb{R}^d)$.* Given $x \in \Omega_D$, let $r > 0$ be such that the ball $B = \{y \in \mathbb{R}^d : |y - x| \leq r\}$ is included in Ω_D, and let $h \in \mathbb{R}^d$ be such that $0 < |h| < r$. Denote

$$\Delta_h \stackrel{\text{def}}{=} (f \diamond \mu)(x + h) - (f \diamond \mu)(x) - h \cdot (\nabla f \diamond \mu)(x) =$$

$$= \int_{\mathring{D}} \big(f(x + h + y) - f(x + y) - h \cdot \nabla f(x + y)\big)\, \mu(y)\, \mathrm{d}y.$$

The bound on the semi-norms of the integral from Theorem 4.15 and the finite increment theorem (Theorem 2.5 (b)) give, for every $\nu \in \mathcal{N}_E$,

$$\|\Delta_h\|_{E;\nu} \leq$$

$$\leq |\mathring{D}| \sup_{y \in D} \|f(x + h + y) - f(x + y) - h \cdot \nabla f(x + y)\|_{E;\nu} \sup_{y \in D} |\mu(y)| \leq$$

$$\leq |\mathring{D}|\, |h| \sup_{|z| \leq |h|} \sup_{y \in D} \|\nabla f(x + y + z) - \nabla f(x + y)\|_{E;\nu} \sup_{y \in D} |\mu(y)|.$$

As y ranges over D, $x + y$ and $x + y + z$ remain inside $B + D$, which is compact (as the sum of compact subsets of $\mathbb{R}^d$, see Theorem A.24) and included in Ω, so ∇f is uniformly continuous on this set by Heine's theorem (Theorem A.34). Hence, for every $\epsilon > 0$, there exists $\eta > 0$ such that, if $|z| \leq \eta$, then

$$\|\nabla f(x + y + z) - \nabla f(x + y)\|_{E;\nu} \leq \frac{\epsilon}{c},$$

where $c = |\mathring{D}| \sup_{y \in D} |\mu(y)|$. Therefore, $|h| \leq \eta$ implies

$$\|\Delta_h\|_{E;\nu} \leq \epsilon\, |h|.$$

Thus, $f \diamond \mu$ satisfies Definition 2.1 of differentiability at the point x, with

$$\nabla(f \diamond \mu)(x) = (\nabla f \diamond \mu)(x).$$

Then, since $\partial_i = \nabla_i$ (Theorem 2.9),

$$\partial_i(f \diamond \mu) = \partial_i f \diamond \mu.$$

The functions $\partial_i f \diamond \mu$ being continuous by Theorem 7.2 (b), it follows that

$$f \diamond \mu \in \mathcal{C}^1(\Omega_D; E).$$

(b) *Case where $f \in \mathcal{C}(\Omega; E)$ and $\mu \in \mathcal{C}^1_D(\mathbb{R}^d)$.* For x, r and h as above, denote

$$\Delta'_h \stackrel{\text{def}}{=} (f \diamond \mu)(x + h) - (f \diamond \mu)(x) + h \cdot (f \diamond \nabla\mu)(x).$$

Since the function $\nabla\mu$ is zero on the open set $\mathbb{R}^d \setminus D$, like μ itself,

$$\Delta'_h = \int_{\mathring{D}} \Big(\big(f(x+h+y) - f(x+y)\big)\,\mu(y) + h \cdot f(x+y)\nabla\mu(y) \Big)\,\mathrm{d}y =$$

$$= \int_{\mathring{D}+h} f(x+y')\,\mu(y'-h)\,\mathrm{d}y' + \int_{\mathring{D}} f(x+y)\,\big(-\mu(y) + h \cdot \nabla\mu(y)\big)\,\mathrm{d}y.$$

Since the integrated functions are zero outside of their domains of integration $\mathring{D}+h$ and $\mathring{D}$, by Theorem 4.17 (a), we can replace these domains by the larger open set $D+\mathring{B}_0$, where $B_0 = \{y \in \mathbb{R}^d : |y| \le r\}$. This gives

$$\Delta'_h = \int_{\mathring{D}+\mathring{B}_0} f(x+y)\,\big(\mu(y-h) - \mu(y) + h \cdot \nabla\mu(y)\big)\,\mathrm{d}y.$$

Using the bound on the semi-norms of the integral from Theorem 4.15, as in (a), and the finite increment theorem (Theorem 2.5 (b)), we obtain, for every $\nu \in \mathcal{N}_E$,

$$\|\Delta'_h\|_{E;\nu} \le$$

$$\le |\mathring{D}+\mathring{B}_0| \sup_{y \in D+B_0} \|f(y)\|_{E;\nu} \sup_{y \in D+B_0} |\mu(y-h) - \mu(y) + h \cdot \nabla(y)| \le$$

$$\le |\mathring{D}+\mathring{B}_0| \sup_{y \in D+B_0} \|f(y)\|_{E;\nu}\,|h| \sup_{|z| \le |h|} \sup_{y \in D+B_0} |\nabla\mu(y+z) - \nabla\mu(y)|.$$

Since $\nabla\mu$ is continuous and hence uniformly continuous on the compact set $D+B_0$ for every $\epsilon > 0$, there exists $\eta > 0$ such that $|h| \le \eta$ implies

$$\|\Delta'_h\|_{E;\nu} \le \epsilon\,|h|.$$

This shows that $f \diamond \mu$ is differentiable at the point x and

$$\nabla(f \diamond \mu)(x) = -(f \diamond \nabla\mu)(x).$$

Thus,

$$\partial_i(f \diamond \mu) = -f \diamond \partial_i\mu.$$

The functions $f \diamond \partial_i\mu$ being continuous again by Theorem 7.2 (b), it follows that

$$f \diamond \mu \in \mathcal{C}^1(\Omega_D; E). \ \square$$

Let us state some continuity properties of weighting in the spaces $\mathcal{C}^m$.

THEOREM 7.5.– *Let Ω be an open subset of $\mathbb{R}^d$, E a Neumann space, D a compact subset of $\mathbb{R}^d$, $0 \le m \le \infty$ and $0 \le \ell \le \infty$.*

Then weighting $\diamond$ is a continuous and hence sequentially continuous bilinear mapping from $\mathcal{C}^m(\Omega; E) \times \mathcal{C}^\ell_D(\mathbb{R}^d)$ into $\mathcal{C}^{m+\ell}(\Omega_D; E)$, from $\mathcal{C}^m_\mathrm{b}(\Omega; E) \times \mathcal{C}^\ell_D(\mathbb{R}^d)$ into $\mathcal{C}^{m+\ell}_\mathrm{b}(\Omega_D; E)$, and from $\mathbf{C}^m_\mathrm{b}(\Omega; E) \times \mathcal{C}^\ell_D(\mathbb{R}^d)$ into $\mathbf{C}^{m+\ell}_\mathrm{b}(\Omega_D; E)$. ▌

Proof. Denote by $\{\| \, \|_{E;\nu} : \nu \in \mathcal{N}_E\}$ the family of semi-norms of E.

Continuity for the spaces $\mathcal{C}^m$. Let $f \in \mathcal{C}^m(\Omega; E)$ and $\mu \in \mathcal{C}^\ell_D(\mathbb{R}^d)$. By iterating the equality $\partial_i(f \diamond \mu) = \partial_i f \diamond \mu$ from Theorem 7.4 (a), it follows that, for every $0 \leq |\alpha| \leq m$ and $0 \leq |\beta| \leq \ell$,

$$\partial^{\alpha+\beta}(f \diamond \mu) = (-1)^{|\beta|}\, \partial^\alpha f \diamond \partial^\beta \mu. \tag{7.6}$$

These functions are continuous by Theorem 7.2 (b), thus

$$f \diamond \mu \in \mathcal{C}^{m+\ell}(\Omega_D; E).$$

If m and ℓ are finite, then, by Definition 2.14 (a) of the semi-norms of $\mathcal{C}^m(\Omega; E)$, for every compact subset $K \subset \Omega_D$ and every $\nu \in \mathcal{N}_E$,

$$\|f \diamond \mu\|_{\mathcal{C}^{m+\ell}(\Omega_D;E);K,\nu} = \sup_{|\gamma| \leq m+\ell} \sup_{x \in K} \|\partial^\gamma(f \diamond \mu)(x)\|_{E;\nu} =$$

$$= \sup_{|\alpha| \leq m,\, |\beta| \leq \ell} \sup_{x \in K} \Big\| \int_{\mathring{D}} \partial^\alpha f(x+y)\, \partial^\beta \mu(y)\, \mathrm{d}y \Big\|_{E;\nu} \leq \tag{7.7}$$

$$\leq |\mathring{D}| \sup_{|\alpha| \leq m} \sup_{z \in K+D} \|\partial^\alpha f(z)\|_{E;\nu} \sup_{|\beta| \leq \ell} \sup_{y \in D} |\partial^\beta \mu(y)| \tag{7.8}$$

(the inequality is given by Theorem 4.15). By Definition 2.14 (b) of the semi-norms of $\mathcal{C}^\ell_{\mathrm{b}}(\mathbb{R}^d)$, since $K + D$ is a compact set included in Ω, this can be rewritten as

$$\|f \diamond \mu\|_{\mathcal{C}^{m+\ell}(\Omega_D;E);K,\nu} \leq |\mathring{D}|\, \|f\|_{\mathcal{C}^m(\Omega;E);K+D,\nu} \|\mu\|_{\mathcal{C}^\ell_{\mathrm{b}}(\mathbb{R}^d)}.$$

By the characterization of continuous bilinear mappings from Theorem A.40, since $\mathcal{C}^\ell_D(\mathbb{R}^d)$ is endowed with the semi-norms of $\mathcal{C}^\ell_{\mathrm{b}}(\mathbb{R}^d)$ by Definition 2.14 (c), it follows from this that $\diamond$ is continuous from $\mathcal{C}^m(\Omega; E) \times \mathcal{C}^\ell_D(\mathbb{R}^d)$ into $\mathcal{C}^{m+\ell}(\Omega_D; E)$.

If m or ℓ is infinite, this inequality still implies continuity because, by Definitions 2.15 and 2.14, $\| \, \|_{\mathcal{C}^\infty(\Omega;E);m,K,\nu} = \| \, \|_{\mathcal{C}^m(\Omega;E);K,\nu}$ and $\| \, \|_{\mathcal{C}^\infty_{\mathrm{b}}(\mathbb{R}^d);\ell,\nu} = \| \, \|_{\mathcal{C}^\ell_{\mathrm{b}}(\mathbb{R}^d);\nu}$.

Continuity for the spaces $\mathcal{C}^m_{\mathrm{b}}$. Let $f \in \mathcal{C}^m_{\mathrm{b}}(\Omega; E)$. By replacing K by Ω_D in (7.7) and $K + D$ by Ω in (7.8), we obtain similarly

$$\|f \diamond \mu\|_{\mathcal{C}^{m+\ell}_{\mathrm{b}}(\Omega_D;E);\nu} \leq |\mathring{D}|\, \|f\|_{\mathcal{C}^m_{\mathrm{b}}(\Omega;E);\nu} \|\mu\|_{\mathcal{C}^\ell_{\mathrm{b}}(\mathbb{R}^d)}, \tag{7.9}$$

which proves that $\diamond$ is continuous from $\mathcal{C}^m_{\mathrm{b}}(\Omega; E) \times \mathcal{C}^\ell_D(\mathbb{R}^d)$ into $\mathcal{C}^{m+\ell}_{\mathrm{b}}(\Omega_D; E)$.

Continuity for the spaces $\mathbf{C}^m_{\mathrm{b}}$. Let $f \in \mathbf{C}^m_{\mathrm{b}}(\Omega; E)$. For every $|\beta| \leq m$, $\partial^\beta f \in \mathbf{C}(\Omega; E)$, so, by Theorem 7.2 (e), $\partial^\beta f \diamond \mu \in \mathbf{C}(\Omega_D; E)$. Furthermore, by the previous case, $\partial^\beta f \diamond \mu$ belongs to $\mathcal{C}_{\mathrm{b}}(\Omega_D; E)$ and therefore to $\mathbf{C}_{\mathrm{b}}(\Omega_D; E)$. Since it is equal to $\partial^\beta(f \diamond \mu)$ by Theorem 7.4 (a), this implies that $f \diamond \mu \in \mathbf{C}^m_{\mathrm{b}}(\Omega_D; E)$.

Since the space $\mathbf{C}^m_{\mathrm{b}}(\Omega; E)$ is endowed with the semi-norms of $\mathcal{C}_{\mathrm{b}}(\Omega; E)$ by Definition 2.14 (d), the inequality (7.9) gives us the continuity of $\diamond$ from $\mathbf{C}^m_{\mathrm{b}}(\Omega; E) \times \mathcal{C}^\ell_D(\mathbb{R}^d)$ into $\mathbf{C}^{m+\ell}_{\mathrm{b}}(\Omega_D; E)$. □

Let us state a variant of the estimate (7.8).

THEOREM 7.6.– *Let $f \in \mathcal{C}_b^m(\Omega; E)$ and $\mu \in \mathcal{C}_D(\mathbb{R}^d)$, where Ω is an open subset of $\mathbb{R}^d$, E is a Neumann space, $m \in \mathbb{N}^*$ and D is a compact subset of $\mathbb{R}^d$. Then, for every semi-norm $\| \ \|_{E;\nu}$ of E,*

$$\|f \diamond \mu\|_{\mathcal{C}_b^m(\Omega_D;E);\nu} \leq \|f\|_{\mathcal{C}_b^m(\Omega;E);\nu} \int_{\mathbb{R}^d} |\mu|. \ \blacksquare$$

Proof. By replacing K by Ω_D in the equality (7.7) with $\ell = 0$ and $\beta = 0$, it follows that

$$\|f \diamond \mu\|_{\mathcal{C}_b^m(\Omega_D;E);\nu} = \sup_{|\alpha| \leq m} \sup_{x \in \Omega_D} \Big\| \int_{\mathring{D}} \partial^\alpha f(x+y)\, \mu(y)\, dy \Big\|_{E;\nu} \leq$$

$$\leq \sup_{|\alpha| \leq m} \sup_{x \in \Omega_D} \int_{\mathring{D}} \|\partial^\alpha f(x+y)\|_{E;\nu}\, |\mu(y)|\, dy \leq \sup_{|\alpha| \leq m} \sup_{z \in \Omega} \|\partial^\alpha f(z)\|_{E;\nu} \int_{\mathbb{R}^d} |\mu|,$$

which is the stated inequality. □

7.4. Local regularization

Let us define the local regularized functions[3] of a continuous function taking values in a Neumann space.

DEFINITION 7.7.– (a) *A* ***regularizing sequence*** *is a sequence of functions $(\rho_n)_{n \in \mathbb{N}}$ such that:*

$$\rho_n \in \mathcal{K}^\infty(\mathbb{R}^d), \quad \rho_n(x) = 0 \textit{ if } |x| \geq \frac{1}{n}, \quad \rho_n \geq 0, \quad \int_{\mathbb{R}^d} \rho_n = 1.$$

(b) *The* ***local regularized functions*** *of a function $f \in \mathcal{C}(\Omega; E)$, where Ω is an open subset of $\mathbb{R}^d$ and E is a Neumann space, are the functions*

$$f \diamond \rho_n \in \mathcal{C}^\infty(\Omega_{D_n}; E),$$

where D_n is the support of ρ_n and $\Omega_{D_n} = \{x \in \mathbb{R}^d : x + D_n \subset \Omega\}$. ■

3 **History of regularization.** Jean LERAY defined the regularized function in 1934 as the convolution $f \star \rho_n$ of a function $f \in L^2(\mathbb{R}^d)$ [55, Formula (1.18) p. 206], which coincides with $f \diamond \check{\rho}_n$, where $\check{\rho}_n(x) = \rho_n(-x)$. He showed that this regularized function is $\mathcal{C}^\infty$ and that regularization commutes with differentiation [55, Lemma 4 p. 208]; in other words, $\partial_i(f \star \rho_n) = \partial_i f \star \rho_n$, which corresponds to our Theorem 7.4 (a).

Justification. There always exists such a regularizing sequence, for example given by

$$\rho_n(x) = cn^d \rho(nx), \tag{7.10}$$

where $\rho \in \mathcal{C}^\infty(\mathbb{R}^d)$ satisfies Theorem 3.19 and $c = 1/\int_{\mathbb{R}^d} \rho$. Indeed, Theorem 6.19 gives $\int_{\mathbb{R}^d} \rho(nx) = n^{-d} \int_{\mathbb{R}^d} \rho(y)\,\mathrm{d}y = n^{-d}/c$. Therefore, local regularized functions exist; they belong to $\mathcal{C}^\infty(\Omega_{D_n}; E)$ by Theorem 7.5. □

Terminology. The functions $f \diamond \rho_n$ are said to be **regularized** because they are regular and converge to f (Theorem 7.9). They are **local** because they are only defined on a subset of Ω when the latter is bounded (indeed, Ω_{D_n} cannot be equal to Ω and cannot contain it, see *Remarks on the domain* Ω_D, p. 149).

Global regularized functions, defined on the whole of Ω, are constructed in Definition 7.13. However, they may differ significantly from f in the neighborhood of the boundary and they do not commute with differentiation. □

Let us state some properties of Ω_{D_n}, and of the open set $\Omega_{1/n}$ obtained by removing from Ω a neighborhood of its boundary of size $1/n$.

THEOREM 7.8.– *Let Ω be an open subset of $\mathbb{R}^d$, $(\rho_n)_{n\in\mathbb{N}}$ a regularizing sequence (Definition 7.7 (a)), D_n the support of ρ_n and*

$$\Omega_{D_n} \stackrel{\text{def}}{=} \{x \in \mathbb{R}^d : x + D_n \subset \Omega\},$$

$$\Omega_{1/n} \stackrel{\text{def}}{=} \Omega_{B(0,1/n)} = \{x \in \mathbb{R}^d : B(x, 1/n) \subset \Omega\}.$$

Then the open sets $(\Omega_{D_n})_{n\in\mathbb{N}}$ cover Ω, the open sets $(\Omega_{1/n})_{n\in\mathbb{N}}$ cover Ω and increase with n, and, for every $n \in \mathbb{N}$,

$$\Omega_{1/n} \subset \Omega_{D_n}.$$

Additionally, for every compact set $K \subset \Omega$, there exists $n_K \in \mathbb{N}$ such that, for every $n \geq n_K$,

$$K \subset \Omega_{1/n} \subset \Omega_{D_n}. \blacksquare$$

Recall that $B(x, r)$ denotes the closed ball $\{y \in \mathbb{R}^d : |y - x| \leq r\}$.

Proof. The open sets $\Omega_{1/n} = \Omega_{B(0,1/n)}$ increase with n by Theorem 7.3 (a). They cover Ω because, if $x \in \Omega$, then Theorem 7.3 (b) applied to the compact set $\{x\}$ gives $r > 0$ such that $\{x\} \subset \Omega_{B(0,r)}$ and therefore $x \in \Omega_{B(0,1/n)}$ if $n \geq 1/r$.

Since $D_n \subset B(0, 1/n)$, Theorem 7.3 (a) gives $\Omega_{B(0,1/n)} \subset \Omega_{D_n}$; therefore, the Ω_{D_n} also cover Ω.

If K is a compact set included in Ω, then Theorem 7.3 (b) gives $r > 0$ such that $K \subset \Omega_{B(0,r)}$ and therefore $K \subset \Omega_{B(0,1/n)} = \Omega_{1/n}$ if $n \geq 1/r$. □

Let us show that the regularized functions converge locally, i.e. outside of an open neighborhood of the boundary of Ω of arbitrary size $1/k$.

THEOREM 7.9.– *Let Ω be an open subset of $\mathbb{R}^d$, E a Neumann space, $0 \leq m \leq \infty$, $(\rho_n)_{n\in\mathbb{N}}$ a regularizing sequence (Definition 7.7 (a)) and $\Omega_{1/n} = \Omega_{B(0,1/n)}$.*

Then, for every $k \in \mathbb{N}$, as $n \to \infty$ and $n \geq k$:

(a) *If $f \in \mathcal{C}^m(\Omega; E)$,*

$$f \diamond \rho_n \to f \ \textit{in}\ \mathcal{C}^m(\Omega_{1/k}; E).$$

(b) *If $f \in \mathbf{C}^m_{\mathrm{b}}(\Omega; E)$,*

$$f \diamond \rho_n \to f \ \textit{in}\ \mathbf{C}^m_{\mathrm{b}}(\Omega_{1/k}; E). \ \blacksquare$$

Case of the entire space. In this case, the *local* regularized function is in fact *global*, since $\mathbb{R}^d_{1/n} = \mathbb{R}^d$. Therefore, if $f \in \mathcal{C}(\mathbb{R}^d; E)$,

$$f \diamond \rho_n \to f \ \text{in}\ \mathcal{C}^\infty(\mathbb{R}^d; E)$$

and, if $f \in \mathbf{C}^m_{\mathrm{b}}(\mathbb{R}^d; E)$,

$$f \diamond \rho_n \to f \ \text{in}\ \mathbf{C}^m_{\mathrm{b}}(\mathbb{R}^d; E). \ \square$$

Proof of Theorem 7.9. This is a special case of Theorem 7.10 with $\omega = \Omega_{1/k}$, since, by Theorem 7.8, $\Omega_{1/k} \subset \Omega_{1/n} \subset \Omega_{D_n}$ for every $n \geq k$. $\square$

Let us show, more generally, that the regularized functions converge on subsets ω of Ω that may extend up to (part of) its boundary.

THEOREM 7.10.– *Let Ω be an open subset of $\mathbb{R}^d$, E a Neumann space, $(\rho_n)_{n\in\mathbb{N}}$ a regularizing sequence, D_n the support of ρ_n, $\Omega_{D_n} \overset{\text{def}}{=} \{x \in \mathbb{R}^d : x + D_n \subset \Omega\}$ and $0 \leq m \leq \infty$. Let ω be an open set included in Ω and $k \in \mathbb{N}$ such that*

$$\omega \subset \Omega_{D_n} \textit{ for every } n \geq k. \tag{7.11}$$

Then, for every $k \in \mathbb{N}$, as $n \to \infty$ and $n \geq k$:

(a) *If $f \in \mathcal{C}^m(\Omega; E)$,*

$$f \diamond \rho_n \to f \ \textit{in}\ \mathcal{C}^m(\omega; E).$$

(b) *If $f \in \mathbf{C}^m_{\mathrm{b}}(\Omega; E)$,*

$$f \diamond \rho_n \to f \ \textit{in}\ \mathbf{C}^m_{\mathrm{b}}(\omega; E). \ \blacksquare$$

Choice of D_n. In this volume, we always choose D_n to be $B(0, 1/n)$, in which case Ω_{D_n} is equal to $\Omega_{1/n}$, namely Ω with a neighborhood of its boundary of size $1/n$ removed, and the local convergence of Theorem 7.9 is sufficient for our purposes.

Theorem 7.10, on the other hand, will be used in subsequent volumes to regularize functions up to part of the boundary, by choosing a regularizing sequence with support in a conic sector $D_n = K \cap B(0, 1/n)$ (as shown in Figure 7.2 on page 149) and $\omega = \Omega_K$. □

Proof of Theorem 7.10. Let $\{\| \|_{E;\nu} : \nu \in \mathcal{N}_E\}$ be the family of semi-norms of E.

(a) *Case where* $f \in \mathcal{C}(\Omega; E)$. By Definition 7.7 (a) of a regularizing sequence and Theorem 4.17,

$$\int_{\mathring{D}_n} \rho_n = \int_{\mathbb{R}^d} \rho_n = 1.$$

Together with Definition 7.1 of the weighting, it therefore follows that, for every $n \geq k$ and $x \in \omega$,

$$(f \diamond \rho_n - f)(x) = \int_{y \in \mathring{D}_n} (f(x+y) - f(x))\, \rho_n(y)\, dy.$$

Let $\nu \in \mathcal{N}_E$. Then (Theorem 4.11 (b), since $\rho_n \geq 0$),

$$\begin{aligned} \|(f \diamond \rho_n - f)(x)\|_{E;\nu} &\leq \int_{y \in \mathring{D}_n} \|f(x+y) - f(x)\|_{E;\nu}\, \rho_n(y)\, dy \leq \\ &\leq \sup_{y \in \mathring{D}_n} \|f(x+y) - f(x)\|_{E;\nu} \int_{\mathring{D}_n} \rho_n \leq \\ &\leq \sup_{z \in \Omega,\ |z-x| \leq 1/n} \|f(z) - f(x)\|_{E;\nu}. \end{aligned} \tag{7.12}$$

Let K be a compact set included in ω. By the strong inclusion theorem (Theorem A.22), there exists $r > 0$ such that $K' = K + B(0, r)$ is included in ω. By Definition 1.3 (a) of the semi-norms of $\mathcal{C}(\omega; E)$, the inequality (7.12) implies that, for $n \geq \sup\{k, 1/r\}$,

$$\begin{aligned} \|f \diamond \rho_n - f\|_{\mathcal{C}(\omega;E);K,\nu} &= \sup_{x \in K} \|(f \diamond \rho_n - f)(x)\|_{E;\nu} \leq \\ &\leq \sup_{x \in K',\ z \in K',\ |z-x| \leq 1/n} \|f(z) - f(x)\|_{E;\nu}. \end{aligned} \tag{7.13}$$

But K' is compact (as a sum of compact subsets of $\mathbb{R}^d$, see Theorem A.24) and included in Ω. Therefore, f is uniformly continuous on K' by Heine's theorem (Theorem A.34). As $n \to \infty$, the right-hand side of the last inequality therefore tends to 0, which proves that

$$f \diamond \rho_n \to f \text{ in } \mathcal{C}(\omega; E).$$

Case where $f \in \mathcal{C}^m(\Omega; E)$, *$m$ finite.* Since $\partial^\beta(f \diamond \rho_n) = \partial^\beta f \diamond \rho_n$ (Theorem 7.4 (a), iterated), by Definition 2.14 (a) of the semi-norms of $\mathcal{C}^m(\omega; E)$, together with the inequality (7.13),

$$\|f \diamond \rho_n - f\|_{\mathcal{C}^m(\omega;E);K,\nu} = \sup_{|\beta|\leq m} \sup_{x\in K} \|\partial^\beta(f \diamond \rho_n - f)(x)\|_{E;\nu} =$$

$$= \sup_{|\beta|\leq m} \sup_{x\in K} \|(\partial^\beta f \diamond \rho_n - \partial^\beta f)(x)\|_{E;\nu} \leq$$

$$\leq \sup_{|\beta|\leq m} \sup_{x\in K',\, z\in K',\, |z-x|\leq 1/n} \|\partial^\beta f(z) - \partial^\beta f(x)\|_{E;\nu}. \quad (7.14)$$

This tends to 0 as above as $n \to \infty$, since here the $\partial^\beta f$ are continuous on Ω and therefore uniformly continuous on K'. Hence,

$$f \diamond \rho_n \to f \text{ in } \mathcal{C}^m(\omega; E).$$

Case where $f \in \mathcal{C}^\infty(\Omega; E)$. The convergence in $\mathcal{C}^\infty(\omega; E)$ follows from the convergence for every finite m, since $\| \, \|_{\mathcal{C}^\infty(\omega;E);m,K,\nu} = \| \, \|_{\mathcal{C}^m(\omega;E);K,\nu}$ by Definitions 2.15 (a) and 2.14 (a) of these semi-norms.

(b) *Case where* $f \in \mathbf{C}^m_{\mathrm{b}}(\Omega; E)$. By replacing the supremum over K in the inequalities (7.13) and (7.14) with the supremum over ω, by Definition 2.14 (b) of the semi-norms of $\mathcal{C}^m_{\mathrm{b}}(\omega; E)$, we obtain

$$\|f \diamond \rho_n - f\|_{\mathcal{C}^m_{\mathrm{b}}(\omega;E);\nu} = \sup_{|\beta|\leq m} \sup_{x\in \omega} \|\partial^\beta(f \diamond \rho_n - f)(x)\|_{E;\nu} =$$

$$\leq \sup_{|\beta|\leq m} \sup_{x\in \omega,\, z\in \Omega,\, |z-x|\leq 1/n} \|\partial^\beta f(z) - \partial^\beta f(x)\|_{E;\nu}.$$

This tends to 0 as $n \to \infty$ because $\partial^\beta f$ are uniformly continuous on Ω by the hypotheses. Since $f \diamond \rho_n$ belongs to $\mathbf{C}^m_{\mathrm{b}}(\omega; E)$ by Theorem 7.5, which is endowed with the semi-norms of $\mathcal{C}^m_{\mathrm{b}}(\omega; E)$ by Definition 2.14 (d), this proves that

$$f \diamond \rho_n \to f \text{ in } \mathbf{C}^m_{\mathrm{b}}(\omega; E). \ \square$$

Let us conclude this section by two estimates on regularized functions. The first is used to prove that $\mathcal{K}^\infty(\Omega)$ is separable (Theorem 7.20). The second will be used in Volume 3 to prove that the velocity of convergence of a sequence of distributions is uniform with respect to test functions.

THEOREM 7.11.– *Let* $f \in \mathcal{K}^m(\Omega)$, *where* Ω *is an open subset of* $\mathbb{R}^d$ *and* $m \in \mathbb{N}$, *and let* $(\rho_n)_{n\in\mathbb{N}}$ *be a regularizing sequence (Definition 7.7 (a)) given by (7.10). Then:*

(a) $$\|f \diamond \rho_n\|_{C_b^m(\Omega_{D_n})} \leq c_{\rho,m} n^{m+d} \|f\|_{C_b(\Omega)},$$

where $c_{\rho,m} = \|\rho\|_{C_b^m(\mathbb{R}^d)} / \int_{\mathbb{R}^d} \rho$.

(b) $$\|f \diamond \rho_n - f\|_{C_b^{m-1}(\Omega_{D_n})} \leq \frac{\sqrt{d}}{n} \|f\|_{C_b^m(\Omega)}. \blacksquare$$

Proof. (a) The inequality (7.9) implies, in particular, that

$$\|f \diamond \rho_n\|_{C_b^m(\Omega_{D_n})} \leq |\mathring{B}(0,1)| \|f\|_{C_b(\Omega)} \|\rho_n\|_{C_b^m(\mathbb{R}^d)},$$

which gives the stated inequality because, by Definition 2.14 (b) of the norm of $C_b^m(\mathbb{R}^d)$,

$$\|\rho_n\|_{C_b^m(\mathbb{R}^d)} = \sup_{0 \leq |\beta| \leq m} \sup_{x \in \mathbb{R}^d} |\partial^\beta \rho_n(x)| \leq$$
$$\leq c n^{m+d} \sup_{0 \leq |\beta| \leq m} \sup_{x \in \mathbb{R}^d} |\partial^\beta \rho(x)| = c n^{m+d} \|\rho\|_{C_b^m(\mathbb{R}^d)},$$

where $c = 1 / \int_{\mathbb{R}^d} \rho$. Indeed, by the construction (7.10), $\rho_n(x) = c n^d \rho(nx)$, and so $\partial_i \rho_n(x) = c n^d n (\partial \rho / \partial x_i)(nx)$ and, by iterating, $\partial^\beta \rho_n(x) = c n^{d+|\beta|} \partial^\beta \rho(nx)$.

(b) Applying the finite increment theorem (Theorem 2.5 (a)) to the extension $\widetilde{f}$ of f by 0, which belongs to $\mathcal{K}^m(\mathbb{R}^d)$ by Theorem 2.20 (a), gives us that, for $|\beta| \leq m-1$ and x and z in $\mathbb{R}^d$,

$$|\partial^\beta f(z) - \partial^\beta f(x)| \leq |z - x| \sup_{x \in \mathbb{R}^d} |\nabla \partial^\beta \widetilde{f}(x)| \leq |z - x| \sqrt{d} \sup_{|\alpha| \leq m} \sup_{x \in \Omega} |\partial^\alpha f(x)|.$$

Together with Definition 2.14 (b) of the semi-norms of $C_b^m(\Omega)$ and the inequality (7.13), it follows that

$$\|f \diamond \rho_n - f\|_{C_b^{m-1}(\Omega_{D_n})} = \sup_{|\beta| \leq m-1} \sup_{x \in \Omega_{D_n}} |\partial^\beta (f \diamond \rho_n - f)(x)| \leq$$
$$\leq \sup_{|\beta| \leq m-1} \sup_{x \in \Omega,\, z \in \Omega,\, |z-x| \leq 1/n} |\partial^\beta f(z) - \partial^\beta f(x)| \leq$$
$$\leq \frac{\sqrt{d}}{n} \sup_{|\alpha| \leq m} \sup_{x \in \Omega} |\partial^\alpha f(x)| = \frac{\sqrt{d}}{n} \|f\|_{C_b^m(\Omega)}. \quad \square$$

7.5. Global regularization

Let us construct a sequence to *localize* the support of functions.

DEFINITION 7.12.– *Let Ω be an open subset of $\mathbb{R}^d$. A* ***localizing sequence*** *is a sequence of functions $(\alpha_n)_{n\in\mathbb{N}}$ such that, for every n:*

$$\alpha_n \in \mathcal{K}^\infty(\Omega), \quad 0 \le \alpha_n \le 1, \quad \alpha_n = \begin{cases} 1 & \text{on } \Omega^n_{3/n}, \\ 0 & \text{outside of } \Omega^n_{2/n}, \end{cases}$$

where $\Omega^n_r \stackrel{\text{def}}{=} \{x \in \Omega : |x| < n,\ B(x,r) \subset \Omega\}$. ▮

Recall that $B(x,r) = \{y \in \mathbb{R}^d : |y - x| \le r\}$.

Let us define a global regularization, namely regularization on the whole of Ω. Unlike the local regularization, it does not commute with differentiation everywhere.

DEFINITION 7.13.– *Let $f \in \mathcal{C}(\Omega; E)$, where Ω is an open subset of $\mathbb{R}^d$ and E is a Neumann space, and let $(\rho_n)_{n\in\mathbb{N}}$ be a regularizing sequence (Definition 7.7) and $(\alpha_n)_{n\in\mathbb{N}}$ a localizing sequence (Definition 7.12).*

The (global) ***regularized functions*** *of f are the functions $R_n f \in \mathcal{K}^\infty(\Omega; E)$ defined, for every $x \in \Omega$, by*

$$(R_n f)(x) \stackrel{\text{def}}{=} (\widetilde{\alpha_n f} \diamond \rho_n)(x),$$

where $\widetilde{\alpha_n f} \in \mathcal{K}(\mathbb{R}^d; E)$ is the extension of $\alpha_n f$ by 0_E outside of Ω. ▮

Justification. The existence of the ρ_n was established in the justification of Definition 7.7 of a regularizing sequence, and the existence of the α_n follows from Urysohn's lemma (Theorem 3.21), since $\Omega^n_{3/n} \Subset \Omega^n_{2/n} \Subset \Omega$.

By Definition 2.15 (e) of $\mathcal{K}^\infty(\Omega; E)$, we need to show that $R_n f$ is infinitely differentiable with compact support. On the one hand, $\alpha_n f \in \mathcal{K}(\Omega; E)$ (Theorem 3.11 (c) with $T(\alpha, f) = \alpha f$, for example), so, by the extension theorem (Theorem 2.20), $\widetilde{\alpha_n f} \in \mathcal{K}(\mathbb{R}^d; E)$, and therefore, by Definition 7.7 (b) of the local regularized functions, $\widetilde{\alpha_n f} \diamond \rho_n \in \mathcal{C}^\infty(\mathbb{R}^d; E)$. Hence, by restriction, we obtain

$$R_n f \in \mathcal{C}^\infty(\Omega; E).$$

On the other hand, by Theorem 7.2 (d),

$$\operatorname{supp}(\widetilde{\alpha_n f} \diamond \rho_n) \subset \operatorname{supp} \widetilde{\alpha_n f} - \operatorname{supp} \rho_n \subset \overline{\Omega^n_{2/n}} + B(0, 1/n) \subset \overline{\Omega^n_{1/n}}.$$

The support of its restriction $R_n f$ is included in $\overline{\Omega^n_{1/n}}$, which is a compact set included in Ω, so, by Theorem 1.18, it is compact. Therefore,

$$R_n f \in \mathcal{K}^\infty(\Omega; E). \ \square$$

Terminology. The functions $R_n f$ are said to be **regularized** because they are regular and converge to f (Theorem 7.15). They are optionally said to be **global** because they are defined on the whole of the domain Ω of f, unlike the *local* regularized functions $f \diamond \rho_n$, which are only defined on a subset Ω_{D_n} of Ω in general (see *Remarks on the domain* Ω_D, p. 149). □

Let us compare the global regularized functions to the local regularized functions.

THEOREM 7.14.– *Let $f \in \mathcal{C}(\Omega; E)$, where Ω is an open subset of $\mathbb{R}^d$ and E is a Neumann space, and let $(\rho_n)_{n\in\mathbb{N}}$ be a regularizing sequence (Definition 7.7) and $(R_n f)_{n\in\mathbb{N}}$ the global regularized functions of f (Definition 7.13). Then, denoting $\Omega^n_r = \{x \in \Omega : |x| < n,\ B(x,r) \subset \Omega\}$:*

(a) $$R_n f = f \diamond \rho_n \ \text{ on } \Omega^n_{4/n}.$$

(b) *If the support of f is compact:*

The support of every $R_n f$ is included in some compact set $K \subset \Omega$

and there exists $n_f \in \mathbb{N}$ such that, for every $n \geq n_f$,

$$R_n f = \widetilde{f} \diamond \rho_n \ \text{ on } \Omega,$$

where $\widetilde{f}$ is the extension of f by 0_E outside of Ω. ▮

Proof. (a) By Definition 7.13 of $R_n f$ and Definition 7.1 of the weighting,

$$(R_n f)(x) = \int_{\mathring{D}_n} (\widetilde{\alpha_n f})(x+y)\,\rho(y)\,dy.$$

If $x \in \Omega^n_{4/n}$ and $y \in \mathring{D}_n$, then $x + y \in \Omega^n_{3/n}$, so, by Definition 7.12 of the localizing functions, $\alpha_n(x+y) = 1$, and thus $(\widetilde{\alpha_n f})(x+y) = f(x+y)$. Therefore,

$$(R_n f)(x) = (f \diamond \rho_n)(x).$$

(b) Suppose that the support of f is compact. It is included in Ω and hence in $\Omega^{n_f}_{3/n_f}$ for n_f sufficiently large (because the $\Omega^n_{3/n}$ are open, increase with n, and cover Ω).

For $n \geq n_f$, we have $\operatorname{supp} f \subset \Omega^n_{3/n}$, so $\alpha_n = 1$ on the support of f. Therefore, $\alpha_n f = f$ on Ω. Hence, $\widetilde{\alpha_n f} \diamond \rho_n = \widetilde{f} \diamond \rho_n$ on $\mathbb{R}^d$. And, by restriction (Definition 7.13),

$$R_n f = \widetilde{f} \diamond \rho_n \text{ on } \Omega.$$

Furthermore, by Theorem 7.2 (d),

$$\operatorname{supp} R_n f \subset \operatorname{supp} \alpha_n f - \operatorname{supp} \rho_n \subset \Omega^{n_f}_{3/n_f} + B(0, 1/n_f) \subset \overline{\Omega^{n_f}_{1/n_f}}.$$

For $n < n_f$, the same property holds because, by Definition 7.12, $\operatorname{supp} \alpha_n \subset \overline{\Omega^n_{2/n}}$, so $\operatorname{supp} R_n f \subset \overline{\Omega^n_{2/n}} + B(0, 1/n) \subset \overline{\Omega^n_{1/n}} \subset \overline{\Omega^{n_f}_{1/n_f}}$. The support of all of the $R_n f$ is thus included in the compact set $\overline{\Omega^{n_f}_{1/n_f}}$, which is included in Ω. □

Let us show that the global regularized functions of a function that belongs to $\mathcal{C}^m(\Omega; E)$ or $\mathcal{K}^m(\Omega; E)$ approximate this function in this space.

THEOREM 7.15.– *Let $f \in \mathcal{C}(\Omega; E)$, where Ω is an open subset of $\mathbb{R}^d$ and E is a Neumann space, $(R_n f)_{n\in\mathbb{N}}$ be the regularized functions given by Definition 7.13 and let $0 \leq m \leq \infty$. Then, as $n \to \infty$:*

(a) *If $f \in \mathcal{C}^m(\Omega; E)$,*

$$R_n f \to f \text{ in } \mathcal{C}^m(\Omega; E).$$

(b) *If $f \in \mathcal{K}^m(\Omega; E)$,*

$$R_n f \to f \text{ in } \mathcal{K}^m(\Omega; E). ▮$$

Proof. Let $\{\| \, \|_{E;\nu} : \nu \in \mathcal{N}_E\}$ be the family of semi-norms of E.

(a) *Case where $f \in \mathcal{C}^m(\Omega; E)$*. Let K be a compact set included in Ω. By Theorem 7.8, there exists $n_K \in \mathbb{N}$ such that $K \subset \Omega_{1/n}$ for every $n \geq n_K$. If $n \geq 4n_K$, then $\Omega_{1/n_K} \subset \Omega^n_{4/n}$, so, by Theorem 7.14 (a), $R_n f = f \diamond \rho_n$ on K. Therefore, by Definition 2.14 of the semi-norms of $\mathcal{C}^m(\Omega; E)$, for every $\nu \in \mathcal{N}_E$,

$$\|R_n f - f\|_{\mathcal{C}^m(\Omega;E);K,\nu} = \sup_{x\in K} \|(R_n f - f)(x)\|_{E;\nu} = \\ = \|f \diamond \rho_n - f\|_{\mathcal{C}^m(\Omega_{1/n_K};E);K,\nu},$$

which tends to 0 by Theorem 7.9 (a) on the convergence of the local regularized functions. This proves that

$$R_n f \to f \text{ in } \mathcal{C}^m(\Omega; E).$$

Case where $f \in \mathcal{C}^\infty(\Omega; E)$. In this case, the convergence follows from the convergence for finite m because, by Definitions 2.15 (a) and 2.14 (a) of the semi-norms, $\| \, \|_{\mathcal{C}^\infty(\omega;E);m,K,\nu} = \| \, \|_{\mathcal{C}^m(\omega;E);K,\nu}$.

(b) *Case where $f \in \mathcal{K}^m(\Omega; E)$*. By Definition 2.14 (e) of $\mathcal{K}^m(\Omega; E)$, the support of f is compact in this case, so by Theorem 7.14 (b), the support of $R_n f$ is included in some compact set K included in Ω, and therefore

$$R_n f \in \mathcal{C}^m_K(\Omega; E).$$

Since the topologies of $\mathcal{C}^m(\Omega; E)$ and $\mathcal{K}^m(\Omega; E)$ coincide on the space $\mathcal{C}^m_K(\Omega; E)$ (Theorem 2.18), the convergence (a) implies the convergence in $\mathcal{K}^m(\Omega; E)$. □

Let us observe that $\mathcal{K}^\infty(\Omega; E)$ is dense in $\mathcal{C}^m(\Omega; E)$ and $\mathcal{K}^m(\Omega; E)$.

THEOREM 7.16.– *Let Ω be an open subset of $\mathbb{R}^d$, E a Neumann space and $0 \leq m \leq \infty$.*

Then $\mathcal{K}^\infty(\Omega; E)$ is sequentially dense and therefore dense in $\mathcal{C}^m(\Omega; E)$ and in $\mathcal{K}^m(\Omega; E)$. ▮

Proof. The stated sequential denseness properties follow from the convergence properties of Theorem 7.15, since, by Definition 7.13, $R_n f \in \mathcal{K}^\infty(\Omega; E)$.

Sequential denseness always implies denseness (Theorem A.14). □

7.6. Partition of unity

Let us construct the **covering by crown-shaped sets** tending to the boundary of an open subset of $\mathbb{R}^d$.

THEOREM 7.17.– *Let Ω be an open subset of $\mathbb{R}^d$ and κ_n the* **crown-shaped set** *defined by*

$$\kappa_n = \Omega^{n+2}_{1/(n+2)} \setminus \overline{\Omega^n_{1/n}},$$

if $n \in \mathbb{N}$ and $\kappa_0 = \Omega^2_{1/2}$, where $\Omega^n_r \overset{\text{def}}{=} \{x \in \Omega : |x| < n,\ B(x,r) \subset \Omega\}$. Then:

(a)
$$\bigcup_{n \in \mathbb{N}^*} \kappa_n = \Omega.$$

(b) *Each κ_n is open, satisfies $\kappa_n \Subset \Omega$ and only intersects with the crown-shaped sets κ_{n-1} and κ_{n+1}.*

(c) *Each compact set $K \subset \Omega$ only intersects with finitely many κ_n.* ▮

Recall that $\overset{\circ}{B}(x,r) = \{y \in \mathbb{R}^d : |y - x| \leq r\}$ and that $\omega \Subset \Omega$ means that ω is bounded and $\overline{\omega} \subset \Omega$.

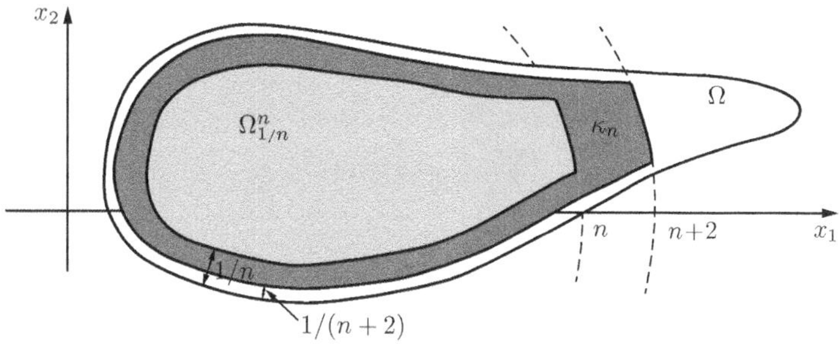

Figure 7.3. *Crown-shaped set κ_n and potato-shaped set $\Omega^n_{1/n}$. The set κ_n is dark gray and $\Omega^n_{1/n}$ is light gray*

Proof. (a) The crown-shaped sets κ_n cover Ω because the $\Omega^n_{1/n}$ form an increasing open covering of Ω. Therefore, for every $x \in \Omega$, there exists a (unique) $n \in \mathbb{N}^*$ such that $x \in \Omega^{n+1}_{1/(n+1)} \setminus \Omega^n_{1/n}$, and this set is included in κ_{n-1} if $n \geq 1$ and in κ_0 if $n = 0$.

(b) Each crown-shaped set κ_n is open because it is the intersection of the open sets $\Omega^{n+2}_{1/(n+2)}$ and $\mathbb{R}^d \setminus \overline{\Omega^n_{1/n}}$ (we checked that each $\Omega^n_{1/n}$ is open in Lemma 1.15).

Furthermore, $\kappa_n \Subset \Omega$ because it is bounded and $\overline{\kappa_n} \subset \overline{\Omega^n_{1/n}} \subset \Omega^{n+1}_{1/(n+1)} \subset \Omega$. Finally, if $m \geq n+2$, then $\kappa_n \subset \Omega^{n+2}_{1/(n+2)} \subset \overline{\Omega^m_{1/m}}$, so

$$\kappa_n \cap \kappa_m = \emptyset.$$

(c) Each compact set K only intersects with finitely many κ_n because, being covered by the increasing open sets $\Omega^n_{1/n}$, it must be covered by one of them, say $\Omega^{n_0}_{1/n_0}$. Therefore, for $n \geq n_0$, we have $K \subset \Omega^n_{1/n}$, which implies that

$$K \cap \kappa_n = \emptyset. \ \square$$

Let us construct a partition of unity subordinate to an open covering of an open subset of $\mathbb{R}^d$.

THEOREM 7.18.– *Let Ω and $\{\omega_i\}_{i\in\mathcal{I}}$ be open subsets of $\mathbb{R}^d$ such that*

$$\Omega = \bigcup_{i\in\mathcal{I}} \omega_i.$$

Then there exists a family $\{\alpha_i\}_{i\in\mathcal{I}}$ of functions, called a ***partition of unity*** *subordinate to the covering $\{\omega_i\}_{i\in\mathcal{I}}$ of Ω, such that:*

(a) *For every $i \in \mathcal{I}$:*

$$\alpha_i \in \mathcal{C}^\infty(\Omega), \quad 0 \leq \alpha_i \leq 1, \quad \operatorname{supp} \alpha_i \subset \omega_i.$$

(b)
$$\sum_{i \in \mathcal{I}} \alpha_i = 1 \ \textit{on } \Omega.$$

(c) *The set of α_i that are not identically zero is countable, and for every compact set $K \subset \Omega$, the set of α_i that are not identically zero on K is finite.*

(d) *If $\omega_i \Subset \Omega$, in other words, if ω_i is bounded and $\overline{\omega_i} \subset \mathring{\Omega}$, then*

$$\alpha_i \in \mathcal{K}^\infty(\omega_i). \ \blacksquare$$

The sum $\sum_{i \in \mathcal{I}} \alpha_i(x)$ is well defined at every point x of Ω because only finitely many $\alpha_i(x)$ are non-zero (by (c) with $K = \{x\}$).

Proof. (a), (b) *and* (c). Let $\{\kappa_n\}_{n \in \mathbb{N}^*}$ be the covering by crown-shaped sets of Ω constructed in Theorem 7.17 and let $Q_n = \overline{\kappa_n}$. Every set Q_n is compact and $\Omega = \bigcup_{n \in \mathbb{N}^*} Q_n$.

Let $n \in \mathbb{N}^*$ be fixed for a while. Every point $x \in Q_n$ belongs to one of the ω_i, so there exists $\epsilon_x > 0$ such that the closed ball $B(x, \epsilon_x) = \{y : |y - x| \leq \epsilon_x\}$ is included in this ω_i. Since the open balls $\mathring{B}(x, \epsilon_x/2)$ cover Q_n, there exists a finite subcovering of balls $\mathring{B}(x_{n,m}, \epsilon_{n,m}/2)$, where $m \in M_n$. We can choose $\epsilon_x < 1/(n+2) - 1/(n+3)$; then

$$\mathring{B}(x_{n,m}, \epsilon_{n,m}) \subset \kappa_{n-1} \cup \kappa_n \cup \kappa_{n+1}. \tag{7.15}$$

For each $m \in M_n$, consider a function $\rho_m \in \mathcal{C}^\infty(\mathbb{R}^d)$ given by Urysohn's lemma (Theorem 3.21) such that

$$\rho_{n,m} = \begin{cases} 1 & \text{on } B(x_{n,m}, \epsilon_{n,m}/2), \\ 0 & \text{outside of } B(x_{n,m}, \epsilon_{n,m}). \end{cases}$$

Now, consider the union of these functions for $n \in \mathbb{N}^*$ and $m \in M_n$. Assign each of them to precisely one of the ω_i containing $\mathring{B}(x_{n,m}, \epsilon_{n,m})$ and

denote by ζ_i the sum of the $\rho_{n,m}$ assigned to ω_i.

For $\ell \in \mathbb{N}^*$, the crown-shaped set κ_ℓ only intersects with $\kappa_{\ell-1}$ and $\kappa_{\ell+1}$ (Theorem 7.17 (b)), so, by (7.15), it only intersects with finitely many $\mathring{B}(x_{n,m}, \epsilon_{n,m})$. Thus, on each κ_ℓ, only finitely many $\rho_{n,m}$ are not identically zero, so each ζ_i is the sum of finitely many of them on this set, and hence is $\mathcal{C}^\infty$ on this set.

Therefore, $\zeta_i \in \mathcal{C}^\infty(\Omega)$. Similarly, on each κ_ℓ, the expression $\sum_{i\in\mathcal{I}} \zeta_i$ is the sum of finitely many $\rho_{n,m}$, so

$$\sum_{i\in\mathcal{I}} \zeta_i \in \mathcal{C}^\infty(\Omega).$$

Moreover, at each point of Ω, one of the $\rho_{n,m}$ is equal to 1, so

$$\sum_{i\in\mathcal{I}} \zeta_i \geq 1.$$

Hence, $\alpha_i = \zeta_i / \sum_{k\in\mathcal{I}} \zeta_k$ has the required properties.

(d) By (a), $\operatorname{supp} \alpha_i \subset \overline{\omega_i}$. If $\omega_i \Subset \Omega$, then $\overline{\omega_i}$ is compact, so by Theorem 1.18, $\operatorname{supp} \alpha_i$ is compact; furthermore, it is included in ω_i. □

Support of α_i. The condition $\operatorname{supp} \alpha_i \subset \omega_i$ of Theorem 7.18 (a) requires α_i to be zero on an open set containing $\Omega \setminus \omega_i$ (Theorem 1.6 (a)), which therefore also contains $\partial\omega_i \setminus \partial\Omega$ (recall that $\partial\Omega$ denotes the boundary of Ω). □

Partition of unity subordinate to covering by crown-shaped sets. Every function of a partition of unity subordinate to the covering by crown-shaped sets of an open set Ω belongs to $\mathcal{K}^\infty(\Omega)$, and its restriction to κ_n satisfies

$$\alpha_n \in \mathcal{K}^\infty(\kappa_n).$$

This is given by Theorem 7.18 (d), since $\kappa_n \Subset \Omega$ by Theorem 7.17 (b). □

Partition of unity composed of functions that are not uniformly continuous. There exist coverings for which any function of any subordinate partition of unity, given by Theorem 7.18, satisfies

$$\alpha_i \notin \mathbf{C}_b(\Omega).$$

This is true even if Ω is bounded. For example, this occurs when Ω is the union of two disjoint open sets ω_1 and ω_2 whose boundaries are not disjoint; in this case, α_1 is equal to 1 on ω_1 and 0 on ω_2, so it *jumps* in the neighborhood of any point shared by the boundaries of these two sets. □

Countable subcoverings. We could restrict Theorem 7.18 to countable coverings because:

Every open covering of an open subset of $\mathbb{R}^d$ has a countable subcovering. (7.16)

Proof. Let $\mathcal{R}$ be an open covering of an open subset Ω of $\mathbb{R}^d$. Any such Ω is a countable union of compact sets, for example the closed potato-shaped sets $\overline{\Omega^n_{1/n}}$ from Theorem 7.17. Since each of these compact sets is covered by a finite subcovering of $\mathcal{R}$, the union of these subcoverings is a countable subcovering of $\mathcal{R}$. □

Let us conclude this section by giving an elementary construction of a continuous function greater than a given function that is finite on any compact set.

THEOREM 7.19.– *Let p be a real function on an open subset Ω of $\mathbb{R}^d$ whose supremum on any compact set included in Ω is finite.*

Then there exists $q \in \mathcal{C}^+(\Omega)$ such that $q \geq p$ on Ω. ▌

Proof. Let $\{\kappa_n\}_{n\in\mathbb{N}^*}$ be the covering by crown-shaped sets of Ω constructed in Theorem 7.17 and $\{\alpha_n\}_{n\in\mathbb{N}^*}$ a subordinate partition of unity (Theorem 7.18). Define $q \in \mathcal{C}^+(\Omega)$ by

$$q(x) = \sum_{n\in\mathbb{N}^*} c_n\, \alpha_n(x), \quad c_n = \sup_{y\in\kappa_{n-1}\cup\kappa_n\cup\kappa_{n+1}} p^+(y),$$

where $p^+(x) = \sup\{p(x), 0\}$.

Let $x \in \kappa_k$. For n not equal to $k-1$, k or $k+1$, the crown-shaped set κ_n does not intersect with κ_k, so $\alpha_n(x) = 0$, since the support of α_n is included in κ_n. Therefore,

$$q(x) = c_{k-1}\alpha_{k-1}(x) + c_k\alpha_k(x) + c_{k+1}\alpha_{k+1}(x).$$

By definition, c_{k-1}, c_k and c_{k+1} are upper bounds for $p^+(x)$, so

$$q(x) \geq p^+(x)\,(\alpha_{k-1} + \alpha_k + \alpha_{k+1})(x) = p^+(x) \sum_{n\in\mathbb{N}^*} \alpha_n(x) = p^+(x),$$

since $\sum_{n\in\mathbb{N}^*} \alpha_n = 1$. Therefore, $q \geq p$ on the union of the κ_k, that is, on Ω. $\Box$

7.7. Separability of $\mathcal{K}^\infty(\Omega)$

Let us show that $\mathcal{K}^\infty(\Omega)$ is sequentially separable and therefore separable.

THEOREM 7.20.– *For every open subset Ω of $\mathbb{R}^d$, there exists a sequence $(\phi_k)_{k\in\mathbb{N}}$ that is sequentially dense, and therefore dense, in $\mathcal{K}^\infty(\Omega)$.* ▮

The sequence constructed in the proof also possesses the following property, which will be used in Volume 3 where $\mathcal{K}^\infty(\Omega)$ is endowed with a stronger topology (that of $\mathcal{D}(\Omega)$).

THEOREM 7.21.– *For every open subset Ω of $\mathbb{R}^d$, there exists a sequence $(\phi_k)_{k\in\mathbb{N}}$ of $\mathcal{K}^\infty(\Omega)$ such that, for every $\varphi \in \mathcal{K}^\infty(\Omega)$, there exists a compact set K included in Ω and a subsequence $(\phi_{\sigma(n)})_{n\in\mathbb{N}}$ converging to φ in $\mathcal{C}^\infty_K(\Omega)$.* ▮

Proof of Theorems 7.20 and 7.21. Given $n \in \mathbb{N}$ and $s \in \mathbb{Z}^d$, denote by $R_{s,n}$ the open cube of side length 2^{1-n} centered at $2^{-n}s$. For fixed n, the set $\{R_{s,n}\}_{s\in\mathbb{Z}^d}$ covers $\mathbb{R}^d$. Let $(\alpha_{s,n})_{s\in\mathbb{Z}^d}$ be a partition of unity subordinate to this covering, as given by Theorem 7.18.

Let $\varphi \in \mathcal{K}^\infty(\Omega)$ and let $\widetilde{\varphi} \in \mathcal{K}^\infty(\mathbb{R}^d)$ be its extension by 0. Define the function $\varphi_n \in \mathcal{K}^\infty(\mathbb{R}^d)$, for every $x \in \mathbb{R}^d$, by

$$\varphi_n(x) = \sum_{s \in \mathbb{Z}^d} \alpha_{s,n}(x)\, a_{s,n}, \tag{7.17}$$

where $a_{s,n} \in \mathbb{Q}$ and satisfies

$$|a_{s,n} - \widetilde{\varphi}(2^{-n}s)| < \frac{1}{n}$$

and $a_{s,n} = 0$ if $\widetilde{\varphi}(2^{-n}s) = 0$.

By the definition of a partition of unity, $\sum_{s\in\mathbb{Z}^d} \alpha_{s,n} = 1$, so

$$\widetilde{\varphi}(x) = \sum_{s\in\mathbb{Z}^d} \alpha_{s,n}(x)\, \widetilde{\varphi}(x).$$

Furthermore, $\alpha_{s,n}(x) = 0$ if $x \notin R_{s,n}$, and $\sum_{s\in\mathbb{Z}^d} |\alpha_{s,n}| = 1$, since $\alpha_{s,n}$ are positive. Therefore,

$$|(\widetilde{\varphi} - \varphi_n)(x)| = \left|\sum_{s\in\mathbb{Z}^d} \alpha_{s,n}(x)\,(\widetilde{\varphi}(x) - a_{s,n})\right| \le \sup_{s\in\mathbb{Z}^d : R_{s,n}\ni x} |(\widetilde{\varphi}(x) - a_{s,n})|.$$

The point x belongs at most to 2^d cubes $R_{s,n}$. On these cubes, for each $i \in [\![1, d]\!]$, we have $2^{-n}(s_i - 1) < x_i < 2^{-n}(s_i + 1)$, and therefore $|x - 2^{-n}s| \le \sqrt{d}\,2^{-n}$. Hence,

$$\begin{aligned}|(\widetilde{\varphi} - \varphi_n)(x)| &\le \sup_{s\in\mathbb{Z}^d} |(\widetilde{\varphi}(2^{-n}s) - a_{s,n})| + \sup_{|y-x|\le\sqrt{d}\,2^{-n}} |\varphi(y) - \varphi(x)| \le \\ &\le \frac{1}{n} + \sup_{|y-x|\le\sqrt{d}\,2^{-n}} |\varphi(y) - \varphi(x)|.\end{aligned}$$

Now, let ρ_k be a regularizing function given by (7.10) on page 158, and $m \in \mathbb{N}$. The inequality (a) from Theorem 7.11 then gives

$$\begin{aligned}\|(\varphi_n - \widetilde{\varphi}) \diamond \rho_k\|_{\mathcal{C}_\mathrm{b}^m(\mathbb{R}^d)} &\le c_{\rho,m}\, k^{m+d}\, \|\varphi_n - \widetilde{\varphi}\|_{\mathcal{C}_\mathrm{b}(\mathbb{R}^d)} \le \\ &\le c_{\rho,m}\, k^{m+d}\left(\frac{1}{n} + \sup_{|y-x|\le\sqrt{d}\,2^{-n}} \widetilde{\varphi}(y) - \widetilde{\varphi}(x)|\right).\end{aligned} \tag{7.18}$$

In addition, the inequality (b) from the same theorem gives

$$\|\widetilde{\varphi} \diamond \rho_k - \widetilde{\varphi}\|_{\mathcal{C}_\mathrm{b}^m(\mathbb{R}^d)} \le \frac{\sqrt{d}}{k}\, \|\widetilde{\varphi}\|_{\mathcal{C}_\mathrm{b}^{m+1}(\mathbb{R}^d)}. \tag{7.19}$$

Choose k sufficiently large that the right-hand side of (7.19) is less than $1/m$. Next, choose n sufficiently large that the right-hand side of (7.18) is less than $1/m$. Then

$$\|\varphi_n \diamond \rho_k - \widetilde{\varphi}\|_{\mathcal{C}_\mathrm{b}^m(\mathbb{R}^d)} \le \frac{2}{m}. \tag{7.20}$$

Denote $Q = \operatorname{supp} \varphi$. The support of φ_n is included in $Q + B(0; \sqrt{d}2^{-n})$, and the support of $\varphi_n \diamond \rho_k$ is therefore included in $Q + B(0; \sqrt{d}2^{-n}) + B(0, 1/k)$. If $2^n > 4\sqrt{d}/r$ and $k > 4/r$, where $r > 0$ is given by the strong inclusion theorem (Theorem A.22) such that $Q + B(0, r) \subset \Omega$, then the support of $\varphi_n \diamond \rho_k$ is included in $K = Q + B(0, r/2)$, which is a compact set included in Ω. Then

$$(\varphi_n \diamond \rho_k)|_\Omega \in \mathcal{C}_K^\infty(\Omega).$$

By Definition 2.15 (b) of the semi-norms of $\mathcal{C}_b^\infty(\Omega)$, (7.20) gives

$$\|(\varphi_n \diamond \rho_k)|_\Omega - \varphi\|_{\mathcal{C}_b^\infty(\Omega);m} = \sup_{0 \le |\beta| \le m} \sup_{x \in \Omega} |(\varphi_n \diamond \rho_k - \varphi)(x)| \le \frac{2}{m}.$$

For every $\epsilon > 0$, choose $m \ge 2/\epsilon$ and then $k = k_m$ and $n = n_m$ such that (7.20) is satisfied and $\phi_m = (\varphi_{n_m} \diamond \rho_{k_m})|_\Omega$ belongs to $\mathcal{C}_K^\infty(\Omega)$. Since the semi-norms $\| \, \|_{\mathcal{C}_b^\infty(\Omega);m}$ increase with m, it follows that, for every $\ell \ge m$,

$$\|\phi_m - \varphi\|_{\mathcal{C}_b^\infty(\Omega);\ell} \le \epsilon.$$

Since the topology of $\mathcal{C}_b^\infty(\Omega)$ coincides with the topology of $\mathcal{K}^\infty(\Omega)$ on $\mathcal{C}_K^\infty(\Omega)$ (Theorem 2.18), this proves that

$$\phi_m \to \varphi \text{ in } \mathcal{C}_K^\infty(\Omega) \text{ and } \mathcal{K}^\infty(\Omega). \tag{7.21}$$

For fixed n, the set of φ_n satisfying (7.17) is countable because it is indexed by $\mathbb{Z}^d \times \mathbb{Q}$ (Theorem A.2 (c) and (d)). The set $\mathcal{E} = \{(\varphi_n \diamond \rho_k)|_\Omega : n \in \mathbb{N},\, k \in \mathbb{N}\}$ is therefore also countable, as is the subset $\mathcal{F}$ of the $(\varphi_n \diamond \rho_k)|_\Omega$ that belong to $\mathcal{K}^\infty(\Omega)$ by Theorem A.2 (a). By Definition A.1 of a countable set, $\mathcal{F}$ can be reordered as a sequence, which is still denoted as $(\phi_k)_{k \in \mathbb{N}}$.

Since $\phi_m \in \mathcal{F}$, the convergence (7.21) completes Theorem 7.21 and shows that $\mathcal{F}$ is sequentially dense in $\mathcal{K}^\infty(\Omega)$. Consequently, $\mathcal{F}$ is dense in $\mathcal{K}^\infty(\Omega)$ because sequential denseness always implies denseness (Theorem A.14). Thus, Theorem 7.20 also is completed. □

Chapter 8

Line Integral of a Vector Field Along a Path

This chapter provides us two essential results to construct primitives:
— The *concentration theorem* (Theorem 8.18) shows that, for any field $q = (q_1, \ldots, q_d)$, the integral $\int_\Omega q \cdot \Psi$ is equal to the integral $\oint_\Gamma q \diamond \rho \cdot d\ell$ around the closed path Γ, where Ψ is a divergence-free *tubular flow* constructed in Theorem 8.18 with support in a tube of axis Γ. Some applications are mentioned in the comment *Utility of the concentration theorem* (p. 186).
— The *theorem on the invariance under homotopy of the line integral of local gradients* (Theorem 8.20) shows that if a field q is of the form $q = \nabla f_B$ on every ball B, then its line integral $\oint_\Gamma q \cdot d\ell$ around a closed path Γ is invariant under homotopy. Some applications are mentioned in the comment *Utility of the invariance theorem...* (p. 187).

We therefore begin by studying the line integral $\oint_\Gamma q \cdot d\ell \stackrel{\text{def}}{=} \int_{t_\text{i}}^{t_\text{e}} (q \circ \Gamma) \cdot \Gamma' \, dt$ (Definition 8.7) of a field $q \in \mathcal{C}(\Omega; E^d)$ along a path $\Gamma \in \mathcal{C}^1([t_\text{i}, t_\text{e}]; \Omega)$. In particular:
— The line integral can be concatenated (Theorem 8.14), i.e. $\oint_\Gamma = \sum_n \oint_{\Gamma_n}$ if $\Gamma = \vec{\cup}_n \Gamma_n$.
— We can reparametrize any concatenation of $\mathcal{C}^1$ paths, or in other words any piecewise $\mathcal{C}^1$ path, as a $\mathcal{C}^1$ path (Theorem 8.4), without changing the line integral (Theorem 8.16).
— The line integral of a gradient around a closed path is zero (Theorem 8.11), i.e. $\oint_\Gamma \nabla f \cdot d\ell = 0$.

8.1. Paths

Let us define paths and closed paths in a separated semi-normed space.

DEFINITION 8.1.– *Let E be a separated semi-normed space and $U \subset E$.*

(a) *A **path** in U is a mapping $\Gamma \in \mathcal{C}([t_\text{i}, t_\text{e}]; U)$, where $[t_\text{i}, t_\text{e}]$ is a closed and bounded interval of $\mathbb{R}$.*

*We say that Γ **joins** the **initial point** $\Gamma(t_\text{i})$ to the **ending point** (or **terminal point**) $\Gamma(t_\text{e})$ in U. The **image** of Γ is the set $[\Gamma] = \{\Gamma(t) : t_\text{i} \leq t \leq t_\text{e}\}$.*

(b) *A **closed path** is a path whose initial point and ending point coincide.*

(c) *We say that a path* **is** C^1, *or* **of class** C^1, *if it is of the form* $\Gamma \in C^1([t_i, t_e]; U)$.

In other words (Definition 2.26), a path is C^1 *if the derivative* Γ', *which is initially only defined on the open set* (t_i, t_e), *has a continuous extension to* $[t_i, t_e]$, *still denoted by* Γ'. ▌

Geometry. The image $[\Gamma]$ of a C^1 path is not necessarily a regular curve or a one-dimensional manifold. It may just be a single point, intersect itself, form angles (between segments, see Theorem 8.4) or cusps, and so on. □

Let us define the "reverse path," where the initial point and ending point are interchanged.

DEFINITION 8.2.– *Let* $\Gamma \in C([t_i, t_e]; E)$ *be a path in a separated semi-normed space* E. *The* **reverse path** *of* Γ *is the path* $\overleftarrow{\Gamma}$ *defined on* $[-t_e, -t_i]$ *by*

$$\overleftarrow{\Gamma}(t) \stackrel{\text{def}}{=} \Gamma(-t). \text{ ▌}$$

Let us concatenate two paths when the ending point of the first is the initial point of the second.

DEFINITION 8.3.– *Let* $\Gamma_1 \in C([t_{i_1}, t_{e_1}]; E)$ *and* $\Gamma_2 \in C([t_{i_2}, t_{e_2}]; E)$ *be two paths in a separated semi-normed space* E *such that*

$$\Gamma_1(t_{e_1}) = \Gamma_2(t_{i_2}).$$

Their **concatenation** *is the path* $\Gamma_1 \overrightarrow{\cup} \Gamma_2$ *defined on* $[t_{i_1}, t_{e_1} + t_{e_2} - t_{i_2}]$ *by:*

$$(\Gamma_1 \overrightarrow{\cup} \Gamma_2)(t) \stackrel{\text{def}}{=} \begin{cases} \Gamma_1(t) & \text{for } t_{i_1} \leq t \leq t_{e_1}, \\ \Gamma_2(t + t_{i_2} - t_{e_1}) & \text{for } t_{e_1} \leq t \leq t_{e_1} + t_{e_2} - t_{i_2}. \end{cases} \text{ ▌}$$

Let us show that we can **reparametrize** any concatenation of C^1 paths to obtain a C^1 path.

THEOREM 8.4.– *Let* $\Gamma = \overrightarrow{\bigcup}_{1 \leq n \leq N} \Gamma_n$ *be the concatenation of finitely many* C^1 *paths in* $\mathbb{R}^d$, *and let* $[t_i, t_e]$ *be its interval of definition.*

Then there exists a bijection $T \in C^1([t_i, t_e])$ *from* $[t_i, t_e]$ *onto itself such that* T' *vanishes at the initial point and at the ending point of each* Γ_n *and is* > 0 *outside of these points. For any such bijection,*

$$\Gamma \circ T \text{ is a } C^1 \text{ path. ▌}$$

Proof. Let $\Lambda \in \mathcal{C}^1([a,b])$ be one of the pieces of Γ and

$$T(t) = a + (b-a)\Big(3\Big(\frac{t-a}{b-a}\Big)^2 - 2\Big(\frac{t-a}{b-a}\Big)^3\Big).$$

Its derivative $T'(t) = 6(t-a)/(b-a) - 6((t-a)/(b-a))^2$ is continuous and > 0 on (a,b), and its extension by 0 is continuous on $[a,b]$.

By Theorem 3.12 (c) on differentiating composite functions, $\Lambda \circ T$ is differentiable and $(\Lambda \circ T)'(t) = (\Lambda' \circ T)(t)\, T'(t)$. This expression tends to 0 as $t \to a$ or $t \to b$, since $(\Lambda' \circ T)(t)$ remains bounded and $T'(t) \to 0$.

By reparametrizing each piece Γ_n of Γ in this way, we obtain a function $\Gamma \circ T$ that is continuous on $[t_\mathrm{i}, t_\mathrm{e}]$, differentiable outside of the points joining different pieces together, and whose derivative tends to 0 at each of these points. By Theorem 2.28 on extending the derivative, it follows that $\Gamma \circ T$ is differentiable at these points and continuously differentiable on $(t_\mathrm{i}, t_\mathrm{e})$. The extension by 0 of its derivative is continuous on the whole of $[t_\mathrm{i}, t_\mathrm{e}]$; in other words, $\Gamma \circ T \in \mathcal{C}^1([t_\mathrm{i}, t_\mathrm{e}]; E)$. □

Let us show that connected open sets are path connected.

THEOREM 8.5.– *Any pair of points of a connected open subset U of separated semi-normed space can be joined by a $\mathcal{C}^1$ path in U.* ▌

Proof. Let E be the space in question, $a \in U$, and X the set of points of U that can be joined to a by a $\mathcal{C}^1$ path in U. It must be proved that $X = U$.

Let us first show that X is open. Let $x \in X$ and $\{\| \; \|_{E;\nu} : \nu \in \mathcal{N}_E\}$ the family of semi-norms of E. By Definition A.7 (b) of an open set, here U, there exists a finite subset N of $\mathcal{N}_E$ and $\epsilon > 0$ such that the ball $B = \{v \in E : \sup_{\nu \in N} \|v - x\|_{E;\nu} \le \epsilon\}$ is included in U. Since x can be joined to a by a path Γ in U of class $\mathcal{C}^1$, every point v of B can be joined to a by the concatenation of Γ and the line segment $[x, v]$. By reparametrizing this concatenation using Theorem 8.4, we obtain a $\mathcal{C}^1$ path that joins v to x. Therefore, $B \subset X$, which shows that X is open in E.

Similarly, its complement $Y = U \setminus X$ is open. Indeed, if we now assume that $x \in Y$, no point of B can be joined to a by a $\mathcal{C}^1$ path in U, otherwise it would also be possible for x, so $B \subset Y$.

The open sets X and Y are disjoint, cover the connected set U and X is non-empty (since it contains a). Therefore, by Definition A.15 of a connected set, $Y = \emptyset$, and $X = U$. □

Path connected. A set is said to be **path connected** if any two of its points can be joined by a path. Theorem 8.5 is slightly stronger, since it gives us $\mathcal{C}^1$ paths. It would also be possible to construct $\mathcal{C}^\infty$ paths, but this is not necessary for our purposes. □

Let us show that, conversely, any two points joined by a path belong to the same connected component.

THEOREM 8.6.– *If two points are connected by a path in a subset U of a separated semi-normed space, then they belong to the same connected component of U.* ▌

Proof. Let E be the space in question and Γ a path in U joining two points a and b. By Definition 8.1, the image $[\Gamma]$ of this path is the image of an interval $[t_\mathrm{i}, t_\mathrm{e}]$ under the continuous mapping Γ. Every interval being connected (Theorem A.16), $[\Gamma]$ is therefore connected (Theorem A.33). Furthermore, it is included in U (by the hypotheses) and contains $\Gamma(t_\mathrm{i})$, that is, a.

By Definition A.15, the connected component generated by a is the largest connected set included in U that contains a. It therefore contains $[\Gamma]$ and certainly also contains $\Gamma(t_\mathrm{e})$, that is, b. □

8.2. Line integral of a field along a path

A **vector field** on a subset of $\mathbb{R}^d$ is any function with d components taking values in a space E, or equivalently any function taking values in the Euclidean product E^d.

Let us define the line integral[1] along a $\mathcal{C}^1$ path of a vector field taking values in a Neumann space.

DEFINITION 8.7.– *Let $q \in \mathcal{C}(\Omega; E^d)$, where $\Omega \subset \mathbb{R}^d$ and E is a Neumann space, and let $\Gamma \in \mathcal{C}^1([t_\mathrm{i}, t_\mathrm{e}]; \Omega)$ be a path in Ω. We denote by Γ' the derivative of Γ.*

---

1 **History of the line integral of a field along a path.** The line integral of a vector field along a path was introduced by Gaspard-Gustave DE CORIOLIS in 1829 [26] to express the work of a force, i.e. the variation of the kinetic energy of a body that moves under the action of this force.
This notion was developed as part of the theory of differential forms established around 1890–1900 by Émile CARTAN [18, Vol. II, pp. 309–396] and Henri POINCARÉ [63, Vol. III, Chapter XXII]; see, for example, [CARTAN, Henri, 19, pp. 215–219] (the son of Émile whom we mentioned above), where the results of sections 8.2 and 8.3 of the present book can be found with the field q "hidden" behind the 0-form ω and the gradient ∇f "hidden" behind the 1-form ω or $\mathrm{d}g$.

French terminology. In French, the line integral is called the *circulation*, a term that English-speakers reserve for the case where the path is closed. The French term *intégrale curviligne*, which is the word-for-word translation of "line integral," is generally reserved for the line integral of a scalar function.

*The **line integral** of q along Γ is the element of E given by*

$$\int_{\Gamma} q \cdot \mathrm{d}\ell \stackrel{\text{def}}{=} \int_{t_i}^{t_e} (q \circ \Gamma) \cdot \Gamma' \, dt. \blacksquare$$

Inconsistent notation! If we want to be consistent with our notation for the Cauchy integral, we need to write either $\int_{t_i}^{t_e} (q \circ \Gamma)(t) \cdot \Gamma'(t)\, \mathrm{d}t$ or $\int_{t_i}^{t_e} (q \circ \Gamma) \cdot \Gamma'$ here. However, we add $\mathrm{d}t$ to the latter anyway to mirror the usage of $\mathrm{d}\ell$. □

Justification of Definition 8.7. For the right-hand side to be defined, by Definition 4.9 of the integral taking values in a Neumann space, the function $(q \circ \Gamma) \cdot \Gamma'$ must be uniformly continuous on (t_i, t_e).

To check this, observe that the composite mapping $q \circ \Gamma$ is continuous (Theorem A.35) on $[t_i, t_e]$, as well as Γ' (extended as in Definition 8.1 (c)). Their product $(q \circ \Gamma) \cdot \Gamma'$ is therefore continuous (Theorem A.35 again) because it is obtained by composing them with the mapping $\cdot$, which is continuous by the inequality (2.2), p. 31. Therefore, by Heine's theorem (Theorem A.34), $(q \circ \Gamma) \cdot \Gamma'$ is uniformly continuous on the compact set $[t_i, t_e]$ and thus on (t_i, t_e).

It is also necessary for $(q \circ \Gamma) \cdot \Gamma'$ to have bounded support. This is the case because $[t_i, t_e]$ is bounded by Definition 8.1 (a) of a path. □

Let us show that the sign of the line integral changes when the path is reversed.

THEOREM 8.8.– *Let $q \in \mathcal{C}(\Omega; E^d)$, where $\Omega \subset \mathbb{R}^d$ and E is a Neumann space, and let Γ be a C^1 path in Ω. Then*

$$\int_{\overleftarrow{\Gamma}} q \cdot \mathrm{d}\ell = -\int_{\Gamma} q \cdot \mathrm{d}\ell. \blacksquare$$

Proof. By Definition 8.2 of the reverse path and Definition 8.7 of the line integral, since $\mathrm{d}(\Gamma(-t))/\mathrm{d}t = -(\mathrm{d}\Gamma/\mathrm{d}t)(-t)$,

$$\int_{\overleftarrow{\Gamma}} q \cdot \mathrm{d}\ell = \int_{-t_e}^{-t_i} q \circ \Gamma(-t) \cdot \frac{\mathrm{d}(\Gamma(-t))}{\mathrm{d}t} \, \mathrm{d}t = -\int_{-t_e}^{-t_i} \left(q \circ \Gamma \cdot \frac{\mathrm{d}\Gamma}{\mathrm{d}t} \right)(-t) \, \mathrm{d}t.$$

The integral being invariant under a symmetry by Theorem 6.18, we obtain

$$\int_{\overleftarrow{\Gamma}} q \cdot \mathrm{d}\ell = -\int_{t_i}^{t_e} \left(q \circ \Gamma \cdot \frac{\mathrm{d}\Gamma}{\mathrm{d}t} \right)(t) \, \mathrm{d}t = -\int_{\Gamma} q \cdot \mathrm{d}\ell. \; \square$$

Let us show that the line integral is invariant under an increasing change of variables.

THEOREM 8.9.– *Let Γ be a $\mathcal{C}^1$ path in a subset Ω of $\mathbb{R}^d$, defined on a bounded interval $[t_{\rm i}, t_{\rm e}]$, and let T be a bijection from a bounded interval $[t'_{\rm i}, t'_{\rm e}]$ onto $[t_{\rm i}, t_{\rm e}]$ such that:*

$$T \in \mathcal{C}^1([t'_{\rm i}, t'_{\rm e}]),\ T' > 0 \text{ on } (t'_{\rm i}, t'_{\rm e}).$$

Then $\Gamma \circ T$ is a $\mathcal{C}^1$ path in Ω and, for every $q \in \mathcal{C}(\Omega; E^d)$, where E is a Neumann space,

$$\int_{\Gamma \circ T} q \cdot \mathrm{d}\ell = \int_{\Gamma} q \cdot \mathrm{d}\ell. \ \blacksquare$$

Proof. Let us first check that $\Gamma \circ T$ is a $\mathcal{C}^1$ path. By Theorem 3.12 (c) on differentiating composite functions, on $(t_{\rm i}, t_{\rm e})$,

$$(\Gamma \circ T)' = (\Gamma' \circ T)\, T'. \qquad (8.1)$$

The right-hand side, and hence the left-hand side, is uniformly continuous because Γ', T and T' are uniformly continuous by the hypotheses, and therefore so are $\Gamma' \circ T$ (Theorem A.35) and its product with T' (Theorem 3.5 (b)). It therefore has (Theorem A.38) a continuous extension on $[t'_{\rm i}, t'_{\rm e}]$. By Definition 8.1 (c) of a $\mathcal{C}^1$ path, this proves that $\Gamma \circ T \in \mathcal{C}^1([t'_{\rm i}, t'_{\rm e}]; \mathbb{R}^d)$.

Let us now prove the invariance of the line integral. Its Definition 8.7 gives, together with (8.1),

$$\int_{\Gamma \circ T} q \cdot \mathrm{d}\ell = \int_{t'_{\rm i}}^{t'_{\rm e}} (q \circ \Gamma \circ T) \cdot (\Gamma \circ T)'\, \mathrm{d}t = \int_{t'_{\rm i}}^{t'_{\rm e}} (((q \circ \Gamma) \cdot \Gamma') \circ T)\, T'\, \mathrm{d}t.$$

Transforming the latter expression with the change of variables formula for an integral from Theorem 6.14 (c), where in this case $|\det[\nabla T]| = T'$ (because $\nabla T = T'$, which is positive by the hypotheses) gives

$$\int_{\Gamma \circ T} q \cdot \mathrm{d}\ell = \int_{t_{\rm i}}^{t_{\rm e}} (q \circ \Gamma) \cdot \Gamma'\, \mathrm{d}t = \int_{\Gamma} q \cdot \mathrm{d}\ell.$$

It remains to be checked that T^{-1} is $\mathcal{C}^1$ because this is assumed by Theorem 6.14 (c). Since $T' > 0$, Theorem A.55 on differentiating the inverse of a function implies that T^{-1} is continuous and differentiable and $(T^{-1})'(t) = 1/(T'(T^{-1}(t)))$. In other words, $(T^{-1})' = \mathcal{Q} \circ T' \circ T^{-1}$, where $\mathcal{Q}(x) = 1/x$. Therefore, $(T^{-1})'$ is continuous, like any composition of continuous mappings (Theorem A.35), since, in addition to T' and T^{-1}, $\mathcal{Q}$ is also continuous (from $(0, \infty)$ into $\mathbb{R}$, by Theorem A.56). □

Independence of the parametrization. By Theorems 8.8 and 8.9, the line integral along a path Γ of class $\mathcal{C}^1$ only depends on its geometry (in other words, its image $[\Gamma]$) and the direction along which it is integrated. □

Line integral along a curve. If the image $[\Gamma]$ of Γ is a rectifiable curve,

$$\oint_{\Gamma} q \cdot \mathrm{d}\ell = \int_{[\Gamma]} q \cdot \tau \, \mathrm{d}\sigma,$$

where $\mathrm{d}\sigma$ is the infinitesimal arc length of $[\Gamma]$ and τ is the oriented unit tangent vector, i.e. $\tau = \Gamma'/|\Gamma'|$ whenever Γ is a $\mathcal{C}^1$ injection such that Γ' does not vanish. □

Let us calculate the line integral along a path consisting of a single point or along a rectilinear path.

THEOREM 8.10.– *Let $q \in \mathcal{C}(\Omega; E^d)$, where $\Omega \subset \mathbb{R}^d$ and E is a Neumann space, and let a and x be two points of Ω such that $[a, x] \subset \Omega$. Then:*

(a) *If $\Gamma_{\{a\}}$ is the path defined on $[0, 1]$ by $\Gamma_{\{a\}}(t) = a$,*

$$\oint_{\Gamma_{\{a\}}} q \cdot \mathrm{d}\ell = 0_E.$$

(b) *If $\Gamma_{\overrightarrow{a,x}}$ is the* ***rectilinear path*** *defined on $[0, 1]$ by $\Gamma_{\overrightarrow{a,x}}(t) = a + t(x - a)$,*

$$\oint_{\Gamma_{\overrightarrow{a,x}}} q \cdot \mathrm{d}\ell = (x - a) \cdot \int_0^1 q(a + t(x - a))\, \mathrm{d}t. \ \blacksquare$$

Proof. Apply Definition 8.7 of the line integral and the equalities $\mathrm{d}a/\mathrm{d}t = 0$ and $\mathrm{d}(a + t(x - a))/\mathrm{d}t = x - a$, respectively. □

Let us calculate the **line integral of a gradient**.

THEOREM 8.11.– *Let $f \in \mathcal{C}^1(\Omega; E)$, where Ω is an open subset of $\mathbb{R}^d$ and E is a Neumann space, and let Γ be a $\mathcal{C}^1$ path in Ω. Then:*

(a) *If a is the initial point of Γ and b is its ending point,*

$$\oint_{\Gamma} \nabla f \cdot \mathrm{d}\ell = f(b) - f(a).$$

(b) *If Γ is a closed path,*

$$\oint_{\Gamma} \nabla f \cdot \mathrm{d}\ell = 0_E. \ \blacksquare$$

Proof. (a) By Theorem 3.12 (a) on changes of variables in a derivative with $\ell = 1$ and $\partial_i = \mathrm{d}/\mathrm{d}t$,

$$(f \circ \Gamma)' = \sum_{j=1}^{d} (\partial_j f \circ \Gamma)\, \Gamma_j' = (\nabla f \circ \Gamma) \boldsymbol{\cdot} \Gamma'.$$

Definition 8.7 of the line integral therefore gives

$$\int_\Gamma \nabla f \boldsymbol{\cdot} \mathrm{d}\ell = \int_{t_\mathrm{i}}^{t_\mathrm{e}} (\nabla f \circ \Gamma) \boldsymbol{\cdot} \Gamma' \, \mathrm{d}t = \int_{t_\mathrm{i}}^{t_\mathrm{e}} (f \circ \Gamma)' \, \mathrm{d}t.$$

With Theorem 6.4 (b) on calculating the integral of a derivative, this gives

$$\int_\Gamma \nabla f \boldsymbol{\cdot} \mathrm{d}\ell = (f \circ \Gamma)(t_\mathrm{e}) - (f \circ \Gamma)(t_\mathrm{i}) = f(b) - f(a).$$

(b) This follows from (a) because the ending point of a closed path coincides with its initial point by Definition 8.1 (b). □

Let us show that the line integral of a vector field depends continuously on the vector field.

THEOREM 8.12.– *Let $\Omega \subset \mathbb{R}^d$, E a Neumann space and Γ a C^1 path in Ω. Then:*

(a) *For every $q \in C(\Omega; E^d)$ and every semi-norm $\| \ \|_{E;\nu}$ of E,*

$$\left\| \int_\Gamma q \boldsymbol{\cdot} \mathrm{d}\ell \right\|_{E;\nu} \le \gamma \, |t_\mathrm{e} - t_\mathrm{i}| \sup_{x \in [\Gamma]} \|q(x)\|_{E^d;\nu},$$

where $[\Gamma] = \{\Gamma(t) : t_\mathrm{i} < t < t_\mathrm{e}\}$ and $\gamma = \sup_{t_\mathrm{i}<t<t_\mathrm{e}} |\Gamma'(t)| < \infty$.

(b) *The mapping $q \mapsto \int_\Gamma q \boldsymbol{\cdot} \mathrm{d}\ell$ is linear and continuous from $C(\Omega; E^d)$ into E.* ▮

Proof. (a) Definition 8.7 of the line integral and the bound on the semi-norms of the integral from Theorem 4.15 give

$$\left\| \int_\Gamma q \boldsymbol{\cdot} \mathrm{d}\ell \right\|_{E;\nu} = \left\| \int_{t_\mathrm{i}}^{t_\mathrm{e}} (q \circ \Gamma) \boldsymbol{\cdot} \Gamma' \, \mathrm{d}t \right\|_{E;\nu} \le$$

$$\le |t_\mathrm{e} - t_\mathrm{i}| \sup_{t_\mathrm{i}<t<t_\mathrm{e}} \|((q \circ \Gamma) \boldsymbol{\cdot} \Gamma')(t)\|_{E;\nu} \le \gamma \, |t_\mathrm{e} - t_\mathrm{i}| \sup_{x \in [\Gamma]} \|q(x)\|_{E^d;\nu},$$

where $\gamma = \sup_{t_\mathrm{i}<t<t_\mathrm{e}} |\Gamma'(t)|$. This quantity is finite since, by Definition 8.1 (c) of a C^1 path, Γ' may be continuously extended to $[t_\mathrm{i}, t_\mathrm{e}]$, and since a continuous function on a compact set is bounded (Theorem A.34).

(b) By Definition 1.3 (a) of the semi-norms of $\mathcal{C}(\Omega; E^d)$, the above inequality can be stated as

$$\left\| \int_\Gamma q \cdot \mathrm{d}\ell \right\|_{E;\nu} \leq c \sup_{x \in [\Gamma]} \|q(x)\|_{E^d;\nu} = c \, \|q\|_{\mathcal{C}(\Omega;E^d);[\Gamma],\nu},$$

which implies the desired continuity by the characterization of continuous linear mappings from Theorem 1.25. □

8.3. Line integral along a concatenation of paths

Let us define the notion of a piecewise $\mathcal{C}^1$ path.

DEFINITION 8.13.– *A **piecewise** $\mathcal{C}^1$ path is a concatenation of finitely many $\mathcal{C}^1$ paths.* ▮

Let us first show that the line integral along a $\mathcal{C}^1$ concatenation of paths is the sum of the line integrals along each piece.

THEOREM 8.14.– *Let $q \in \mathcal{C}(\Omega; E^d)$, where $\Omega \subset \mathbb{R}^d$ and E is a Neumann space, and let Γ, $\Gamma_1, \ldots,$ and Γ_N be $\mathcal{C}^1$ paths in Ω such that*

$$\Gamma = \vec{\bigcup_{1 \leq n \leq N}} \Gamma_n.$$

Then

$$\int_\Gamma q \cdot \mathrm{d}\ell = \sum_{1 \leq n \leq N} \int_{\Gamma_n} q \cdot \mathrm{d}\ell. \text{ ▮}$$

Proof. By Definition 8.7 of the line integral, we have to show that

$$\int_{t_\mathrm{i}}^{t_\mathrm{e}} (q \circ \Gamma) \cdot \Gamma' \, \mathrm{d}t = \sum_{1 \leq n \leq N} \int_{t_{\mathrm{i}_n}}^{t_{\mathrm{e}_n}} (q \circ \Gamma) \cdot \Gamma' \, \mathrm{d}t.$$

It follows from the additivity with respect to the interval of integration (Theorem 6.2) since, by Definition 8.3 of concatenation, $t_\mathrm{i} = t_{\mathrm{i}_1} \ldots < t_{\mathrm{e}_n} = t_{\mathrm{i}_{n+1}} < \ldots t_{\mathrm{e}_N} = t_\mathrm{e}$. □

Let us now extend this property to piecewise $\mathcal{C}^1$ paths that are not necessarily $\mathcal{C}^1$ as a whole by using it as the definition of the line integral along such a path.

DEFINITION 8.15.– *Let $q \in C(\Omega; E^d)$, where $\Omega \subset \mathbb{R}^d$ and E is a Neumann space, and consider a piecewise C^1 path in Ω*

$$\Gamma = \overrightarrow{\bigcup_{1 \le n \le N}} \Gamma_n.$$

The ***line integral*** *of q along Γ is here the element of E defined by*

$$\fint_{\Gamma} q \cdot \mathrm{d}\ell \stackrel{\text{def}}{=} \sum_{1 \le n \le N} \fint_{\Gamma_n} q \cdot \mathrm{d}\ell. \blacksquare$$

Justification. The notation $\fint_\Gamma$ is admissible because, when Γ is C^1, it has the same line integral as in Definition 8.7 by the additivity property of Theorem 8.14.

This definition is still admissible for piecewise C^1 paths because the line integral does not depend on how Γ is partitioned into C^1 pieces, even though there are infinitely many possible partitions. Indeed, it is equal to the line integral of the *minimal* partition, which is unique. More precisely, define $t_1 = t_\mathrm{i}$, then define t_{i+1} inductively as the largest real number such that the restriction Λ_i of Γ to $[t_i, t_{i+1}]$ is a C^1 path. Repeat until $t_{I+1} = t_\mathrm{e}$. Then

$$\Gamma = \overrightarrow{\bigcup_{1 \le i \le I}} \Lambda_i.$$

This partition, called the *minimal* partition, only depends on Γ and not on the original partition into Γ_n. We indeed find the line integral of this minimal partition from the above definition, since, for every $i \in [\![1, I]\!]$, there exists $n_i \in \mathbb{N}$ such that $\Lambda_i = \overrightarrow{\bigcup}_{n_i \le n < n_{i+1}} \Gamma_n$, and therefore, again by the additivity property of Theorem 8.14,

$$\sum_{1 \le n \le N} \fint_{\Gamma_n} q \cdot \mathrm{d}\ell = \sum_{1 \le i \le I} \sum_{n_i \le n < n_{i+1}} \fint_{\Gamma_n} q \cdot \mathrm{d}\ell = \sum_{1 \le i \le I} \fint_{\Lambda_i} q \cdot \mathrm{d}\ell. \square$$

Let us show reparametrizing a piecewise C^1 path as a C^1 path does not change the line integral.

THEOREM 8.16.– *Let $q \in C(\Omega; E^d)$, where $\Omega \subset \mathbb{R}^d$ and E is a Neumann space, and consider a piecewise C^1 path in Ω*

$$\Gamma = \overrightarrow{\bigcup_{1 \le n \le N}} \Gamma_n.$$

Let T be a reparametrization of Γ as a C^1 path, that is, let T be a bijection from the interval on which Γ is defined onto itself as given by Theorem 8.4. Then

$$\fint_{\Gamma \circ T} q \cdot \mathrm{d}\ell = \fint_{\Gamma} q \cdot \mathrm{d}\ell$$

and, for every $n \in [\![1, N]\!]$,

$$\int_{\Gamma_n \circ T} q \cdot d\ell = \int_{\Gamma_n} q \cdot d\ell. \blacksquare$$

Proof. For each piece Γ_n of Γ, Theorem 8.9 on changes of variables in the line integral along a $\mathcal{C}^1$ path gives, since T is $\mathcal{C}^1$ on $[t_{i_n}, t_{e_n}]$ and $T' > 0$ on (t_{i_n}, t_{e_n}),

$$\int_{\Gamma_n \circ T} q \cdot d\ell = \int_{\Gamma_n} q \cdot d\ell.$$

Since $\Gamma \circ T$ is the concatenation of the $\Gamma_n \circ T$, Definition 8.15 of the line integral along a piecewise $\mathcal{C}^1$ path gives

$$\int_{\Gamma \circ T} q \cdot d\ell = \sum_n \int_{\Gamma_n \circ T} q \cdot d\ell = \sum_n \int_{\Gamma_n} q \cdot d\ell = \int_{\Gamma} q \cdot d\ell. \square$$

8.4. Tubular flow and the concentration theorem

Let us construct a divergence-free test field with support in a tubular neighborhood of a path[2].

By definition, the **divergence**[3] of ψ is $\nabla \cdot \psi = \partial_1 \psi_1 + \cdots \partial_d \psi_d$.

THEOREM 8.17.– *Let $\mathcal{T} = [\Gamma] + B$, known as a **tube**, where $\Gamma \in \mathcal{C}^1([t_i, t_e]; \mathbb{R}^d)$ is a closed path in $\mathbb{R}^d$, $[\Gamma] = \{\Gamma(t) : t_i \leq t \leq t_e\}$ is its image and B is a compact subset of $\mathbb{R}^d$. Additionally, let $\rho \in \mathcal{C}^\infty_B(\mathbb{R}^d)$.*

*We define $\Psi \in \mathcal{C}^\infty_{\mathcal{T}}(\mathbb{R}^d; \mathbb{R}^d)$, known as the **tubular flow**, by*

$$\Psi(x) \stackrel{\text{def}}{=} \int_{t_i}^{t_e} \rho(x - \Gamma(t))\, \Gamma'(t)\, dt.$$

It satisfies

$$\nabla \cdot \Psi = 0. \blacksquare$$

2 **History of the construction of a tubular flow.** The divergence-free field Ψ from Theorem 8.17 was obtained by Jacques SIMON in 1993 [70, Lemma, p. 1170] by constructing a concentrated incompressible flow $\vec{\delta}_\Gamma$ and then regularizing it, as explained in the comment *Underlying idea: the concentrated flow* on the next page.
The concentrated incompressible field was also constructed by Stanislav Konstantinovitch SMIRNOV in 1993 [76, p. 842] to conversely decompose any incompressible field ψ into an integral $\psi = \int_\mu \vec{\delta}_{\Gamma_\mu} d\mu$ of concentrated fields.

3 **History of the divergence.** The term *divergence* was introduced by William Kingdon CLIFFORD in 1878 [24].

Terminology. We speak of a **tubular flow** because Ψ is the velocity field of an incompressible flow (meaning that the divergence is zero) with support in the tube $\mathcal{T}$ of axis $[\Gamma]$ (see Figure 8.1). This flow is stationary outside of $\mathcal{T}$ and the flux through any section S of $\mathcal{T}$ with the same orientation as Γ is equal to 1. □

Utility of tubular flows. Constructing such a flow is a key step in our construction of primitives, through the *concentration theorem* (Theorem 8.18), as it is explained in the comment *Utility* ..., p. 186. □

Underlying idea: the concentrated flow. The function Ψ is the regularized function $\vec{\delta}_\Gamma \diamond \rho$ of the distribution $\vec{\delta}_\Gamma \in \mathcal{D}'(\mathbb{R}^d;\mathbb{R}^d)$ of support $[\Gamma]$ defined, for every $\phi \in \mathcal{D}(\mathbb{R}^d;\mathbb{R}^d)$, by

$$\langle \vec{\delta}_\Gamma, \phi \rangle = \oint_\Gamma \phi \cdot \mathrm{d}\ell.$$

This distribution represents a "concentrated" incompressible flow on $[\Gamma]$. If $[\Gamma]$ is a regular curve, then, at each point of Γ, the "concentrated vector" $\vec{\delta}_\Gamma$ is "equal" to the tangent vector with the same orientation as Γ.

This distribution $\vec{\delta}_\Gamma$ is divergence free, namely $\nabla \cdot \vec{\delta}_\Gamma = 0$, because, for every $\varphi \in \mathcal{D}(\mathbb{R}^d)$,

$$\langle \nabla \cdot \vec{\delta}_\Gamma, \varphi \rangle = -\langle \vec{\delta}_\Gamma, \nabla\varphi \rangle = -\oint_\Gamma \nabla\varphi \cdot \mathrm{d}\ell = 0,$$

since the line integral of a gradient around a closed path is always zero (Theorem 8.11 (b)). Therefore, the field $\Psi = \vec{\delta}_\Gamma \diamond \rho$ is also divergence free, since

$$\nabla \cdot \Psi = \nabla \cdot (\vec{\delta}_\Gamma \diamond \rho) = (\nabla \cdot \vec{\delta}_\Gamma) \diamond \rho = 0. \ \square$$

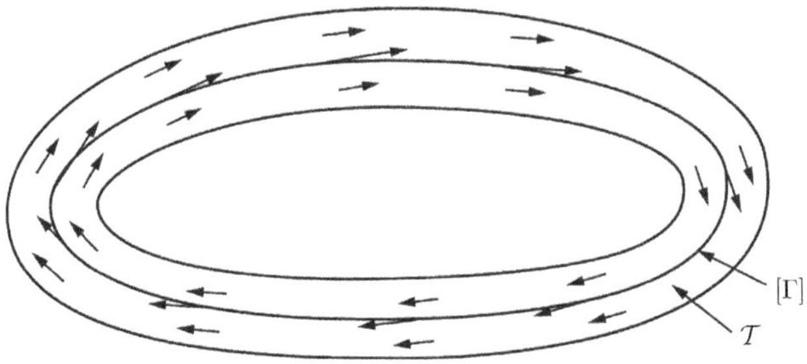

Figure 8.1. *Divergence-free tubular flow in the tube $\mathcal{T}$ of axis $[\Gamma]$*

Proof of Theorem 8.17. Regularity of Ψ. Its definition can be written as

$$\Psi(x) = L(R(x)),$$

where, for every $g \in \mathcal{C}(\mathbb{R}^d)$,

$$L(g) \stackrel{\text{def}}{=} \int_{t_i}^{t_e} g(-\Gamma(t))\,\Gamma'(t)\,\mathrm{d}t,$$

and where $R(x)(y) = \rho(x+y)$, i.e. $R(x) = \tau_{-x}\rho$, where τ_x is a translation.

By the bound on the semi-norms of the integral from Theorem 4.15 and Definition 1.3 (b) of the semi-norms (in this case just the norm) of $\mathcal{C}_b(\mathbb{R}^d)$,

$$|L(g)| \leq |t_e - t_i| \sup_{y \in \mathbb{R}^d} |g(y)| \sup_{t_i \leq t \leq t_e} |\Gamma'(t)| = \gamma \, \|g\|_{\mathcal{C}_b(\mathbb{R}^d)},$$

where γ only depends on Γ. By the characterization of continuous linear mappings from Theorem 1.25, this implies that $L \in \mathcal{L}(\mathcal{C}_b(\mathbb{R}^d); \mathbb{R}^d)$. But $R \in \mathcal{C}^\infty(\mathbb{R}^d; \mathcal{C}_b(\mathbb{R}^d))$ by the differentiability properties of the translation from Theorem 3.18 (d), since $\rho \in \mathcal{K}^\infty(\mathbb{R}^d)$ by the hypotheses. The composite mapping $L \circ R$, namely Ψ, therefore (Theorem 3.2) belongs to $\mathcal{C}^\infty(\mathbb{R}^d; \mathbb{R}^d)$.

Support of Ψ. If $x \notin [\Gamma] + B$, then, for every $t \in [t_i, t_e]$, we have $x - \Gamma(t) \notin B$, so $\rho(x - \Gamma(t)) = 0$, and hence $\Psi(x) = 0$. The support of Ψ is therefore included in the tube $\mathcal{T} = [\Gamma] + B$, which is compact (as a sum of compact subsets of $\mathbb{R}^d$, see Theorem A.24).

Divergence of Ψ. Let $x \in \mathbb{R}^d$. Since each mapping L_i is continuous and linear from $\mathcal{C}_b(\mathbb{R}^d)$ into $\mathbb{R}$, it commutes with the partial derivative ∂_i by Theorem 3.1, so

$$\sum_{i=1}^d \partial_i \Psi_i(x) = \sum_{i=1}^d \partial_i (L_i(R(x))) = \sum_{i=1}^d L_i(\partial_i(R(x))) =$$

$$= \int_{t_i}^{t_e} \sum_{i=1}^d \partial_i \rho(x - \Gamma(t)) \, \Gamma_i'(t) \, \mathrm{d}t = \int_{t_i}^{t_e} \nabla r(\Gamma(t)) \cdot \Gamma'(t) \, \mathrm{d}t,$$

where $r(y) = -\rho(x - y)$. The right-hand side is Definition 8.7 of the line integral of ∇r around the closed path Γ, so, by Theorem 8.11 (b), it is zero. In other words,

$$(\nabla \cdot \Psi)(x) = \oint_\Gamma \nabla r \cdot \mathrm{d}\ell = 0. \; \square$$

Let us show that, for any field q, the integral $\int_\Omega q \cdot \Psi$ is equal to the "concentrated" integral along Γ, $\oint_\Gamma q \diamond \rho \cdot \mathrm{d}\ell$. We call this result the **concentration theorem**[4].

THEOREM 8.18.– *Let* $q \in \mathcal{C}(\Omega; E^d)$, *where* Ω *is an open subset of* $\mathbb{R}^d$ *and* E *is a Neumann space.*

Let $\mathcal{T} = [\Gamma] + B$ *be a tube included in* Ω, *where* Γ *is a closed* $\mathcal{C}^1$ *path in* Ω *and* B *is a compact subset of* $\mathbb{R}^d$, $\rho \in \mathcal{C}_B^\infty(\mathbb{R}^d)$, *and* $\Psi \in \mathcal{C}_\mathcal{T}^\infty(\mathbb{R}^d; \mathbb{R}^d)$ *is the tubular flow given by Theorem 8.17. Then*

$$\int_\Omega q(x) \cdot \Psi(x) \, \mathrm{d}x = \oint_\Gamma q \diamond \rho \cdot \mathrm{d}\ell. \; \blacksquare$$

4 **History of the concentration theorem.** Theorem 8.18 was established for a Banach space E by Jacques SIMON in 1993 [72, p. 207, last equality].

Utility of the concentration theorem. Theorem 8.18 is a *key step* of our proof of the *orthogonality theorem* (Theorem 9.2), namely the construction of a primitive of a field q taking values in a Neumann space that is orthogonal to divergence-free test fields. Indeed, the concentration theorem is used to deduce the condition $\oint_\Gamma q \cdot d\ell = 0_E$ for any closed path Γ from the orthogonality condition $\int_\Omega q \cdot \psi = 0_E$, enabling us to explicitly construct a primitive, (see the equality (9.4), p. 195). □

Proof of Theorem 8.18. By permuting the variables with Theorem 6.5, the definition of Ψ gives

$$\int_\Omega q(x) \cdot \Psi(x)\, dx = \int_\Omega \sum_{i=1}^{d} q_i(x) \Big(\int_{t_i}^{t_e} \rho(x - \Gamma(t))\, \Gamma_i'(t)\, dt \Big)\, dx =$$

$$= \int_{t_i}^{t_e} \sum_{i=1}^{d} \Big(\int_\Omega q_i(x)\, \rho(x - \Gamma(t))\, dx \Big) \Gamma_i'(t)\, dt.$$

In other words, together with the expression of the weighted function from Theorem 7.2 (c) and Definition 8.7 of the line integral,

$$\int_\Omega q(x) \cdot \Psi(x)\, dx = \int_{t_i}^{t_e} \sum_{i=1}^{d} (q_i \diamond \rho)(\Gamma(t))\, \Gamma_i'(t)\, dt =$$

$$= \int_{t_i}^{t_e} (q \diamond \rho)(\Gamma(t)) \cdot \Gamma'(t)\, dt = \oint_\Gamma q \diamond \rho \cdot d\ell. \ \square$$

8.5. Invariance under homotopy of the line integral of a local gradient

Let us define the notion of homotopy.

DEFINITION 8.19.– *Let U be a subset of a separated semi-normed space.*

Two closed paths Γ and Γ_* in U defined on the same interval $[t_i, t_e]$ are ***homotopic*** *in U if we can transform one into the other by means of a continuous deformation. In other words, if there exists $H \in C([t_i, t_e] \times [0, 1]; U)$ such that, for every $t \in [t_i, t_e]$ and $s \in [0, 1]$,*

$$H(t, 0) = \Gamma(t), \quad H(t, 1) = \Gamma_*(t), \quad H(t_i, s) = H(t_e, s).$$

The ***image*** *of H is the set $[H] = \{H(t, s) : t_i \le t \le t_e,\ 0 \le s \le 1\}$.* ▌

Let us show that, if a field is locally a gradient, its line integral around closed paths is invariant under homotopy. We call this result the **theorem on the invariance under homotopy of the line integral of a local gradient**[5].

5 **History of the theorem on the invariance under homotopy of the line integral of a local gradient.** We are not familiar with the origin of Theorem 8.20. It is a classical result of the theory of differential

THEOREM 8.20.– *Let $q \in \mathcal{C}(\Omega; E^d)$, where Ω is an open subset of $\mathbb{R}^d$ and E is a Neumann space such that, for every open ball $B \Subset \Omega$, there exists $f_B \in \mathcal{C}^1(B; E)$ satisfying:*

$$\nabla f_B = q \text{ on } B.$$

Then, if Γ and Γ_* are two closed $\mathcal{C}^1$ paths that are homotopic in Ω,

$$\int_{\Gamma} q \cdot \mathrm{d}\ell = \int_{\Gamma_*} q \cdot \mathrm{d}\ell. \blacksquare$$

Utility of the theorem on the invariance under homotopy of the line integral of a local gradient. Theorem 8.20 is a key step in proving existence results for primitives on a simply connected open set using the *gluing theorem for local primitives* (Theorem 9.4):

— Primitive of a field of $\mathcal{C}^1$ functions satisfying Poincaré's condition (Theorem 9.10) or of a merely continuous field satisfying a weaker version of this condition (Theorem 9.11).

— Stream function of a two-dimensional divergence-free field (Theorem 9.12). □

Proof of Theorem 8.20. Intermediate closed paths. After reparametrizing Γ and Γ_* with Theorem 8.9 if necessary, we may assume that they are defined on $[0,1]$. Let H be a homotopy between Γ and Γ_* in Ω; in other words, let $H \in \mathcal{C}([0,1] \times [0,1]; \Omega)$ such that, for every t and s in $[0,1]$,

$$H(t,0) = \Gamma(t), \quad H(t,1) = \Gamma_*(t), \quad H(0,s) = H(1,s).$$

Define $N+1$ closed paths $\Gamma_n \in \mathcal{C}([0,1]; \Omega)$, where $n \in [\![0, N]\!]$, by

$$\Gamma_n(t) = H\Big(t, \frac{n}{N}\Big).$$

Split each Γ_n into N pieces $\Gamma_n^m \in \mathcal{C}([m/N, (m+1)/N]; \Omega)$, where $m \in [\![0, N-1]\!]$, defined by $\Gamma_n(t) = H(t, n/N)$, so (see Figure 8.2),

$$\Gamma_n = \Gamma_n^0 \,\vec{\cup}\, \Gamma_n^1 \,\vec{\cup} \ldots \vec{\cup}\, \Gamma_n^{N-1}.$$

Finally, define the intermediate points a_n^m, where $n \in [\![0, N]\!]$ and $m \in [\![0, N]\!]$, by

$$a_n^m = H\Big(\frac{m}{N}, \frac{n}{N}\Big)$$

and denote by $T_n^m = \Gamma_{\overrightarrow{a_n^m, a_{n+1}^m}}$ the *transverse* rectilinear path joining a_n^m to a_{n+1}^m.

forms taking values in a Banach space. It can be seen, for example, in [CARTAN, Henri, 19, Theorem 3.7.3, p. 229], where the field q is "hidden" behind the 1-form ω and the existence of f_B such that $\nabla f_B = q$ corresponds to the hypothesis that "ω is closed".

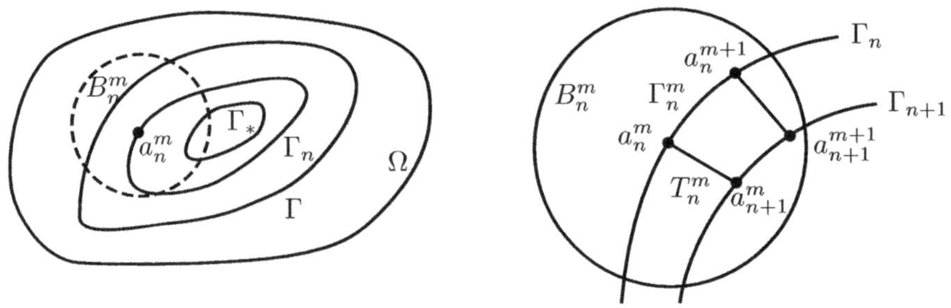

Figure 8.2. *Intermediate closed paths*

The image $[H] = \{H(t, s) : 0 \le t \le 1,\ 0 \le s \le 1\}$ is compact as the image of a compact set, here $[0, 1] \times [0, 1]$, under a continuous mapping (Theorem A.33). Thus, by the strong inclusion theorem (Theorem A.22), there exists $\delta > 0$ such that

$$[H] + B(0, \delta) \subset \Omega.$$

Choose N sufficiently large that $|t - t'| \le 1/N$ and $|s - s'| \le 1/N$ imply that $|H(t, s) - H(t', s')| \le \delta/3$, and let B_n^m be the open ball of center a_n^m and radius $2\delta/3$. Then

the paths Γ_n^m, Γ_{n+1}^m, T_n^m and T_n^{m+1} are included in B_n^m.

Invariance of the line integral along the Γ_n. By the hypotheses, there exists a function $f_n^m \in \mathcal{C}^1(B_n^m; E)$ such that

$$q = \nabla f_n^m \text{ on } B_n^m.$$

The formula for the line integral of a gradient (Theorem 8.11 (a)) gives

$$\int_{\Gamma_{n+1}^m} q \cdot d\ell - \int_{\Gamma_n^m} q \cdot d\ell = \int_{\Gamma_{n+1}^m} \nabla f_n^m \cdot d\ell - \int_{\Gamma_n^m} \nabla f_n^m \cdot d\ell =$$

$$= \left(f_n^m(a_{n+1}^{m+1}) - f_n^m(a_{n+1}^m)\right) - \left(f_n^m(a_n^{m+1}) - f_n^m(a_n^m)\right) =$$

$$= \int_{T_n^{m+1}} q \cdot d\ell - \int_{T_n^m} q \cdot d\ell.$$

Summing over m from 0 to $N - 1$, we obtain

$$\int_{\Gamma_{n+1}} q \cdot d\ell - \int_{\Gamma_n} q \cdot d\ell = \int_{T_n^N} q \cdot d\ell - \int_{T_n^0} q \cdot d\ell.$$

The right-hand side of the equation is zero because the paths T_n^0 and T_n^N coincide. Indeed, T_n^0 joins the initial points a_n^0 and a_{n+1}^0 of Γ_n and Γ_{n+1}, while T_n^N joins their ending points a_n^N and a_{n+1}^N, and these ending points coincide with the initial points, since

$$a_n^0 = H\Big(0, \frac{n}{N}\Big) = H\Big(1, \frac{n}{N}\Big) = a_n^N,$$

and similarly $a_{n+1}^0 = a_{n+1}^N$. Therefore,

$$\oint_{\Gamma_{n+1}} q \cdot \mathrm{d}\ell - \oint_{\Gamma_n} q \cdot \mathrm{d}\ell = 0_E.$$

This holds for each n, so

$$\oint_{\Gamma_N} q \cdot \mathrm{d}\ell = \oint_{\Gamma_0} q \cdot \mathrm{d}\ell.$$

This proves the stated result, since $\Gamma_0 = \Gamma$ and $\Gamma_N = \Gamma_*$. □

Stokes' formula. Theorem 8.20 on the invariance under homotopy is a (non-elementary) variant of Stokes' formula[6]

$$\int_{\partial H} \sigma = \int_H \mathrm{d}\sigma,$$

where σ is an exterior differential k-form and H is a $k+1$-chain. Details may be found, for example, in [BOURBAKI, 15, § 11.3.4, p. 49], when σ takes values in a Banach space E.

Indeed, the hypothesis $\nabla f_B = q$ gives $\partial_i q_j = \partial_i \partial_j f_B = \partial_j \partial_i f_B = \partial_j q$, so the differential 1-form $\sigma = \Sigma_j q_j \, \mathrm{d}x_j$ satisfies

$$\mathrm{d}\sigma = \Sigma_{ij} \partial_i q_j \, \mathrm{d}x_i \wedge \mathrm{d}x_j = \Sigma_{i<j} (\partial_i q_j - \partial_j q_i) \, \mathrm{d}x_i \wedge \mathrm{d}x_j = 0_E.$$

Since a homotopy H between Γ and Γ_* is an oriented 2-chain with boundary $\partial H = \overrightarrow{\Gamma} \cup \overleftarrow{\Gamma}_*$, it follows that

$$\oint_{\Gamma} q \cdot \mathrm{d}\ell - \oint_{\Gamma_*} q \cdot \mathrm{d}\ell = \oint_{\partial H} \sigma = \oint_H \mathrm{d}\sigma = 0_E. \ \square$$

6 **History of Stokes' formula.** We did not find any specific references about the origin of this formula. Attributed to Sir George Gabriel STOKES, it was supposedly discovered by Mikhail Vasilyevitch OSTROGRADSKY around 1820 and then rediscovered by Lord KELVIN. It can also be found associated with names such as Carl Friedrich GAUSS and George GREEN in various forms, among which is the formula of Theorem 10.8.

Chapter 9

Primitives of Continuous Functions

The purpose of this chapter is to determine the conditions under which a continuous field $q = (q_1, \ldots, q_d)$ admits a primitive f, that is, a function f satisfying $\nabla f = q$.

We begin by explicitly constructing a primitive when $\int_\Gamma q \cdot d\ell = 0$ for every closed path Γ in Ω (Theorem 9.1) by integrating q along paths. From this, we deduce that it is sufficient for q to be orthogonal to the divergence-free test fields, in other words, to have $\int_\Omega q \cdot \psi = 0$ for every ψ such that $\nabla \cdot \psi = 0$ (Theorem 9.2), using a tubular flow as a test field and its integral concentration property. This is the *orthogonality theorem*. These conditions are in fact both necessary and sufficient.

We then show that, when Ω is simply connected, it suffices that there exists a primitive on every ball $B \subset \Omega$ (Theorem 9.4), thanks to the theorem on the invariance under homotopy of the line integral. This is the *primitive gluing theorem*. We therefore construct such local primitives:
— When q is C^1 and $\partial_i q_j = \partial_j q_i$ (for every i and j), by integrating q along line segments (Theorem 9.5). This is *Poincaré's theorem*.
— When q is merely continuous and $\int_\Omega q_j \partial_i \varphi = \int_\Omega q_i \partial_j \varphi$ for every test function φ (Theorem 9.7), by regularization. This condition is a weak version of the Poincaré condition.

Thus, when Ω is simply connected, there exists a primitive whenever either the Poincaré condition or its weak version are satisfied (Theorems 9.10 and 9.11). These conditions are necessary and sufficient.

We compare these conditions in Theorem 9.14. Finally, we show that, if the primitive exists, it is unique up to an additive constant on each connected component Ω_m of Ω and that, by fixing its values at a point of each Ω_m, we obtain a continuous mapping $q \mapsto f$ (Theorem 9.18).

9.1. Explicit primitive of a field with line integral zero

Let us explicitly construct a **primitive** q^* of a field $q = (q_1, \ldots, q_d)$, that is, a function q^* satisfying $\nabla q^* = q$, whenever the line integral of q is zero around every closed path[1].

1 **History of the explicit construction of a primitive in Theorem 9.1.** We do not know the origin of this result, which is a classical result of the theory of differential forms taking values in a Banach space (see, for example, [Henri CARTAN, 19, Theorem 3.4.3, p. 220], where q is "hidden" behind the differential form ω).

THEOREM 9.1.– *Let $q \in \mathcal{C}(\Omega; E^d)$, where Ω is an open subset of $\mathbb{R}^d$ and E is a Neumann space, such that, for every closed path Γ in Ω of class $\mathcal{C}^1$,*

$$\oint_\Gamma q \cdot \mathrm{d}\ell = 0_E. \tag{9.1}$$

On each connected component Ω_m of Ω, choose a point a_m. Then:

(a) *For each m and each $x \in \Omega_m$, the element of E defined by*

$$q^*(x) \stackrel{\text{def}}{=} \oint_{\Gamma(a_m,x)} q \cdot \mathrm{d}\ell$$

is independent of the path $\Gamma(a_m, x)$ of class $\mathcal{C}^1$ joining a_m to x in Ω_m (such a path always exists).

(b) *We have $q^* \in \mathcal{C}^1(\Omega; E)$ and*

$$\nabla q^* = q.$$

(c) *If the line segment $[a_m, x]$ is included in Ω,*

$$q^*(x) = (x - a_m) \cdot \int_0^1 q(a_m + t(x - a_m))\,\mathrm{d}t. \blacksquare$$

Optimality of Theorem 9.1 (b). The condition (9.1) is necessary and sufficient for q to have a primitive because, if $q = \nabla q^*$, then $\oint_\Gamma q \cdot \mathrm{d}\ell = 0$, since the line integral of a gradient around a closed path is always zero (Theorem 8.11 (b)). □

Inconsistent notation. Here, we have denoted the primitive by q^*, but elsewhere it is always denoted by f. This is intentional to highlight that q^* is a special, explicit primitive, whereas f is arbitrary. □

Proof of Theorem 9.1. (a) Since each connected component Ω_m of Ω is connected and open (Theorem A.16), each of its points x is joined (Theorem 8.5) to the point a_m by a path $\Gamma(a_m, x)$ in Ω_m, and therefore in Ω, of class $\mathcal{C}^1$.

Let us check that $\oint_{\Gamma(a_m,x)} q \cdot \mathrm{d}\ell$ does not depend on the path Γ joining a_m to x. Let Γ and Γ_* be two such paths. The concatenation $\Gamma \vec{\cup} \overleftarrow{\Gamma}_*$ of Γ and of the reverse path of Γ_* is a piecewise closed $\mathcal{C}^1$ path. By Definition 8.15 of the line integral along a concatenation, and since the sign of the line integral changes along the reverse path by Theorem 8.8,

$$\oint_{\Gamma \vec{\cup} \overleftarrow{\Gamma}_*} q \cdot \mathrm{d}\ell = \oint_\Gamma q \cdot \mathrm{d}\ell + \oint_{\overleftarrow{\Gamma}_*} q \cdot \mathrm{d}\ell = \oint_\Gamma q \cdot \mathrm{d}\ell - \oint_{\Gamma_*} q \cdot \mathrm{d}\ell.$$

Reparametrizing this path using Theorem 8.4, we obtain, by Theorem 8.16, a closed $\mathcal{C}^1$ path with the same line integral, which is zero by the hypotheses. Therefore, we indeed have

$$\int_{\Gamma} q \cdot \mathrm{d}\ell = \int_{\Gamma_*} q \cdot \mathrm{d}\ell.$$

(b) We need to show that q^* is continuously differentiable and $\partial_i q^* = q_i$. Let $x \in \Omega$, $\eta > 0$ such that the ball $\{y \in \mathbb{R}^d : |y - x| \leq \eta\}$ is included in Ω, and let s be a non-zero real number such that $|s| \leq \eta$.

Again by Definition 8.15 and Theorem 8.8, the definition of q^* gives

$$q^*(x + s\mathbf{e}_i) - q^*(x) = \int_{\Gamma(a, x+s\mathbf{e}_i)} q \cdot \mathrm{d}\ell - \int_{\Gamma(a,x)} q \cdot \mathrm{d}\ell = \int_{\Lambda} q \cdot \mathrm{d}\ell,$$

where $\Lambda = \overleftarrow{\Gamma(a, x)} \vec{\cup} \Gamma(a, x + s\mathbf{e}_i)$. Since the path Λ joins x to $x + s\mathbf{e}_i$ and the line integral is independent of the path joining these two points by (a), this equation holds when Λ is chosen to be the rectilinear path $\Gamma_{\overrightarrow{x, x+s\mathbf{e}_i}}$. The formula for the line integral along such a path from Theorem 8.10 (b) gives

$$q^*(x + s\mathbf{e}_i) - q^*(x) = s\mathbf{e}_i \cdot \int_0^1 q(x + ts\mathbf{e}_i)\,\mathrm{d}t = s \int_0^1 q_i(x + ts\mathbf{e}_i)\,\mathrm{d}t.$$

Therefore,

$$q^*(x + s\mathbf{e}_i) - q^*(x) - s q_i(x) = s \int_0^1 (q_i(x + ts\mathbf{e}_i) - q_i(x))\,\mathrm{d}t.$$

For every semi-norm $\| \, \|_{E;\nu}$ of E, the bound on the semi-norms of the Cauchy integral from Theorem 4.15 gives

$$\|q^*(x + s\mathbf{e}_i) - q^*(x) - s q_i(x)\|_{E;\nu} \leq |s| \sup_{0 \leq r \leq s} \|q_i(x + r\mathbf{e}_i) - q_i(x)\|_{E;\nu}.$$

For every $\epsilon > 0$, we can choose η such that the right-hand side is $\leq \epsilon\,|s|$ because q_i is continuous, so by the characterization of the partial derivatives (2.7) from Definition 2.8,

$$\partial_i q^*(x) = q_i(x).$$

Since its partial derivatives are continuous, q^* indeed belongs to $\mathcal{C}^1(\Omega; E)$ by Theorem 2.10.

(c) This is again the formula for the line integral along a rectilinear path from Theorem 8.10 (b). □

Connected components of an open subset of $\mathbb{R}^d$. Let us observe that the number of points a_m that must be chosen in Theorem 9.1 is countable (and possibly finite) since:

Every open subset U of $\mathbb{R}^d$ has, at most, countably many connected components. (9.2)

Proof. By Theorem A.16, the connected components of U are pairwise disjoint and each of them is open and therefore contains a point of $\mathbb{Q}^d$. Therefore, the set of them is countable as the image of a countable set (Theorem A.2 (b)), here a subset of $\mathbb{Q}^d$ (which is countable by Theorem A.2 (a), (d) and (c)). □

9.2. Primitive of a field orthogonal to the divergence-free test fields

Let us show that a field $q = (q_1, \ldots, q_d)$ has a primitive f whenever it is "orthogonal" to every divergence-free **test field** $\psi = (\psi_1, \ldots, \psi_d)$. This is the **orthogonality theorem**[2].

THEOREM 9.2.– *Let $q \in \mathcal{C}(\Omega; E^d)$, where Ω is an open subset of $\mathbb{R}^d$ and E is a Neumann space, such that:*

$$\int_\Omega q \cdot \psi = 0_E, \quad \forall \psi \in \mathcal{K}^\infty(\Omega; \mathbb{R}^d) \text{ such that } \nabla \cdot \psi = 0. \tag{9.3}$$

Then there exists $f \in \mathcal{C}^1(\Omega; E)$ such that

$$\nabla f = q. \blacksquare$$

Optimality of Theorem 9.2. The condition (9.3) is necessary and sufficient for q to have a primitive, since, if $q = \nabla f$, then $\nabla \cdot \psi = 0$ implies

$$\int_\Omega q \cdot \psi = \int_\Omega \sum_{i=1}^d \partial_i f \, \psi_i = -\int_\Omega f \sum_{i=1}^d \partial_i \psi_i = -\int_\Omega f \, \nabla \cdot \psi = 0_E. \ \square$$

2 **History of the existence of primitives for fields that are orthogonal to the divergence-free test fields. Real values.** Theorem 9.2 is a special case of the *orthogonality theorem for distributions* given in Vol. 3, which follows from the *cohomology theorem* of Georges DE RHAM in the case $E = \mathbb{R}$. The latter showed, in 1955, [28, Theorem 17', p. 114] that a *current* T is homologous to 0 if and only if $T(\psi) = 0$ for every form ψ that is $\mathcal{C}^\infty$, closed, and has compact support (currents generalize differential forms on a manifold in the same way that distributions generalize functions; for a differential form, this result means that *every closed differential form is exact*).
Jacques-Louis LIONS observed, in 1969, [56, p. 69] that the orthogonality theorem for real distributions, and therefore for continuous functions, follows from the result of DE RHAM by considering the current $T = q_1 \mathrm{d}x_1 + \cdots + q_n \mathrm{d}x_n$ (a good explanation of the transition from differential forms to primitives is given for functions in [RUDIN, 66, § 10.42 and 10.43, pp. 262–264]).
In light of the importance of this result when solving the Navier–Stokes equations, various more direct and elementary proofs have been given for specific examples of distributions or real functions: by Olga LADYZHENSKAYA in 1963 [50, Theorem 1, p. 28] for $q \in (L^2(\Omega))^d$; by Luc TARTAR in 1978 [78] for $q \in (H^{-1}(\Omega))^d$; and by Jacques SIMON in 1993 [70] for every $q \in (\mathcal{D}'(\Omega))^d$.

Vector values. Jacques SIMON proved Theorem 9.2 for a Banach space E in 1993 [71, Theorem 5 (i), p. 4]. Here, we use the same method, which is based on the concentration theorem (Theorem 8.18). (The proof given by Georges DE RHAM [28] does not seem to extend to this case, since it uses the reflexivity properties of spaces of currents.)

Orthogonality. Generalizing the notion of orthogonality with respect to a scalar product, we can say that a field q satisfying (9.3) is *orthogonal to* $\mathcal{K}^\infty_{\rm div}(\Omega;\mathbb{R}^d) = \{\psi \in \mathcal{K}^\infty(\Omega;\mathbb{R}^d) : \nabla \cdot \psi = 0\}$ with respect to the bilinear mapping $(q,\psi) \mapsto \int_\Omega q \cdot \psi$ from $\mathcal{C}(\Omega;E^d) \times \mathcal{K}^\infty(\Omega;\mathbb{R}^d)$ into E.

We can say that the space $\mathcal{C}_\nabla(\Omega;E^d)$ of fields that are gradients is *the orthogonal complement of the space* $\mathcal{K}^\infty_{\rm div}(\Omega;\mathbb{R}^d)$ since the condition (9.3) is equivalent to $q \in \mathcal{C}_\nabla(\Omega;E^d)$. This can be denoted as

$$\mathcal{C}_\nabla(\Omega;E^d) = (\mathcal{K}^\infty_{\rm div}(\Omega;\mathbb{R}^d))^\perp. \ \square$$

Proof of Theorem 9.2. Let Γ be a closed path $\mathcal{C}^1$ in Ω. Since its image $[\Gamma]$ is compact, by the separation theorem (Theorem A.22), there exists $r > 0$ such that the tube $\mathcal{T} = [\Gamma] + B(0,r)$ is included in Ω. Let n_Γ be an integer $\geq 1/r$.

For every $n \geq n_\Gamma$, let $\Psi_n \in \mathcal{C}^\infty_{\mathcal{T}_n}(\mathbb{R}^d;\mathbb{R}^d)$, where $\mathcal{T}_n = [\Gamma] + B(0,1/n)$, be the tubular flow given by Theorem 8.17, related to the regularizing function ρ_n given by Definition 7.7 (a). It satisfies $\nabla \cdot \Psi_n = 0$ and its restriction belongs to $\mathcal{K}^\infty(\Omega;\mathbb{R}^d)$ by Theorem 2.16 (c), so the hypothesis (9.3) gives

$$\int_\Omega q \cdot \Psi_n = 0_E.$$

The concentration theorem (Theorem 8.18) then gives

$$\oint_\Gamma q \diamond \rho_n \cdot {\rm d}\ell = \int_\Omega q \cdot \Psi_n = 0_E. \tag{9.4}$$

Now, $[\Gamma] \subset \Omega_{B(0,r)} \subset \Omega_{B(0,1/n_\Gamma)}$ by Theorem 7.3 and therefore $q \diamond \rho_n \to q$ in $\mathcal{C}(\Omega_{B(0,1/n_\Gamma)};E)$ by Theorem 7.9 (a). Therefore,

$$\oint_\Gamma q \diamond \rho_n \cdot {\rm d}\ell \to \oint_\Gamma q \cdot {\rm d}\ell,$$

since the line integral depends continuously on q (Theorem 8.12 (b)) and hence sequentially continuously on q (Theorem A.29). In the limit,

$$\oint_\Gamma q \cdot {\rm d}\ell = 0_E.$$

This implies the existence of f such that $\nabla f = q$ by Theorem 9.1. $\square$

9.3. Gluing of local primitives on a simply connected open set

Let us define the notion of simply connected set.

DEFINITION 9.3.– *A subset U of a separated semi-normed space is said to be* ***simply connected*** *if every closed path of U is homotopic in U to a closed path consisting of a single point.* ▌

Simple connectedness versus connectedness. The term "simply connected" is unfortunately not defined in the same way by all authors. For some, it requires connectedness, which is not the case here.

To require connectedness, it suffices to replace the condition "every closed path is homotopic to **a** point" with "every closed path is homotopic to **every** point of U." □

Simple connectedness in $\mathbb{R}^d$ versus the presence of "holes":
— The space $\mathbb{R}^d$ is simply connected (this can be verified by choosing $T(t,s) = (1-s)\,\Gamma(t)$).
— In $\mathbb{R}$, every open set is simply connected, even if it has holes.
— In $\mathbb{R}^2$, an open set is simply connected if and only if it does not have holes. For instance, the *crown* $\{x \in \mathbb{R}^2 : 1 < |x| < 2\}$ is connected but not simply connected.
— In $\mathbb{R}^d$, $d \geq 3$, simply connected open sets can have holes. For instance, the set $\{x \in \mathbb{R}^3 : 1 < |x| < 2\}$ is both connected and simply connected. □

Simple connectedness of star-shaped sets. In a separated semi-normed space:

$$\text{Every star-shaped set is connected and simply connected.} \tag{9.5}$$

Proof. Let U be a set that is star shaped with respect to a point a, that is for every $z \in U$, the line segment $[a, z]$ is included in U.

It is simply connected because every closed path Γ is homotopic in U to the path consisting of the single point $\{a\}$ via the homotopy $H(t,s) = sa + (1-s)\Gamma(t)$.

It is connected because if it was covered by two disjoint non-empty open sets $\mathcal{O}_1$ and $\mathcal{O}_2$, then a would belong to one of them, say $\mathcal{O}_1$, and $\mathcal{O}_2$ would contain a point z of U. The sets $\mathcal{U}_i = \{s \in \mathbb{R} : a + s(z-a) \in \mathcal{O}_i\}$ would then form an open covering of the interval $[0,1]$, which contradicts its connectedness (Theorem A.16). □

Let us show that, on a simply connected open set, if a field is locally a gradient, then it is a gradient. In other words, if the field has local primitives, then it has a global primitive. We call this the **local primitive gluing theorem**[3].

THEOREM 9.4.– *Let $q \in \mathcal{C}(\Omega; E^d)$, where E is a Neumann space and*

$$\Omega \text{ is a simply connected open subset of } \mathbb{R}^d,$$

such that, for every open ball $B \Subset \Omega$, there exists $f_B \in \mathcal{C}^1(B; E)$ such that

$$\nabla f_B = q \text{ on } B.$$

Then there exists $f \in \mathcal{C}^1(\Omega; E)$ such that

$$\nabla f = q. \ \blacksquare$$

3 **History of the local primitive gluing theorem.** We are not familiar with the origin of Theorem 9.4, which is a classical result from the theory of differential forms taking values in a Banach space. It can be seen, for example, in [Henri CARTAN, 19, Theorem 3.8.1, p. 230], where q is "hidden" behind the *closed differential* 1*-form* ω, where *closed* means that it is *locally a gradient* [19, Definition, p. 222].

Proof. Let Γ be a closed path in Ω of class C^1. By Definition 9.3 of a simply connected open set, Γ is homotopic in Ω to a closed path Γ_* consisting of a single point. Since q is locally a gradient by the hypotheses, its line integral around a closed path is invariant under homotopy by Theorem 8.20, so

$$\oint_\Gamma q \cdot d\ell = \oint_{\Gamma_*} q \cdot d\ell.$$

Since the line integral around a closed path consisting of a single point is zero (Theorem 8.10 (a)),

$$\oint_\Gamma q \cdot d\ell = 0_E.$$

This holds for every Γ, which guarantees the existence of a primitive f by Theorem 9.1. □

Optimality of Theorem 9.4. The existence of local primitives is clearly a necessary condition for the existence of a global primitive. This condition is therefore necessary and sufficient.

On an arbitrary open set, the result no longer holds in general. An example of a function with local primitives but no global primitive is given in Theorem 9.15, in two dimensions, and in Theorem 9.16, in arbitrary dimensions. □

Caution. Theorem 9.4 is a *gluing theorem for* **certain** *local primitives*, but **not all** local primitives. Indeed, *it may be impossible to glue together the given local primitives* f_B because they may differ by a constant. However, when Ω is simply connected, we can add a constant to each of them to glue them together. □

9.4. Explicit primitive on a star-shaped set: Poincaré's theorem

A subset U of a vector space is said to be **star shaped with respect to the point** a if, for every $u \in U$, it contains the line segment $[a, u] = \{a + t(u - a) : 0 \le t \le 1\}$. A set is said to be **star shaped** if it is star shaped with respect to one of its points.

Let us explicitly construct a primitive of a continuously differentiable vector field $q = (q_1, \dots, q_d)$ such that $\partial_i q_j = \partial_j q_i$ for every i and j on a star-shaped open set, a result that is weaker (see Theorem 9.14 (e)) than the conditions (9.1) and (9.3), considered earlier for an arbitrary open set. This is called **Poincaré's theorem**[4].

THEOREM 9.5.– *Let $q \in C^1(\Omega; E^d)$, where E is a Neumann space and*

Ω is an open subset of $\mathbb{R}^d$ that is star shaped with respect to a point a,

4 **History of Poincaré's theorem.** Theorem 9.5 was established by Henri POINCARÉ in 1899 [63, p. 10] in the real-valued case.

such that, for every i and j in $[\![1, d]\!]$,

$$\partial_i q_j = \partial_j q_i.$$

Then the function defined, for every $x \in \Omega$, by

$$q^*(x) \overset{\text{def}}{=} (x - a) \cdot \int_0^1 q(a + t(x - a))\,\mathrm{d}t$$

satisfies $q^* \in \mathcal{C}^2(\Omega; E)$ and

$$\nabla q^* = q. \blacksquare$$

Proof. For every $x \in \Omega$ and $j \in [\![1, d]\!]$, we denote

$$Q_j(x) = \int_0^1 q_j(a + t(x - a))\,\mathrm{d}t.$$

Suppose for now that differentiation under the integral sign is admissible, so that $Q_j \in \mathcal{C}^1(\Omega; E)$ and

$$\partial_i Q_j(x) = \int_0^1 t\,\partial_i q_j(a + t(x - a))\,\mathrm{d}t. \tag{9.6}$$

Then $q^*(x) = \sum_{j=1}^d (x_j - a_j) Q_j(x)$, so Theorem 3.6 and the Leibniz rule give $q^* \in \mathcal{C}^1(\Omega; E)$ and

$$\begin{aligned}\partial_i q^*(x) &= Q_i(x) + \sum_{j=1}^d (x_j - a_j)\partial_i Q_j(x) = \\ &= \int_0^1 q_i(a + t(x - a)) + t \sum_{j=1}^d (x - a)_j \partial_i q_j(a + t(x - a))\,\mathrm{d}t.\end{aligned}$$

Now, apply the hypothesis $\partial_i q_j = \partial_j q_i$ and observe that, by Theorem 3.12 (a) on changes of variables in a derivative with $T(t) = a + t(x - a)$,

$$\frac{\mathrm{d}}{\mathrm{d}t}\big(q_i(a + t(x - a))\big) = \sum_{j=1}^d \partial_j q_i(a + t(x - a))\,(x - a)_j,$$

since $\mathrm{d}T_j/\mathrm{d}t(t) = (x - a)_j$. We therefore obtain

$$\partial_i q^*(x) = \int_0^1 q_i(a + t(x - a)) + t\,\frac{\mathrm{d}}{\mathrm{d}t}\big(q_i(a + t(x - a))\big)\,\mathrm{d}t.$$

By the Leibniz rule once again, together with the expression of the integral of a derivative from Theorem 6.4 (a), we finally obtain

$$\partial_i q^*(x) = \int_0^1 \frac{\mathrm{d}}{\mathrm{d}t}\big(t\, q_i(a + t(x-a))\big)\, \mathrm{d}t = q_i(x).$$

This proves that

$$\nabla q^* = q.$$

Hence, $\partial_j \partial_i q^* = \partial_j q_i \in \mathcal{C}(\Omega; E)$, and so $q^* \in \mathcal{C}^2(\Omega; E)$.

We still need to check (9.6). Let B be an open ball such that $\overline{B} \subset \Omega$. Theorem 4.27 on differentiating under the integral sign on $B \times (0,1)$ with $f(x,t) = q_j(a+t(x-a))$ and $g_i(x,t) = t\, \partial_i q_j(a + t(x-a))$ then gives (9.6) on each B and therefore on the whole of Ω. Indeed, the hypotheses of this theorem are satisfied:
— The functions f and g_i are uniformly continuous and bounded because they are uniformly continuous and bounded on the compact set $\overline{B} \times [0,1]$ by Heine's theorem (Theorem A.34), since they are continuous on this set.
— For any fixed t, the mapping $x \mapsto f(x,t)$ is differentiable and $\partial_i f(x,t) = g_i(x,t)$. This is an elementary fact, which completes the proof of (9.6), and hence the proof of Theorem 9.5. □

9.5. Explicit primitive under the weak Poincaré condition

Before we consider the weak Poincaré condition, let us observe that every open set Ω that is star shaped, with respect to a point a, is the union of the subsets $\Omega^{*a}_{1/n}$ of the $\Omega_{1/n}$ that are star shaped with respect to a.

THEOREM 9.6.– *Let Ω be an open subset of $\mathbb{R}^d$ that is star shaped with respect to a point a and, for every $n \in \mathbb{N}$, let*

$$\Omega^{*a}_{1/n} \stackrel{\text{def}}{=} \{x \in \Omega : [a,x] \subset \Omega_{1/n}\},$$

where $[a,x] = \{a + t(x-a) : 0 \leq t \leq 1\}$ and $\Omega_{1/n} = \{x \in \Omega : B(x, 1/n) \subset \Omega\}$.

Then $\Omega^{*a}_{1/n}$ is a star shaped open set and

$$\Omega = \bigcup_{n \in \mathbb{N}} \Omega^{*a}_{1/n}. \ \blacksquare$$

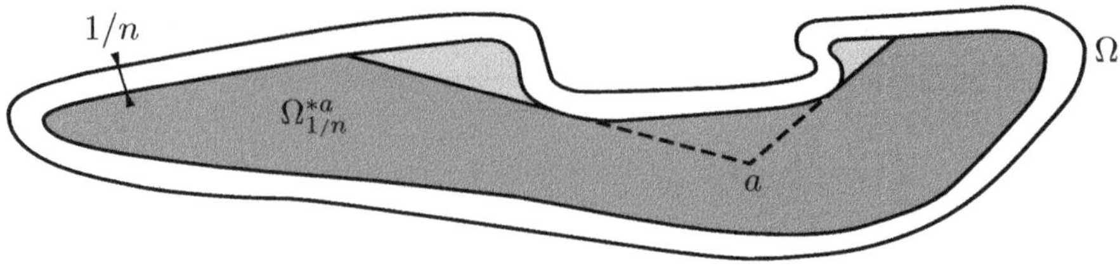

Figure 9.1. *Subset $\Omega^{*a}_{1/n}$ of $\Omega_{1/n}$ that is star shaped with respect to a. $\Omega^{*a}_{1/n}$ is dark gray and $\Omega_{1/n}$ is the union of the light gray and dark gray regions*

Proof of Theorem 9.6. The set $\Omega^{*a}_{1/n}$ is star shaped because, if $x \in \Omega^{*a}_{1/n}$, then, for all $y \in [a, x]$, we have $[a, y] \subset [a, x] \subset \Omega_{1/n}$, so $y \in \Omega^{*a}_{1/n}$ and hence $[a, x] \subset \Omega^{*a}_{1/n}$.

Let us show that this set is open. Let $x \in \Omega^{*a}_{1/n}$. Then $[a, x]$ is a compact set included in $\Omega_{1/n}$, which is open (Theorem 7.2 (a)), so the strong inclusion theorem (Theorem A.22) gives $r > 0$ such that $[a, x] + B(0, r) \subset \Omega_{1/n}$. If $y \in B(x, r)$,

$$[a, y] = \{a + t(x - a) + t(y - x) : 0 \leq t \leq 1\} \subset [a, x] + B(0, r) \subset \Omega_{1/n},$$

since $|t(y - x)| \leq tr \leq r$, so $y \in \Omega^{*a}_{1/n}$. This proves that $\Omega^{*a}_{1/n}$ is open.

Finally, Ω is the union of the $\Omega^{*a}_{1/n}$ because, if $x \in \Omega$, then $[a, x] \subset \Omega$ so, again by Theorem A.22, there exists $r > 0$ such that $[a, x] + B(0, r) \subset \Omega$; then $[a, x] \subset \Omega_{1/n}$, that is, $x \in \Omega^{*a}_{1/n}$, whenever $n \geq 1/r$. □

If q is merely continuous, Poincaré's condition $\partial_i q_j = \partial_j q_i$ is no longer meaningful in the "classical" sense, but we can state a "weak" formulation that still guarantees the existence of an explicit primitive on a star-shaped open set as follows:

THEOREM 9.7.– *Let $q \in \mathcal{C}(\Omega; E^d)$, where E is a Neumann space and*

Ω is an open subset of $\mathbb{R}^d$ that is star shaped with respect to a point a,

such that, for every i and j in $[\![1, d]\!]$ and every $\varphi \in \mathcal{K}^\infty(\Omega)$,

$$\int_\Omega q_j \, \partial_i \varphi = \int_\Omega q_i \, \partial_j \varphi. \tag{9.7}$$

Then the function defined by

$$q^*(x) \stackrel{\text{def}}{=} (x - a) \cdot \int_0^1 q(a + t(x - a)) \, dt$$

satisfies $q^* \in \mathcal{C}^1(\Omega; E)$ and

$$\nabla q^* = q. \blacksquare$$

Weak Poincaré condition. We name *weak Poincaré condition* the equality (9.7) because, when $q \in \mathcal{C}^1(\Omega; E^d)$, it is equivalent to Poincaré's condition $\partial_i q_j = \partial_j q_i$ as we will see in Theorem 9.9. □

Proof of Theorem 9.7. We will proceed in four steps.

1. *Regularization.* Let $q \diamond \rho_n \in \mathcal{C}^\infty(\Omega_{1/n}; E^d)$ be a regularized function of q given by Definition 7.7, where the support of ρ_n is included in the ball $B(0, 1/n)$ and $\Omega_{1/n} = \Omega_{B(0,1/n)}$, namely the set $\{x \in \mathbb{R}^d : B(x, 1/n) \subset \Omega\}$. Let us show that, for every i and j,

$$\partial_i(q_j \diamond \rho_n) - \partial_j(q_i \diamond \rho_n) = 0_E \text{ on } \Omega_{1/n}. \tag{9.8}$$

The expression of the derivative of a weighting from Theorem 7.4 (b) and the second expression of the weighting itself from Theorem 7.2 (c) give

$$\partial_i(q \diamond \rho_n)(x) = -(q \diamond \partial_i \rho_n)(x) = -\int_\Omega q(y)\, \partial_i \rho_n(y - x)\, dy.$$

Hence,

$$\partial_i(q_j \diamond \rho_n)(x) - \partial_j(q_i \diamond \rho_n)(x) = \int_\Omega q_i(y)\, \partial_j \rho_n(y - x) - q_j(y)\, \partial_i \rho_n(y - x)\, dy.$$

The right-hand side is zero by the weak Poincaré condition (9.7) applied to φ defined by $\varphi(y) = \rho_n(y - x)$, which establishes (9.8).

2. *Primitive on star shaped subsets.* Let $\Omega^{*a}_{1/n}$ be the subset of $\Omega_{1/n}$ that is star shaped with respect to a. This set is open and star shaped (Theorem 9.6). By Poincaré's theorem (Theorem 9.5), the property (9.8) implies that the function defined on $\Omega^{*a}_{1/n}$ by

$$q^*_n(x) = (x - a) \cdot \int_0^1 (q \diamond \rho_n)(a + t(x - a))\, dt \tag{9.9}$$

is a primitive of $q \diamond \rho_n$. In other words, for every i,

$$\partial_i q^*_n = q_i \diamond \rho_n \text{ on } \Omega^{*a}_{1/n}.$$

3. *Convergence.* Let $k \in \mathbb{N}$ and $n \geq k$. Then $\Omega^{*a}_{1/k} \subset \Omega^{*a}_{1/n} \subset \Omega_{1/n}$ and the regularizing property from Theorem 7.10 (a) gives, as $n \to \infty$,

$$q_i \diamond \rho_n \to q_i \text{ on } \mathcal{C}(\Omega^{*a}_{1/k}; E).$$

That is,

$$\partial_i q^*_n \to q_i \text{ on } \mathcal{C}(\Omega^{*a}_{1/k}; E).$$

Moreover the expression (9.9) of q_n^* implies, as we will verify in Lemma 9.8,

$$q_n^* \to q^* \text{ on } \mathcal{C}(\Omega_{1/k}^{*a}; E). \tag{9.10}$$

The completeness property of $\mathcal{C}^1(\Omega_{1/k}^{*a}; E)$ from Theorem 2.23 then shows that $q^* \in \mathcal{C}^1(\Omega_{1/k}^{*a}; E)$ and

$$\partial_i q^* = q_i \text{ on } \Omega_{1/k}^{*a}.$$

4. *Gluing.* Since the $(\Omega_{1/k}^{*a})_{k\geq 1}$ cover Ω (Theorem 9.6), it follows that q^* belongs to $\mathcal{C}^1(\Omega; E)$ and $\nabla q^* = q$ on the whole of Ω. □

We still need to establish the convergence (9.10), which corresponds to the following property.

LEMMA 9.8.– *Let $q \in \mathcal{C}(\Omega; E^d)$, where Ω is a star-shaped open subset of $\mathbb{R}^d$ with respect to a point a and E is a Neumann space, and for every $x \in \Omega$, let*

$$q^*(x) \stackrel{\text{def}}{=} (x-a) \cdot \int_0^1 q(a + t(x-a))\,\mathrm{d}t.$$

Then the mapping $q \mapsto q^*$ is linear and continuous, and therefore sequentially continuous, from $\mathcal{C}(\Omega; E^d)$ into $\mathcal{C}(\Omega; E)$. ▮

Proof. Let $\{\| \,\|_{E;\nu} : \nu \in \mathcal{N}_E\}$ be the family of semi-norms of E. Definition 1.3 (a) of the semi-norms of $\mathcal{C}(\Omega; E)$, together with the inequality (2.2), p. 31, and the bound on the semi-norms of the integral from Theorem 4.17 (b), implies that, for every compact set $K \subset \Omega$ and every $\nu \in \mathcal{N}_E$,

$$\|q^*\|_{\mathcal{C}(\Omega;E);K,\nu} = \sup_{x\in K} \|q^*(x)\|_{E;\nu} \leq c \sup_{x\in D} \|q(x)\|_{E^d;\nu} = c\,\|q\|_{\mathcal{C}(\Omega;E^d);D,\nu},$$

where $c = \sup_{x\in K} |x-a|$ and $D = \{a + t(x-a) : x \in K,\ 0 \leq t \leq 1\}$. The final term is well defined (Definition 1.3 (a)) because D is compact in $\mathbb{R}^d$ (since it is closed and bounded) and included in Ω (since this set is star shaped).

By the characterization of continuous linear mappings from Theorem 1.25, this shows that the mapping $q \mapsto q^*$ is continuous. It is therefore sequentially continuous, like any continuous mapping (Theorem A.29). □

Continuity taking values in $\mathcal{C}^1(\Omega; E)$. It follows from Lemma 9.8 and Theorem 9.7 that, if Ω is star shaped, the mapping $q \mapsto q^*$ is continuous from $\mathcal{C}(\Omega; E^d)$ into $\mathcal{C}^1(\Omega; E)$. Indeed, $q \mapsto \partial_i q^*$ is also continuous from $\mathcal{C}(\Omega; E^d)$ into $\mathcal{C}(\Omega; E)$ because $\partial_i q^* = q_i$.

This property is generalized to an arbitrary open set Ω in Theorem 9.18. □

Let us check that the hypothesis (9.7) of Theorem 9.7 is a weak version of Poincaré's condition $\partial_i q_j = \partial_j q_i$.

THEOREM 9.9.– *Let $q \in \mathcal{C}^1(\Omega; E^d)$, where Ω is an open subset of $\mathbb{R}^d$ and E is a Neumann space. Then*

$$\partial_i q_j = \partial_j q_i,$$

where i and j belong to $[\![1, d]\!]$ if and only if, for every $\varphi \in \mathcal{K}^\infty(\Omega)$,

$$\int_\Omega q_j\, \partial_i \varphi = \int_\Omega q_i\, \partial_j \varphi. \blacksquare$$

Proof. Direct implication. If $\partial_i q_j = \partial_j q_i$, then, applying the formula of integration by parts from Theorem 6.12 twice, we obtain

$$\int_\Omega q_i\, \partial_j \varphi = -\int_\Omega \partial_j q_i\, \varphi = -\int_\Omega \partial_i q_j\, \varphi = \int_\Omega q_j\, \partial_i \varphi.$$

Converse. Let $\varphi \in \mathcal{K}^\infty(\Omega)$. Again applying the formula of integration by parts from Theorem 6.12, we obtain $\int_\Omega q_i\, \partial_j \varphi = -\int_\Omega \partial_j q_i\, \varphi$ and $\int_\Omega q_j\, \partial_i \varphi = -\int_\Omega \partial_i q_j\, \varphi$. Therefore, by subtraction,

$$\int_\Omega q_i\, \partial_j \varphi - q_j\, \partial_i \varphi = \int_\Omega (\partial_i q_j - \partial_j q_i)\, \varphi.$$

If this vanishes for every φ, then $\partial_i q_j = \partial_j q_i$ by the weak vanishing property from Theorem 6.13. $\square$

9.6. Primitives on a simply connected open set

Let us show that, on a simply connected open set, Poincaré's condition $\partial_i q_j = \partial_j q_i$ guarantees the existence of a primitive for a continuously differentiable field.

THEOREM 9.10.– *Let $q \in \mathcal{C}^1(\Omega; E^d)$, where E is a Neumann space and*

Ω is a simply connected open subset of $\mathbb{R}^d$,

such that, for every i and j in $[\![1, d]\!]$,

$$\partial_i q_j = \partial_j q_i.$$

Then there exists $f \in \mathcal{C}^2(\Omega; E)$ such that

$$\nabla f = q. \blacksquare$$

Proof. By Poincaré's theorem (Theorem 9.5), the hypothesis $\partial_i q_j = \partial_j q_i$ implies the existence of a primitive on every ball $B \subset \Omega$. This implies the existence of a primitive on the whole of Ω by Theorem 9.4 on gluing together local primitives, since Ω is simply connected. $\square$

Optimality of Theorem 9.10. The condition $\partial_i q_j = \partial_j q_i$ is necessary and sufficient for a continuously differentiable field q to have a primitive because, if $q = \nabla f$, then

$$\partial_i q_j = \partial_i \partial_j f = \partial_j \partial_i f = \partial_j q_i.$$

When Ω is simply connected, this condition is therefore necessary and sufficient.

For an arbitrary open set Ω, it is necessary but not always sufficient (Theorem 9.16). □

Let us remain in the case of a simply connected open set and show that, if q is merely continuous, the weak Poincaré condition also guarantees the existence of a primitive.

THEOREM 9.11.– *Let $q \in \mathcal{C}(\Omega; E^d)$, where E is a Neumann space and*

$$\Omega \text{ is a simply connected open subset of } \mathbb{R}^d,$$

such that, for every i and j in $[\![1, d]\!]$ and all $\varphi \in \mathcal{K}^\infty(\Omega)$,

$$\int_\Omega q_j \, \partial_i \varphi = \int_\Omega q_i \, \partial_j \varphi. \tag{9.11}$$

Then there exists $f \in \mathcal{C}^1(\Omega; E)$ such that

$$\nabla f = q. \blacksquare$$

Proof. By Theorem 9.7, the hypothesis (9.11) implies the existence of a primitive on every ball $B \subset \Omega$. This implies the existence of a primitive on the whole of Ω by Theorem 9.4 on gluing together local primitives, since Ω is simply connected. □

Optimality of Theorem 9.11. When Ω is simply connected, the condition (9.11) is necessary and sufficient for q to have a primitive because, if $q = \nabla f$, then, for every $\varphi \in \mathcal{K}^\infty(\Omega)$,

$$\int_\Omega q_j \, \partial_i \varphi = \int_\Omega \partial_j f \, \partial_i \varphi = -\int_\Omega f \, \partial_j \partial_i \varphi = -\int_\Omega f \, \partial_i \partial_j \varphi = \int_\Omega \partial_i f \, \partial_j \varphi = \int_\Omega q_i \, \partial_j \varphi.$$

For an arbitrary open set Ω, it is necessary but not always sufficient (Theorem 9.14 (e)). □

Let us show that, on a simply connected open subset of $\mathbb{R}^2$, every divergence-free field $v = (v_1, v_2)$ derives from a stream function. This is known as **Haar's lemma**[5].

THEOREM 9.12.– *Let $v \in \mathcal{C}^1(\Omega; E^2)$, where Ω is a simply connected open subset of $\mathbb{R}^2$ and E is a Neumann space, such that*

$$\partial_1 v_1 + \partial_2 v_2 = 0_E.$$

*Then there exists a **stream function** $f \in \mathcal{C}^2(\Omega; E)$ such that:*

$$v_1 = \partial_2 f, \quad v_2 = -\partial_1 f. \blacksquare$$

5 **History of Haar's lemma.** Alfréd HAAR showed Theorem 9.12 with $E = \mathbb{R}$ between 1926 [43] and 1929 [44].

Proof. The field $q = (-v_2, v_1)$ satisfies

$$\partial_1 q_2 - \partial_2 q_1 = \partial_1 v_1 + \partial_2 v_2 = 0_E,$$

so, by Theorem 9.10, it has a primitive f such that

$$\partial_1 f = q_1 = -v_2 \text{ and } \partial_2 f = q_2 = v_1. \ \square$$

Abbreviated formulation of Theorem 9.12. Denoting by $^\perp$ a rotation of $\pi/2$ in the counterclockwise direction, and so $\nabla^\perp = (\partial_2, -\partial_1)$, the result of Theorem 9.12 can be stated as follows:

$$\textit{If } \Omega \textit{ is simply connected and } \nabla \cdot v = 0_E, \textit{ then there exists } f \textit{ such that } \nabla^\perp f = v. \qquad (9.12)$$

$\square$

Uniqueness. The stream function f obtained in Theorem 9.12 is unique up to an additive constant on each connected component of Ω by Theorem 9.17 (b) (since ∇f is unique). $\square$

Weak existence condition. By Theorem 9.11, there exists a stream function $f \in \mathcal{C}^1(\Omega; E)$ whenever the divergence of the field $v = (v_1, v_2)$ is zero in the following weak sense: for every $\varphi \in \mathcal{K}^\infty(\Omega)$,

$$\int_\Omega v_1 \partial_1 \varphi + v_2 \partial_2 \varphi = 0_E. \ \square$$

Stream functions in arbitrary dimensions. In three or more dimensions, constructing a stream function associated with a divergence-free function becomes much more complex. Details may be found, for example, in [GIRAULT–RAVIART, 40, Chapter I, § 3.3].

9.7. Comparison of the existence conditions for a primitive

Let us introduce the subspace of fields with a primitive.

DEFINITION 9.13.– *Let Ω be an open subset of $\mathbb{R}^d$ and E a Neumann space. We denote*

$$\mathcal{C}_\nabla(\Omega; E^d) \stackrel{\text{def}}{=} \{q \in \mathcal{C}(\Omega; E^d) : \exists f \in \mathcal{C}^1(\Omega; E) \textit{ such that } \nabla f = q\},$$

which is a vector space that we endow with the semi-norms of $\mathcal{C}(\Omega; E^d)$. ▮

Let us compare the conditions used in previous sections to obtain the existence of a primitive.

THEOREM 9.14.– *Let $q \in \mathcal{C}(\Omega; E^d)$, where Ω is an open subset of $\mathbb{R}^d$ and E is a Neumann space. Then:*

(a) $q \in C_\nabla(\Omega; E^d) \quad \Leftrightarrow \quad \exists f \in C^1(\Omega; E)$ *such that* $\nabla f = q$

$$\Leftrightarrow \quad \int_\Omega q \cdot \psi = 0_E, \quad \forall \psi \in \mathcal{K}^\infty(\Omega; \mathbb{R}^d) \text{ such that } \nabla \cdot \psi = 0$$

$$\Leftrightarrow \quad \oint_\Gamma q \cdot \mathrm{d}\ell = 0_E, \quad \forall \Gamma \text{ closed } C^1 \text{ path in } \Omega.$$

(b) *If* Ω *is simply connected,*

$$q \in C_\nabla(\Omega; E^d) \quad \Leftrightarrow \quad \exists f \in C^1(\Omega; E) \text{ such that } \nabla f = q$$

$$\Leftrightarrow \quad \int_\Omega q \cdot \psi = 0_E, \quad \forall \psi \in \mathcal{K}^\infty(\Omega; \mathbb{R}^d) \text{ such that } \nabla \cdot \psi = 0$$

$$\Leftrightarrow \quad \oint_\Gamma q \cdot \mathrm{d}\ell = 0_E, \quad \forall \Gamma \text{ closed } C^1 \text{ path in } \Omega$$

$$\Leftrightarrow \quad \oint_\Gamma q \cdot \mathrm{d}\ell = \oint_{\Gamma_*} q \cdot \mathrm{d}\ell, \quad \forall \Gamma \text{ and } \Gamma_* \text{ homotopic closed } C^1 \text{ paths}$$

$$\Leftrightarrow \quad \int_\Omega q_j \partial_i \varphi = \int_\Omega q_i \partial_j \varphi, \quad \forall i,\ \forall j,\ \forall \varphi \in \mathcal{K}^\infty(\Omega)$$

$$\Leftrightarrow \quad \forall \text{ ball } B \Subset \Omega,\ \exists f_B \in C^1(B; E) \text{ such that } \nabla f_B = q \text{ on } B.$$

(c) $\forall$ *ball* $B \Subset \Omega$, $\exists f_B \in C^1(B; E)$ *such that* $\nabla f_B = q$ *on* B

$$\Leftrightarrow \quad \int_\Omega q_j \partial_i \varphi = \int_\Omega q_i \partial_j \varphi, \quad \forall i,\ \forall j,\ \forall \varphi \in \mathcal{K}^\infty(\Omega)$$

$$\Leftrightarrow \quad \oint_\Gamma q \cdot \mathrm{d}\ell = \oint_{\Gamma_*} q \cdot \mathrm{d}\ell, \quad \forall \Gamma \text{ and } \Gamma_* \text{ homotopic closed } C^1 \text{ paths.}$$

(d) *If* $q \in C^1(\Omega; E^d)$,

$$\partial_i q_j = \partial_j q_i \quad \Leftrightarrow \quad \int_\Omega q_j \partial_i \varphi = \int_\Omega q_i \partial_j \varphi, \quad \forall \varphi \in \mathcal{K}^\infty(\Omega).$$

(e) *Having* $q \in C_\nabla(\Omega; E^d)$ *implies the properties of (c), but there exists open sets* Ω *for which the converse is false.* ▮

Case of a field with compact support. If the support of q is compact and included in Ω, the equivalences in (b) hold even when Ω is not simply connected.

Indeed, if q has local primitives on Ω, then its extension by 0_E has local primitives on the whole of $\mathbb{R}^d$, which is simply connected, so it has a primitive on the whole of $\mathbb{R}^d$ whose restriction is a primitive of q on Ω. Therefore, the properties in (c) are equivalent to those in (a). □

Proof of Theorem 9.14. (a) *First equivalence.* For every $f \in \mathcal{C}^1(\Omega; E)$ and every $\psi \in \mathcal{K}^\infty(\Omega; \mathbb{R}^d)$, the formula of integration by parts from Theorem 6.12 gives

$$\int_\Omega \nabla f \cdot \psi = \sum_{i=1}^d \int_\Omega \partial_i f \, \psi_i = -\sum_{i=1}^d \int_\Omega f \, \partial_i \psi_i = -\int_\Omega f \, \nabla \cdot \psi.$$

If $\nabla f = q$, we therefore have $\int_\Omega q \cdot \psi = -\int_\Omega f \, \nabla \cdot \psi = 0_E$ whenever $\nabla \cdot \psi = 0$. The converse is given by the orthogonality theorem (Theorem 9.2).

Second equivalence. If $q = \nabla f$, its line integral around closed paths is zero (Theorem 8.11 (b)). The converse is given by Theorem 9.1.

(c) *First equivalence.* If $\int_\Omega q_j \partial_i \varphi = \int_\Omega q_i \partial_j \varphi$ for all i, j and φ, and if B is a ball included in Ω, then Poincaré's theorem (Theorem 9.11) shows that, by restricting to the φ with support in B, q has a primitive f_B on this set.

Consider now the converse (which is straightforward when Ω is simply connected, see the comment *Optimality of Theorem 9.11*, p. 204, but here Ω is an arbitrary open set).

Let $\varphi \in \mathcal{K}^\infty(\Omega)$ with support K. Each point x of K is contained in an open ball B_x included in Ω, so, since K is compact, we can take a finite subcovering $(B_m)_{m \in M}$, where B_m denotes B_{x_m}. Let $(\alpha_m)_{m \in M}$ be a partition of unity subordinate to the covering by the B_m of their union ω, given by Theorem 7.18. The support of $q_j \partial_i \varphi$ is included in the support of φ and therefore certainly included in the open set ω, so its integral may be restricted to ω by Theorem 4.17 (a), that is,

$$\int_\Omega q_j \partial_i \varphi = \int_\omega q_j \partial_i \varphi.$$

Since $\sum_{m \in M} \alpha_m = 1$ on ω, it follows that

$$\int_\Omega q_j \partial_i \varphi = \int_\omega q_j \partial_i \Big(\sum_{m \in M} \alpha_m \varphi \Big) = \sum_{m \in M} \int_\omega q_j \partial_i (\alpha_m \varphi). \tag{9.13}$$

Suppose now that q has a primitive on each ball, and let f_m be a primitive on B_m. Since the support of $\alpha_m \varphi$ is included in B_m, together with Theorem 6.12 on integration by parts, it follows that

$$\int_\omega q_j \partial_i (\alpha_m \varphi) = \int_{B_m} \partial_j f_m \partial_i (\alpha_m \varphi) = -\int_{B_m} f_m \partial_j \partial_i (\alpha_m \varphi).$$

The derivatives commute by Schwarz's theorem (Theorem 2.12), so we can exchange i and j in this formula, and therefore in (9.13), which gives

$$\int_\Omega q_j \partial_i \varphi = \int_\Omega q_i \partial_j \varphi.$$

Second equivalence. If $q = \nabla f_B$ on each ball B included in Ω, the homotopy invariance theorem (Theorem 8.20) gives $\oint_\Gamma q \cdot d\ell = \oint_{\Gamma_*} q \cdot d\ell$ for every pair of homotopic closed paths Γ and Γ_* in Ω.

Conversely, suppose that this property is satisfied and let B be a ball included in Ω. Every closed path Γ of B is homotopic in this set to a closed path Γ_* consisting of a single point. Since the line integral is zero around Γ_* (Theorem 8.10 (a)), it is also zero around Γ, and therefore Theorem 9.1 gives f_B such that $q = \nabla f_B$ on B.

(b) If Ω is simply connected, $\nabla f = q$ is equivalent to $\nabla f_B = q$ on each B by Theorem 9.4 on gluing together local primitives, and therefore the properties of (a) are equivalent to those of (c).

(d) This is Theorem 9.9.

(e) If $\nabla f = q$ on Ω, then this certainly also holds on B. The converse is false, as shown by the examples that we will construct in Theorems 9.15, for $d = 2$, and 9.16, for arbitrary $d \geq 2$. □

9.8. Fields with local primitives but no global primitive

Let us show that there exist open sets on which the existence of local primitives does not guarantee the existence of a global primitive. Let us begin with an example in two dimensions with real values.

THEOREM 9.15.– *Let $\Omega = \{x \in \mathbb{R}^2 : |x| > 1\}$, and let $q \in C^\infty(\Omega; \mathbb{R}^2)$ be the field defined for every $x \in \Omega$ by $q(x) = (-x_2, x_1)/|x|^2$.*

For every ball B included in Ω, there exists $f_B \in C^\infty(B)$ such that $\nabla f_B = q$ on B. But there does not exist a function $f \in C^1(\Omega)$ such that $\nabla f = q$ on the whole of Ω. ▌

Proof. In polar coordinates, $\nabla = \mathbf{e}_r \partial_r + (\mathbf{e}_\theta / r)\partial_\theta$ and $q(\theta, r) = \mathbf{e}_\theta / r$, so

$$\nabla\theta = q \text{ except at } \theta = 0.$$

Indeed, θ is discontinuous on the half-line $D = \{(r, \theta) : \theta = 0\}$, since it is equal to 0 on one side and 2π on the other.

The field q does not have a primitive, as this primitive would necessarily be continuous (Theorem 2.10) and its restriction to $\Omega \setminus D$ would be of the form $\theta + c$ (Theorem 2.7), which is a contradiction.

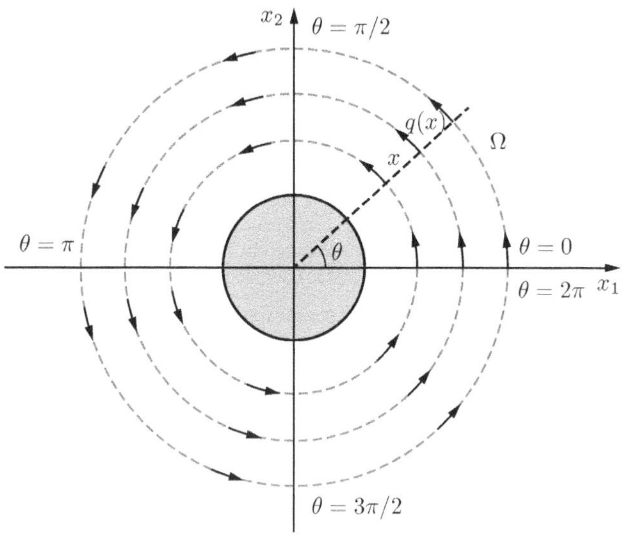

Figure 9.2. *Field q with local primitive θ but no global primitive. The set Ω is the exterior of the dashed disk*

Nevertheless, $\nabla\theta = q$ on every ball B because q is also a gradient on any ball that intersects with D, which can be seen by choosing another half-line as the origin of θ. □

Let us show that such open sets also exist in arbitrary dimensions, $d \geq 2$.

THEOREM 9.16.– *Let $d \geq 2$, and let E be a Neumann space that is not just $\{0_E\}$.*

Then there exists an open subset Ω of $\mathbb{R}^d$ and a field $q \in \mathcal{C}^\infty(\Omega; E^d)$ such that: for every ball B included in Ω, there exists a function $f_B \in \mathcal{C}^\infty(B; E)$ such that $\nabla f_B = q$ on B, but there does not exist a function $f \in \mathcal{C}^1(\Omega; E)$ such that $\nabla f = q$ on the whole of Ω. ▌

Proof. Let Ω_2 be the open subset of $\mathbb{R}^2$ and $\mathbf{q}$ the field given in Theorem 9.15, and let $u \in E$, $u \neq 0_E$. In two dimensions, the field defined on Ω_2 by $q(x) = \mathbf{q}(x)\,u$ is as required. In dimensions higher than two, the field defined on $\Omega_2 \times \mathbb{R}^{d-2}$ by $q(x_1, \ldots, x_d) = (\mathbf{q}_1(x_1, x_2), \mathbf{q}_2(x_1, x_2), 0, \ldots, 0)\,u$ is as required. □

Is simple connectedness necessary for gluing together local primitives? Recall that simple connectedness is sufficient for gluing together local primitives, that is, to guarantee that any field with local primitives has a global primitive by Theorem 9.4 and therefore to guarantee that a field satisfying Poincaré's condition, whether the strong or the weak variant, has a primitive.

Conversely, it is necessary if $d = 1$ or 2, but this is no longer the case when $d \geq 3$, although simple connectedness remains necessary for $d = 3$ if certain regularity conditions are imposed on the open set. These results, which were communicated to me by Pierre DREYFUSS and Nicolas DEPAUW, appeal to difficult ideas from algebraic topology, presented in [DREYFUSS, 29]. An overview is given below.

Case where $d = 1$. Every open subset of $\mathbb{R}$ is simply connected and therefore has the local primitive gluing property.

Case where $d = 2$. Every open subset Ω of $\mathbb{R}^2$ that is not simply connected has at least one *hole*; in other words, there exists a point $z \notin \Omega$ enclosed by a closed path Γ in Ω. This set therefore does not have the local primitive gluing property, since the field q introduced in Theorem 9.15, after being translated in such a way that z is at the origin, is locally a gradient on Ω but globally not a gradient.

Case where $d = 3$. The exterior of the *Alexander horned sphere* (presented and studied in [HATCHER, 45, p. 171]) has the local primitive gluing property but is not simply connected.

However, for an open subset of $\mathbb{R}^3$ that is bounded and locally on one side of the graph of a continuous function, the local primitive gluing property implies simple connectedness.

Case where $d \geq 4$. The local primitive gluing property of an open subset of $\mathbb{R}^d$ does not imply simple connectedness, even for bounded open sets that are locally on one side of the graph of a continuous function. □

9.9. Uniqueness of primitives

Let us show that primitives are unique up to an additive constant on each connected component of the domain.

THEOREM 9.17.– *Let $q \in \mathcal{C}_\nabla(\Omega; E^d)$, where Ω is an open subset of $\mathbb{R}^d$ and E is a Neumann space. Then:*

(a) *The field q has infinitely many primitives.*

(b) *The set of all primitives may be deduced from a given primitive by adding arbitrary constants to each connected component Ω_m of Ω.*

(c) *Given a point $a_m \in \Omega_m$ and $c_m \in E$ for each connected component Ω_m of Ω, there exists one and only one primitive f such that: for every m,*

$$f(a_m) = c_m. \blacksquare$$

Reminder. We denote by $\mathcal{C}_\nabla(\Omega; E^d)$ the space of continuous fields with a primitive (Definition 9.13). □

Proof. (a) Every function $f + c$, where $c \in E$, is a primitive of q.

(b) If f and f' are two primitives, $\nabla(f' - f) = 0$, so by Theorem 2.7, $f' - f$ is constant on each Ω_m. Conversely, $f + c$ is a primitive if c is constant on each Ω_m.

(c) If g is a primitive, then the function defined on each Ω_m by $f = g - g(a_m) + c_m$ is a primitive such that $f(a_m) = c_m$ for every m. This is the only one because, if f' is another primitive satisfying this property, then $f - f'$ is zero at a_m and is constant on Ω_m by (b) for every m, and is therefore zero on the whole of Ω. □

Case of a connected open subset. If Ω is connected, then its only connected component is Ω itself, which simplifies the statement of parts (b) and (c) of Theorem 9.17. □

9.10. Continuous primitive mapping

Let us construct a continuous linear primitive mapping.

THEOREM 9.18.– *Let $q \in \mathcal{C}_\nabla(\Omega; E^d)$, where Ω is an open subset of $\mathbb{R}^d$ and E is a Neumann space. For each connected component Ω_m of Ω, let $a_m \in \Omega_m$. Finally, let $f \in \mathcal{C}^1(\Omega; E)$ be the unique function such that*

$$\nabla f = q, \quad f(a_m) = 0_E, \ \forall m.$$

Then the mapping $q \mapsto f$ is linear and continuous and therefore sequentially continuous from $\mathcal{C}_\nabla(\Omega; E^d)$ into $\mathcal{C}^1(\Omega; E)$.

This mapping coincides with the mapping $q \mapsto q^*$ given by Theorem 9.1. ▮

Reminder. The space $\mathcal{C}_\nabla(\Omega; E^d)$ of continuous fields that have a primitive is endowed with the semi-norms of $\mathcal{C}(\Omega; E^d)$ by Definition 9.13. □

Notation. We could denote this mapping by ∇^{-1}, that is, $\nabla^{-1} q \stackrel{\text{def}}{=} f$, or alternatively $\nabla^{-1} \stackrel{\text{def}}{=} q^*$. □

Before giving the proof, let us state a consequence of sequential continuity.

THEOREM 9.19.– *Consider a sequence $(f_n)_{n\in\mathbb{N}}$ and a function f of $\mathcal{C}^1(\Omega; E)$, where Ω is an open subset of $\mathbb{R}^d$ and E is a Neumann space. For each connected component Ω_m of Ω, let $a_m \in \Omega_m$. Suppose that, as $n \to \infty$,*

$$\nabla f_n \to \nabla f \text{ in } \mathcal{C}(\Omega; E^d)$$

and for every m,

$$f_n(a_m) \to f(a_m) \text{ in } E.$$

Then

$$f_n \to f \text{ in } \mathcal{C}^1(\Omega; E). \ ▮$$

Proof of Theorem 9.19. Let g_n and g be the functions of $\mathcal{C}^1(\Omega; E)$ defined on each Ω_m by $g_n = f_n - f_n(a_m)$ and $g = f - f(a_m)$. Theorem 9.18 shows that $g_n \to g$ in $\mathcal{C}^1(\Omega; E)$, which implies that $f_n \to f$ in $\mathcal{C}^1(\Omega; E)$. □

Proof of Theorem 9.18. Let $y \in \Omega$, Ω_m the connected component of Ω containing y, Γ a piecewise $\mathcal{C}^1$ path joining a_m to y in Ω_m (such a path exists by Theorem 8.5), and $\epsilon > 0$ such that the ball $B(y,\epsilon) = \{x \in \mathbb{R}^d : |x-y| \le \epsilon\}$ is included in Ω.

Let $x \in \mathring{B}(y,\epsilon)$ and Λ the rectilinear path joining y to x, i.e. the path defined on $[0,1]$ by $\Lambda(t) = y + t(x-y)$. The path $\Gamma \overrightarrow{\cup} \Lambda$ joins a_m to x in Ω_m, so the expression of the line integral of a gradient (Theorem 8.11 (a)) and the expression of their concatenation (Definition 8.15) give

$$f(x) = f(a_m) + \int_{\Gamma \overrightarrow{\cup} \Lambda} \nabla f \cdot d\ell = \int_{\Gamma} q \cdot d\ell + \int_{\Lambda} q \cdot d\ell.$$

Let $\{\| \ \|_{E;\nu} : \nu \in \mathcal{N}_E\}$ be the family of semi-norms of E. The bound on the semi-norms of the line integral from Theorem 8.12 (a) gives, for every $\nu \in \mathcal{N}_E$,

$$\|f(x)\|_{E;\nu} \le (\gamma_\Gamma + \gamma_\Lambda) \sup_{z \in [\Gamma] \cup B(y,\epsilon)} \|q(z)\|_{E^d;\nu},$$

where $[\Gamma]$ is the image of Γ, $\gamma_\Gamma = \sup_{t_i \le t \le t_e} |\Gamma'(t)|$ and $\gamma_\Lambda = \sup_{0 \le t \le 1} |\Lambda'(t)|$, and so $\gamma_\Lambda = |x-y| \le \epsilon$.

Consider now a compact set $K \subset \Omega$. The open sets $\mathring{B}(y,\epsilon)$ cover this set, so there exists (Definition A.17 (a)) a finite subcovering $\mathcal{R}$. Thus,

$$\sup_{x \in K} \|f(x)\|_{E;\nu} \le c \sup_{z \in D} \|q(z)\|_{E^d;\nu},$$

where $c = \sup_{\mathcal{R}} \gamma_\Gamma + \epsilon$ is finite and $D = \bigcup_{\mathcal{R}} [\Gamma] \cup B(y,\epsilon)$. But D is compact, as a finite union of closed and bounded sets (which means that it is itself closed and bounded (Theorem A.10) in $\mathbb{R}^d$ and therefore compact by the Borel–Lebesgue theorem (Theorem A.23 (b))). And it is included in Ω. Therefore, by Definition 1.3 (a) of the semi-norms of $\mathcal{C}(\Omega; E)$, the above inequality can be written as

$$\|f\|_{\mathcal{C}(\Omega;E);K,\nu} \le c\,\|q\|_{\mathcal{C}(\Omega;E^d);D,\nu}.$$

By Definition 2.14 (a) of the semi-norms of $\mathcal{C}^1(\Omega; E)$, since $K \subset D$ and $\partial_i f = q_i$,

$$\|f\|_{\mathcal{C}^1(\Omega;E);K,\nu} = \sup\{\|f\|_{\mathcal{C}(\Omega;E);K,\nu}, \sup_{1 \le i \le d} \|\partial_i f\|_{\mathcal{C}(\Omega;E);K,\nu}\} \le$$
$$\le \sup\{c, 1\} \|q\|_{\mathcal{C}(\Omega;E^d);D,\nu}.$$

By the characterization of continuous linear mappings from Theorem 1.25, this proves that the mapping $q \mapsto f$ is continuous. It is therefore sequentially continuous, like any continuous mapping (Theorem A.29). □

Chapter 10

Additional Results: Integration on a Sphere

The purpose of this chapter is to present a few results about surface integrals that we will use in Volume 3 to construct distributions associated with singular functions. The main result is Stokes' formula (Theorem 10.8).

These properties are classical for functions taking values in a Banach space or a complete semi-normed space, but we need to extend them to functions taking values in a Neumann space, which is new. The proofs, made as elementary as possible, are unsurprising.

This chapter does not undertake a general study of surface integration, which we will do in Volume 5 for integrable distributions on a locally Lipschitz graph.

10.1. Surface integration on a sphere

Let us define the surface integral on a sphere of a continuous function taking values in a Neumann space.

DEFINITION 10.1.– *Let $f \in \mathcal{C}(S_r; E)$, where $S_r = \{x \in \mathbb{R}^d : |x| = r\}$, $d \geq 2$, $r > 0$, and E is a Neumann space. The* ***surface integral*** $\int_{S_r} f \,\mathrm{d}s \in E$ *is defined by*

$$\int_{S_r} f \,\mathrm{d}s \stackrel{\text{def}}{=} \lim_{\epsilon \to 0} \frac{1}{\epsilon} \int_{C_{r,r+\epsilon}} f\Big(\frac{rx}{|x|}\Big) \,\mathrm{d}x,$$

where $C_{r,b} = \{x \in \mathbb{R}^d : r < |x| < b\}$. ▮

Let us observe that, since the sphere S_r is compact, every continuous function is uniformly continuous and bounded on it by Heine's theorem (Theorem A.34), and the topologies on the spaces of these functions coincide (Theorem 1.10 (e)). In other words,

$$\mathcal{C}(S_r; E) \mathrel{\underset{\leftrightarrow}{=}} \mathbf{C}_{\mathrm{b}}(S_r; E). \tag{10.1}$$

To justify Definition 10.1, we will use the following additivity property, which is a higher dimensional analog of the additivity of the integral with respect to the interval of integration (recall that we only defined the Cauchy integral on open sets).

THEOREM 10.2.– *Let $0 \leq a \leq r \leq b \leq \infty$, and let E be a Neumann space. Then:*

(a) *If $f \in \mathcal{B}(C_{a,b}; E)$, where $C_{a,b} = \{x \in \mathbb{R}^d : a < |x| < b\}$,*

$$\int_{C_{a,b}} f = \int_{C_{a,r}} f + \int_{C_{r,b}} f.$$

(b) *If $f \in \mathcal{B}(\mathring{B}_b; E)$, where $\mathring{B}_b = \{x \in \mathbb{R}^d : |x| < b\}$,*

$$\int_{\mathring{B}_b} f = \int_{\mathring{B}_a} f + \int_{C_{a,b}} f. \blacksquare$$

Proof. (a) Since $C_{a,b} = (C_{a,r} \cup C_{r,b}) \cup S_r$ and the sphere S_r is negligible (Theorem 5.9), we have (Theorems 6.1 and 4.21 (b))

$$\int_{C_{a,b}} f = \int_{C_{a,r} \cup C_{r,b}} f = \int_{C_{a,r}} f + \int_{C_{r,b}} f.$$

(b) This similarly follows from the fact that $\mathring{B}_b = (\mathring{B}_a \cup C_{a,b}) \cup S_a$. □

Justification of Definition 10.1. Denote by $\widetilde{f}$ the extension of f that is constant on half-lines from 0, that is, suppose that $\widetilde{f}(x) = f(rx/|x|)$ for every $x \neq 0$. Then

$$\int_{S_r} f \, \mathrm{ds} \stackrel{\mathrm{def}}{=} \lim_{\epsilon \to 0} \frac{1}{\epsilon} \int_{C_{r,r+\epsilon}} \widetilde{f}. \tag{10.2}$$

The integral on the right-hand side is well defined according to Definition 4.9 of the Cauchy integral because $\widetilde{f}$ is uniformly continuous with bounded support, that is,

$$\widetilde{f} \in \mathcal{B}(C_{r,r+\epsilon}; E).$$

Indeed, as the composition of the continuous mappings $x \mapsto rx/|x|$ from $\mathbb{R}^d \setminus \{0\}$ into S_r and f from S_r into E, the function $\widetilde{f}$ is continuous from $\mathbb{R}^d \setminus \{0\}$ into E (Theorem A.35). It is therefore uniformly continuous on the compact set $\overline{C_{r,r+\epsilon}}$ by Heine's theorem (Theorem A.34) and therefore is certainly continuous on $C_{r,r+\epsilon}$. Additionally, its support is bounded because it is included in $C_{r,r+\epsilon}$.

The limit of the right-hand side of (10.2) exists because, as we will check later, there exists some $c_f \in E$ such that

$$\int_{C_{a,b}} \widetilde{f} = (b^d - a^d)\, c_f \tag{10.3}$$

and because $\mathrm{d}r^d/\mathrm{d}r = dr^{d-1}$ (Theorem A.56), so, by the characterization of the derivative from Theorem A.52,

$$\frac{1}{\epsilon}\int_{C_{r,r+\epsilon}} \widetilde{f} = \frac{(r+\epsilon)^d - r^d}{\epsilon} c_f \to dr^{d-1} c_f. \tag{10.4}$$

We still need to check (10.3). Let $0 < \eta < 1$. By Theorem 10.2 (a) and the effect of a homothety on an integral (Theorem 6.19),

$$\int_{C_{a,b}} \widetilde{f} = \int_{C_{\eta b,b}} \widetilde{f} - \int_{C_{\eta a,a}} \widetilde{f} + \int_{C_{\eta a,\eta b}} \widetilde{f} = (b^d - a^d)\int_{C_{\eta,1}} \widetilde{f} + \eta^d \int_{C_{a,b}} \widetilde{f}. \tag{10.5}$$

Given $(\eta_n)_{n\in\mathbb{N}} \to 0$, the sequence formed by the $\int_{C_{\eta_n,1}} \widetilde{f}$ is Cauchy, since, for every semi-norm $\| \; \|_{E;\nu}$ of E and $0 < \eta' \leq \eta$, by Theorem 10.2 (a), by the bound on the semi-norms of the integral from Theorem 4.15 and the measure of a crown (Theorem 5.4 (b)),

$$\Big\| \int_{C_{\eta',1}} \widetilde{f} - \int_{C_{\eta,1}} \widetilde{f} \Big\|_{E;\nu} = \Big\| \int_{C_{\eta',\eta}} \widetilde{f} \Big\|_{E;\nu} \leq v_d \eta^d \sup_{x\in S_r} \|f(x)\|_{E;\nu}.$$

Therefore, since E is sequentially complete, $\int_{C_{\eta_n,1}} \widetilde{f}$ has a limit c_f. In the limit, (10.5) implies (10.3) because we also have $\eta_n^d \to 0$. □

Value of c_f. In the limit, (10.4) implies $\int_{S_r} f = dr^{d-1} c_f$. □

Case where $d = 1$. In this case, S_r is just the points $+r$ and $-r$, the function f is just the specification of $f(+r)$ and $f(-r)$, and $\int_{S_r} f\,\mathrm{d}s = f(+r) + f(-r)$. □

10.2. Properties of the integral on a sphere

Let us state some fundamental properties of the surface integral.

THEOREM 10.3.– *Let $f \in \mathcal{C}(S_r; E)$, where $S_r = \{x \in \mathbb{R}^d : |x| = r\}$, $d \geq 2$, $r > 0$, and E is a Neumann space, and let $\| \; \|_{E;\nu}$ be a semi-norm of E. Then, denoting by $v_d = |\mathring{B}_{\mathbb{R}^d}|$ the measure of the open unit ball:*

(a) *The mapping $f \mapsto \int_{S_r} f\,\mathrm{d}s$ is continuous and linear from $\mathcal{C}(S_r; E)$ into E.*

(b)
$$\left\| \int_{S_r} f\,\mathrm{d}s \right\|_{E;\nu} \leq d\upsilon_d r^{d-1} \sup_{x \in S_r} \|f(x)\|_{E;\nu}.$$

(c)
$$\int_{S_r} 1\,\mathrm{d}s = d\upsilon_d r^{d-1}. \blacksquare$$

Proof. (b) By Definition 10.1 of the integral on a sphere,

$$\left\| \int_{S_r} f\,\mathrm{d}s \right\|_{E;\nu} = \left\| \lim_{\epsilon \to 0} \frac{1}{\epsilon} \int_{C_{r,r+\epsilon}} f\left(\frac{rx}{|x|}\right) \mathrm{d}x \right\|_{E;\nu} =$$
$$= \lim_{\epsilon \to 0} \frac{1}{\epsilon} \left\| \int_{C_{r,r+\epsilon}} f\left(\frac{rx}{|x|}\right) \mathrm{d}x \right\|_{E;\nu},$$

since the semi-norms $\| \; \|_{E;\nu}$ are continuous on E (because they are contractions, see Theorem A.5).

This implies the stated inequality because, by the bound on the semi-norms of the integral from Theorem 4.15 and the inequality $|C_{r,r+\epsilon}| \leq \epsilon d\upsilon_d (r+\epsilon)^{d-1}$ satisfied by the measure of a crown (Theorem 5.4 (b)),

$$\frac{1}{\epsilon} \left\| \int_{C_{r,r+\epsilon}} f\left(\frac{rx}{|x|}\right) \mathrm{d}x \right\|_{E;\nu} \leq \frac{|C_{r,r+\epsilon}|}{\epsilon} \sup_{x \in C_{r,r+\epsilon}} \left\| f\left(\frac{rx}{|x|}\right) \right\|_{E;\nu} \leq$$
$$\leq d\upsilon_d (r+\epsilon)^{d-1} \sup_{x \in S_r} \|f(x)\|_{E;\nu}.$$

(a) If $g \in \mathcal{C}(S_r; E)$, the linearity of the Cauchy integral (Theorem 4.11 (a)) implies

$$\int_{C_{r,r+\epsilon}} (f+g)\left(\frac{rx}{|x|}\right) \mathrm{d}x = \int_{C_{r,r+\epsilon}} f\left(\frac{rx}{|x|}\right) \mathrm{d}x + \int_{C_{r,r+\epsilon}} g\left(\frac{rx}{|x|}\right) \mathrm{d}x,$$

so, in the limit, $\int_{S_r} f + g\,\mathrm{d}s = \int_{S_r} f\,\mathrm{d}s + \int_{S_r} g\,\mathrm{d}s$. For $t \in \mathbb{R}$, we can similarly show that $\int_{S_r} tf\,\mathrm{d}s = t\int_{S_r} g\,\mathrm{d}s$. This proves that the mapping $f \mapsto \int_{S_r} f\,\mathrm{d}s$ is linear.

By the characterization of continuous linear mappings from Theorem 1.25, its continuity then follows from the inequality, for every semi-norm of E,

$$\left\| \int_{S_r} f\,\mathrm{d}s \right\|_{E;\nu} \leq d\upsilon_d r^{d-1} \|f\|_{\mathcal{C}(S_r;E);S_r,\nu}.$$

This is a restatement of the inequality (b) because, since S_r is compact by the Borel–Lebesgue theorem (Theorem A.23 (b)), $\sup_{x \in S_r} \|f(x)\|_{E;\nu} = \|f\|_{\mathcal{C}(S_r;E);S_r,\nu}$, by Definition 1.3 (a) of the semi-norms of $\mathcal{C}(S_r; E)$.

(c) Definition 10.1 of the surface integral gives, together with the expression of the measure of an open set and a crown from Theorems 4.12 (b) and 5.4 (b),

$$\int_{S_r} 1 \, \mathrm{d}s = \lim_{\epsilon \to 0} \frac{1}{\epsilon} \int_{C_{r,r+\epsilon}} 1 = \lim_{\epsilon \to 0} \frac{|C_{r,r+\epsilon}|}{\epsilon} = \lim_{\epsilon \to 0} \frac{\upsilon_d((r+\epsilon)^d - r^d)}{\epsilon},$$

which is equal to $\upsilon_d \, \mathrm{d}r^d / \, \mathrm{d}r = d\upsilon_d r^{d-1}$ by the characterization of the derivative and the expression of the derivative of a power from Theorems A.52 and A.56. □

Value of υ_2**.** Since the number π is defined as the ratio of the circumference of a circle to its diameter, that is, $\pi \overset{\text{def}}{=} \int_{S_r} 1 \, \mathrm{d}s / 2r$ when $d = 2$, Theorem 10.3 (c) gives $\upsilon_2 = \pi$. □

Let us show that the integral depends continuously on the radius of the sphere.

THEOREM 10.4.– *Let* $f \in \mathbf{C}_{\mathrm{b}}(C_{a,b}; E)$*, where* $C_{a,b} = \{x \in \mathbb{R}^d : a < |x| < b\}$*,* $d \geq 2$*,* $0 \leq a < b < \infty$*, and E is a Neumann space. Then*

$$\Big[r \mapsto \int_{S_r} f \, \mathrm{d}s\Big] \in \mathbf{C}_{\mathrm{b}}((a,b); E). \; ❚$$

Proof. Let $a < r \leq s < b$. Applying the homothety $y = sx/r$ yields (Theorem 6.19)

$$\int_{C_{s,s+\epsilon s/r}} f\Big(\frac{sy}{|y|}\Big) \, \mathrm{d}y = \Big(\frac{s}{r}\Big)^d \int_{C_{r,r+\epsilon}} f\Big(\frac{sx}{|x|}\Big) \, \mathrm{d}x.$$

Dividing by $\epsilon s/r$ and taking the limit as $\epsilon \to 0$, it follows, by Definition 10.1 of the surface integral for radius s, that

$$\int_{S_s} f \, \mathrm{d}s = \Big(\frac{s}{r}\Big)^{d-1} \lim_{\epsilon \to 0} \frac{1}{\epsilon} \int_{C_{r,r+\epsilon}} f\Big(\frac{sx}{|x|}\Big) \, \mathrm{d}x.$$

Multiplying by $(r/s)^{d-1}$ and subtracting $\int_{S_r} f$, we obtain, again by Definition 10.1, now for radius r,

$$\Big(\frac{r}{s}\Big)^{d-1} \int_{S_s} f \, \mathrm{d}s - \int_{S_r} f \, \mathrm{d}s = \lim_{\epsilon \to 0} \frac{1}{\epsilon} \int_{C_{r,r+\epsilon}} \Big(f\Big(\frac{sx}{|x|}\Big) - f\Big(\frac{rx}{|x|}\Big)\Big) \, \mathrm{d}x.$$

Let us estimate the integral on the right-hand side. For every semi-norm of E, the bound on the semi-norms of the integral from Theorem 4.15 gives

$$\Big\| \frac{1}{\epsilon} \int_{C_{r,r+\epsilon}} \Big(f\Big(\frac{sx}{|x|}\Big) - f\Big(\frac{rx}{|x|}\Big)\Big) \, \mathrm{d}x \Big\|_{E;\nu} \leq \frac{|C_{r,r+\epsilon}|}{\epsilon} \sup_{|z-y|=|s-r|} \|f(z) - f(y)\|_{E;\nu},$$

since $|sx/|x| - rx/|x|| = |s-r|$. Together with the bound $|C_{r,r+\epsilon}| \leq \epsilon d\upsilon_d(r+\epsilon)^{d-1}$ on the measure of a crown from Theorem 5.4 (b), this gives, in the limit,

$$\Big\| \Big(\frac{r}{s}\Big)^{d-1} \int_{S_s} f \, \mathrm{d}s - \int_{S_r} f \, \mathrm{d}s \Big\|_{E;\nu} \leq d\upsilon_d r^{d-1} \sup_{|z-y|=|s-r|} \|f(z) - f(y)\|_{E;\nu}.$$

Moreover, the bound on the semi-norms of the surface integral from Theorem 10.3 (b) gives

$$\left\|\left(\left(\frac{r}{s}\right)^{d-1}-1\right)\int_{S_s} f\,\mathrm{d}s\right\|_{E;\nu} \leq c_\nu d v_d (r^{d-1}-s^{d-1}),$$

where $c_\nu = \sup_{x\in C_{a,b}} \|f(x)\|_{E;\nu}$. The latter is finite because, since f is uniformly continuous by the hypotheses and $C_{a,b}$ is precompact (Theorem A.23 (b)), $f(C_{a,b})$ is also precompact (Theorem A.33) and therefore bounded (Theorem A.19 (a)).

Since $|r^{d-1}-s^{d-1}| \leq (d-1)b^{d-2}|r-s|$ (Theorem A.56), the last two inequalities imply

$$\left\|\int_{S_s} f\,\mathrm{d}s - \int_{S_r} f\,\mathrm{d}s\right\|_{E;\nu} \leq d v_d b^{d-1} \sup_{|z-y|=|s-r|} \|f(z)-f(y)\|_{E;\nu} + c\,|s-r|,$$

where $c = c_\nu d v_d (d-1) b^{d-2}$. By Definition 1.2 (b) of uniform continuity, the uniform continuity of f therefore implies the uniform continuity of the function $r \mapsto \int_{S_r} f\,\mathrm{d}s$. □

10.3. Radial calculation of integrals

Let us show that the integral of a continuous function on a crown is differentiable with respect to its outer radius and its derivative is the surface integral.

THEOREM 10.5.– *Let $f \in \mathbf{C}_b(C_{a,b}; E)$, where $C_{a,b} = \{x \in \mathbb{R}^d : a < |x| < b\}$, $d \geq 2$, $0 \leq a < b < \infty$, and E is a Neumann space. Then:*

(a)
$$\left[r \mapsto \int_{C_{a,r}} f\right] \in C^1([a,b]; E).$$

(b)
$$\frac{\mathrm{d}}{\mathrm{d}r}\int_{C_{a,r}} f = \int_{S_r} f\,\mathrm{d}s.$$

(c)
$$\int_{C_{a,b}} f = \int_a^b \left(\int_{S_r} f\,\mathrm{d}s\right)\mathrm{d}r. \blacksquare$$

Proof. (b) The integral $\int_{C_{a,r}} f$ makes sense according to Definition 4.9 of the Cauchy integral because the function f belongs to $\mathcal{B}(C_{a,b}; E)$; in other words, it is uniformly continuous with bounded support (Theorem 4.8 (b)), since $C_{a,b}$ is bounded.

Let $a < r < r + \epsilon \leq b$. As $\epsilon \to 0$,

$$\frac{1}{\epsilon} \int_{C_{r,r+\epsilon}} \Big(f(x) - f\Big(\frac{rx}{|x|}\Big)\Big) \, \mathrm{d}x \to 0_E. \tag{10.6}$$

Indeed, for every semi-norm $\| \ \|_{E;\nu}$ of E, the inequalities satisfied by the integral and the measure of a crown from Theorems 4.15 and 5.4 (b) give

$$\frac{1}{\epsilon} \Big\| \int_{C_{r,r+\epsilon}} \Big(f(x) - f\Big(\frac{rx}{|x|}\Big)\Big) \, \mathrm{d}x \Big\|_{E;\nu} \leq$$

$$\leq \frac{|C_{r,r+\epsilon}|}{\epsilon} \sup_{r<|x|<r+\epsilon} \Big\| f(x) - f\Big(\frac{rx}{|x|}\Big) \Big\|_{E;\nu} \leq$$

$$\leq dv_d b^{d-1} \sup_{|x-y|<\epsilon} \|f(x) - f(y)\|_{E;\nu},$$

which tends to 0 because f is uniformly continuous (Definition 1.2 (b)) by the hypotheses.

By Definition 10.1 of the surface integral, (10.6) gives

$$\frac{1}{\epsilon} \int_{C_{r,r+\epsilon}} f \to \int_{S_r} f \, \mathrm{d}s.$$

In other words, by the additivity property from Theorem 10.2 (a),

$$\frac{1}{\epsilon} \Big(\int_{C_{a,r+\epsilon}} f - \int_{C_{a,r}} f \Big) \to \int_{S_r} f \, \mathrm{d}s.$$

Similarly, this holds for $a < r - \epsilon < r < b$. By Definition 2.1 (b) of differentiability in dimension one (or by its characterization from Theorem A.52), this proves that the mapping $r \mapsto \int_{C_{a,r}} f$ is differentiable on (a, b) with derivative $\int_{S_r} f \, \mathrm{d}s$.

(a) Denote $F(r) = \int_{C_{a,r}} f$. By the characterization of a C^1 function on a closed set, here $[a, b]$, from Theorem 2.27, we need first to check that F is uniformly continuous and bounded on $[a, b]$. Since differentiability only implies continuity on the open set (a, b) (Theorem 2.10), we shall adopt a different approach. For $a \leq s \leq r \leq b$ and every semi-norm of E, we have, again by Theorems 10.2 (a), 4.15, and 5.4 (b),

$$\Big\| \int_{C_{a,r}} f - \int_{C_{a,s}} f \Big\|_{E;\nu} = \Big\| \int_{C_{s,r}} f \Big\|_{E;\nu} \leq c_\nu |C_{s,r}| \leq c_\nu d v_d r^{d-1} (r - s),$$

where $c_\nu = \sup_{x\in C_{a,b}} \|f(x)\|_{E;\nu}$. This establishes the uniform continuity and boundedness of F on the whole of $[a,b]$.

Second, we need to check that the derivative $\mathrm{d}F/\mathrm{d}r$ is uniformly continuous and bounded. This is indeed the case because, by (b), it is equal to $\int_{S_r} f\,\mathrm{d}s$, which is uniformly continuous and bounded (Theorem 10.4).

(c) The calculation of the integral of a derivative from Theorem 6.4 (b) gives, together with (b),

$$\int_a^b \Big(\int_{S_r} f\,\mathrm{d}s\Big)\,\mathrm{d}r = \int_a^b \frac{\mathrm{d}F}{\mathrm{d}r}\,\mathrm{d}r = F(b) - F(a) = \int_{C_{a,b}} f,$$

because $F(a) = \int_\emptyset f = 0_E$. □

Let us calculate the integral of powers of $|x|$ on a crown.

THEOREM 10.6.– *Let $C_{a,b} = \{x \in \mathbb{R}^d : a < |x| < b\}$, where $0 < a < b < \infty$ and $d \geq 2$, $s \in \mathbb{R}$, and $\upsilon_d = |\mathring{B}_{\mathbb{R}^d}|$ is the measure of the open unit ball. Then*

$$\int_{C_{a,b}} |x|^s\,\mathrm{d}x = d\upsilon_d \frac{b^{s+d} - a^{s+d}}{s+d}.\ \blacksquare$$

Proof. We have (Theorems 10.5 (c), 10.3 (c), A.56, and 6.4 (a))

$$\int_{C_{a,b}} |x|^s\,\mathrm{d}x = \int_a^b \Big(\int_{S_r} r^s\,\mathrm{d}s\Big)\,\mathrm{d}r = \int_a^b d\upsilon_d r^{s+d-1}\,\mathrm{d}r =$$

$$= \int_a^b \frac{d\upsilon_d}{s+d}\frac{\mathrm{d}r^{s+d}}{\mathrm{d}r}\,\mathrm{d}r = d\upsilon_d\frac{b^{s+d} - a^{s+d}}{s+d}.\ \square$$

10.4. Surface integral as an integral of dimension $d - 1$

Let us express a surface integral on a sphere of $\mathbb{R}^d$ as an integral on a ball of $\mathbb{R}^{d-1}$. This result is used in the next section to prove Stokes' formula.

THEOREM 10.7.– *Let $f \in \mathcal{C}(S_r; E)$, where $S_r = \{x \in \mathbb{R}^d : |x| = r\}$, $r > 0$, $d \geq 2$, and E is a Neumann space. Furthermore, let $\mathring{B}_r^* = \{x_* \in \mathbb{R}^{d-1} : |x_*| < r\}$ and*

$$\nu_d(x) = \frac{x_d}{|x|}, \quad \Upsilon_r(x_*) = \sqrt{r^2 - |x_*|^2}.$$

Then

$$\int_{S_r} f\nu_d\,\mathrm{d}s = \int_{\mathring{B}_r^*} f(x_*, \Upsilon_r(x_*)) - f(x_*, -\Upsilon_r(x_*))\,\mathrm{d}x_*.\ \blacksquare$$

Proof. We will proceed in several steps.

1. *Expression in terms of the extension of f independent of x_d.* Let us begin by showing that

$$\int_{S_r} f\,\mathrm{d}s = \lim_{\epsilon\to 0}\Big(\frac{1}{\epsilon}\int_{A_+}\bar{f} + \frac{1}{\epsilon}\int_{A_-}\bar{f}\Big), \tag{10.7}$$

where

$$A_+ = \{(x_*, x_d) : |x_*| < r,\ \Upsilon_r(x_*) < x_d < \Upsilon_{r+\epsilon}(x_*)\},$$
$$A_- = \{(x_*, x_d) : |x_*| < r,\ -\Upsilon_{r+\epsilon}(x_*) < x_d < -\Upsilon_r(x_*)\},$$

and where $\bar{f}$ is the extension by f that is constant on vertical lines, namely

$$\bar{f} = \begin{cases} f(x_*, \Upsilon_r(x_*)) & \text{on } A_+, \\ f(x_*, -\Upsilon_r(x_*)) & \text{on } A_-. \end{cases}$$

To do this, decompose $C_{r,r+\epsilon} = A_+ \cup A_- \cup L \cup N$, where (see Figure 10.1),

$$L = \{(x_*, x_d) : r < |x_*| < r+\epsilon,\ |x_d| < \Upsilon_{r+\epsilon}(x_*)\},$$
$$N = \{(x_*, x_d) : |x_*| = r,\ |x_d| < \Upsilon_{r+\epsilon}(x_*)\}.$$

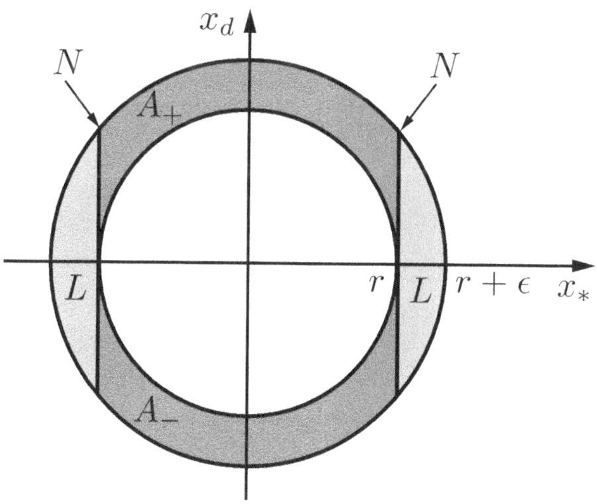

Figure 10.1. *Decomposition of the crown $C_{r,r+\epsilon}$. A is dark gray and L is light gray*

Since N is the product $S_r^* \times (-\sqrt{2\epsilon r+\epsilon^2}, \sqrt{2\epsilon r+\epsilon^2})$, it is negligible in $\mathbb{R}^d$ by Theorem 5.8, because the sphere S_r^* is negligible (Theorem 5.9) in $\mathbb{R}^{d-1}$.

Definition 10.1 of the surface integral therefore gives, together with additivity, and because negligible subsets do not contribute to the integral (Theorems 4.21 (b) and 6.1),

$$\int_{S_r} f \, \mathrm{d}s = \lim_{\epsilon \to 0} \frac{1}{\epsilon} \int_{C_{r,r+\epsilon}} \widetilde{f} = \lim_{\epsilon \to 0} \Big(\frac{1}{\epsilon} \int_{A+} \widetilde{f} + \frac{1}{\epsilon} \int_{A-} \widetilde{f} + \frac{1}{\epsilon} \int_{L} \widetilde{f} \Big),$$

where $\widetilde{f}(x) = f(rx/|x|)$.

The limit of the term involving the lateral part L is zero because, for any semi-norm $\| \, \|_{E;\nu}$ of E, by the bound on the semi-norms of the integral from Theorem 4.15,

$$\Big\| \frac{1}{\epsilon} \int_L \widetilde{f} \Big\|_{E;\nu} \leq \frac{|L|}{\epsilon} \sup_{x \in L} \|\widetilde{f}(x)\|_{E;\nu} = \frac{|L|}{\epsilon} \sup_{x \in S_r} \|f(x)\|_{E;\nu},$$

which tends to 0 as $\epsilon \to 0$, since f is bounded (by (10.1)) and

$$|L| \leq c\epsilon^{3/2}.$$

Indeed, $L \subset C^*_{r,r+\epsilon} \times (-\sqrt{2\epsilon r + \epsilon^2}, \sqrt{2\epsilon r + \epsilon^2})$ thus $|L| \leq |C^*_{r,r+\epsilon}| 2\sqrt{2\epsilon r + \epsilon^2}$ (Theorems 4.3 and 4.4 (a)), and $|C^*_{r,r+\epsilon}| \leq \epsilon(d-1)v_{d-1}(r+\epsilon)^{d-2}$ (Theorem 5.4 (b)), and so $c \leq 2(d-1)v_{d-1}(2r)^{d-2}\sqrt{3r}$ if $\epsilon \leq r$.

The property (10.7) follows from the additional fact that the limit of the two other terms does not change when we replace $\widetilde{f}$ by $\overline{f}$, since

$$\Big\| \frac{1}{\epsilon} \int_{A_+} \widetilde{f} - \overline{f} \Big\|_{E;\nu} \leq \frac{|A_+|}{\epsilon} \sup_{x \in A_+} \|\widetilde{f}(x) - \overline{f}(x)\|_{E;\nu} =$$

$$= \frac{|A_+|}{\epsilon} \sup_{x \in A_+} \|f(\widetilde{x}) - f(\overline{x})\|_{E;\nu} \to 0,$$

where $\widetilde{x} = rx/|x|$ and $\overline{x} = (x_*, \Upsilon_r(x_*))$. Indeed, $|A_+| \leq |C_{r,r+\epsilon}| \leq c\epsilon$ by Theorem 5.4 (b), and the supremum tends to 0 because f is uniformly continuous (again by (10.1)) and $\sup_{x \in A_+} |\widetilde{x} - \overline{x}| \to 0$. The latter convergence follows from the fact that, for $x \in A_+$, we have $|\widetilde{x} - x| = |r| - |x| \leq \epsilon$ and

$$|x - \overline{x}| = |x_d - \Upsilon_r(x_*)| \leq \sqrt{(r+\epsilon)^2 - |x_*|^2} - \sqrt{r^2 - |x_*|^2} \leq \sqrt{\epsilon^2 + 2\epsilon r}.$$

2. *Integration with respect to* x_d. By separating the variables of the integral (Theorem 6.10) and integrating with respect to x_d,

$$\int_{A_+} \overline{\nu_d f} = \int_{\mathring{B}^*_r} \Big(\int_{\Upsilon_r(x_*)}^{\Upsilon_{r+\epsilon}(x_*)} \overline{\nu_d f}(x_*, x_d) \, \mathrm{d}x_d \Big) \, \mathrm{d}x_* =$$

$$= \int_{\mathring{B}^*_r} (\Upsilon_{r+\epsilon}(x_*) - \Upsilon_r(x_*)) \frac{\Upsilon_r(x_*)}{r} f(x_*, \Upsilon_r(x_*)) \, \mathrm{d}x_*,$$

since, on A_+, the term $\overline{\nu_d f}$ does not depend on x_d and

$$\overline{\nu_d}(x_*, x_d) = \nu_d(x_*, \Upsilon_r(x_*)) = \frac{\Upsilon_r(x_*)}{r}.$$

Similarly,

$$\int_{A_-} \overline{\nu_d f} = -\int_{\mathring{B}_r^*} (\Upsilon_{r+\epsilon}(x_*) - \Upsilon_r(x_*)) \frac{\Upsilon_r(x_*)}{r} f(x_*, -\Upsilon_r(x_*)) \, dx_*,$$

since $\overline{\nu_d}(x_*, x_d) = \nu_d(x_*, -\Upsilon_r(x_*)) = -\Upsilon_r(x_*)/r$ on A_-.

Therefore,

$$\frac{1}{\epsilon}\int_{A_+} \overline{\nu_d f} + \frac{1}{\epsilon}\int_{A_-} \overline{\nu_d f} - \int_{\mathring{B}_r^*} f(x_*, \Upsilon_r(x_*)) - f(x_*, -\Upsilon_r(x_*)) \, dx_* = \int_{\mathring{B}_r^*} \Phi,$$

where

$$\Phi(x_*) = \Big(\frac{\Upsilon_{r+\epsilon}(x_*) - \Upsilon_r(x_*)}{\epsilon} \frac{\Upsilon_r(x_*)}{r} - 1\Big)\big(f(x_*, \Upsilon_r(x_*)) - f(x_*, -\Upsilon_r(x_*))\big).$$

3. *Estimation.* For every semi-norm $\| \ \|_{E;\nu}$ of E, we therefore have (Theorems 4.11 (b), 4.14 (a), and 4.11 (a))

$$\Big\|\frac{1}{\epsilon}\int_{A_+} \overline{\nu_d f} + \frac{1}{\epsilon}\int_{A_-} \overline{\nu_d f} - \int_{\mathring{B}_r^*} f(x_*, \Upsilon_r(x_*)) - f(x_*, -\Upsilon_r(x_*)) \, dx_*\Big\|_{E;\nu} \leq$$

$$\leq \int_{\mathring{B}_r^*} \|\Phi\|_{E;\nu} \leq c\delta_\epsilon,$$

where $c = 2\sup_{x \in S_r} \|f(x)\|_{E;\nu}$ and

$$\delta_\epsilon = \int_{\mathring{B}_r^*} \Big|\frac{\Upsilon_{r+\epsilon} - \Upsilon_r}{\epsilon}\frac{\Upsilon_r}{r} - 1\Big|.$$

As we will check in step 4, we have $\delta_\epsilon \to 0$ as $\epsilon \to 0$. Together with (10.7) applied to $f\nu_d$, we therefore obtain, in the limit,

$$\Big\|\int_{S_r} f\nu_d \, ds - \int_{\mathring{B}_r^*} f(x_*, \Upsilon(x_*)) - f(x_*, -\Upsilon(x_*)) \, dx_*\Big\|_{E;\nu} = 0.$$

This holds for every semi-norm of E, which establishes the stated equality, provided that we verify that $\delta_\epsilon \to 0$.

4. *Limit of de_ϵ.* Denote $\varphi(r) = (r^2 - |x_*|^2)^{1/2}$. Observe that (Theorems 3.12 (c) and A.56) φ is differentiable at every point $r > |x_*|$ and

$$\varphi'(r) = \frac{1}{2}\frac{dr^2}{dr}(r^2 - |x_*|^2)^{-1/2} = r(r^2 - |x_*|^2)^{-1/2} > 0.$$

Similarly, φ' is differentiable and

$$\varphi''(r) = (r^2 - |x_*|^2)^{-1/2} - r^2(r^2 - |x_*|^2)^{-3/2} = -|x_*|^2(r^2 - |x_*|^2)^{-3/2} < 0.$$

Therefore, φ' is decreasing (Theorem A.53) and the finite increment theorem (Theorem A.54) gives

$$\varphi'(r+\epsilon) \leq \frac{\varphi(r+\epsilon) - \varphi(r)}{\epsilon} \leq \varphi'(r).$$

Hence,

$$0 \leq 1 - \Big(\frac{\Upsilon_{r+\epsilon} - \Upsilon_r}{\epsilon}\frac{\Upsilon_r}{r}\Big)(x_*) = 1 - \frac{\varphi(r+\epsilon) - \varphi(r)}{\epsilon}\frac{1}{\varphi'(r)} \leq 1 - \frac{\varphi'(r+\epsilon)}{\varphi'(r)}$$

and

$$\delta_\epsilon \leq \int_{\mathring{B}_r^*} 1 - \frac{\varphi'(r+\epsilon)}{\varphi'(r)} = \int_{\mathring{B}_r^*} 1 - \Big(\frac{1 - |x_*|/r^2}{1 - |x_*|/(r+\epsilon)^2}\Big)^{1/2}.$$

The integrated function tends to 0 as $\epsilon \to 0$ (but not uniformly, because it is equal to 1 if $|x_*| = r$). Since it increases with $|x_*|$, denoting it by ϕ, we have the inequality (Theorems 10.2 (b) and 4.14 (a)), for $0 < s < r$,

$$\delta_\epsilon = \int_{\mathring{B}_r^*} \phi = \int_{C_{s,r}^*} \phi + \int_{\mathring{B}_s^*} \phi \leq |C_{s,r}^*| + |\mathring{B}_r^*|\Big(1 - \Big(\frac{1 - s^2/r^2}{1 - s^2/(r+\epsilon)^2}\Big)^{1/2}\Big).$$

Hence, $\delta_\epsilon \to 0$ as $\epsilon \to 0$ because, given $\eta > 0$, we obtain $\delta_\epsilon \leq \eta$ for every $\epsilon \leq \epsilon_0$ by choosing $s < r$ such that $|C_{s,r}^*| \leq (d-1)v_{d-1}r^{d-2}(r-s) \leq \eta/2$ (Theorems 5.4 (b)), then by choosing ϵ_0 such that the other term on the right-hand side is $\leq \eta/2$. □

10.5. A Stokes formula

Let us state **Stokes' formula** on a crown[1] for a function taking values in a Neumann space.

THEOREM 10.8.– *Let $f \in \mathcal{C}^1(\overline{C_{a,b}}; E)$, where $C_{a,b} = \{x \in \mathbb{R}^d : a < |x| < b\}$, $d \geq 2$, $0 < a < b < \infty$, and E is a Neumann space. Then, for every $i \in [\![1, d]\!]$,*

$$\int_{C_{a,b}} \partial_i f = \int_{S_b} f n_i \, \mathrm{d}s + \int_{S_a} f n_i \, \mathrm{d}s,$$

where n is the outward unit normal vector of $C_{a,b}$ on S_a and S_b, namely

$$n_i(x) = \sigma\frac{x_i}{|x|}, \text{ where } \sigma \text{ is equal to } 1 \text{ on } S_b \text{ and } -1 \text{ on } S_a. \blacksquare$$

1 **History of Stokes' formula.** See footnote 6, Chapter 8, p. 189.

Proof. Decompose $C_{a,b} = (A \cup B \cup C) \cup N$, where

$$A_+ = \{(x_*, x_d) : |x_*| < a,\ \Upsilon_a(x_*) < x_d < \Upsilon_b(x_*)\},$$
$$A_- = \{(x_*, x_d) : |x_*| < a,\ -\Upsilon_b(x_*) < x_d < -\Upsilon_a(x_*)\},$$
$$L = \{(x_*, x_d) : a < |x_*| < b,\ -\Upsilon_b(x_*) < x_d < \Upsilon_b(x_*)\},$$
$$N = \{(x_*, x_d) : |x_*| = a,\ -\Upsilon_b(x_*) < x_d < \Upsilon_b(x_*)\},$$

with $x_* \in \mathbb{R}^{d-1}$, $\Upsilon_a(x_*) = \sqrt{a^2 - |x_*|^2}$, and $\Upsilon_b(x_*) = \sqrt{b^2 - |x_*|^2}$ (this is the decomposition shown in Figure 10.1, with $r = a$ and $r + \epsilon = b$).

The set N is negligible because it is included in $\omega_\epsilon = C^*_{a-\epsilon,a+\epsilon} \times (-a, a)$ for every $\epsilon > 0$, and $|\omega_\epsilon| = 2(d-1)v_{d-1}(a+\epsilon)^{d-2}\epsilon 2a$ (Theorems 4.3, 4.4 (a), and 5.4 (b)), which tends to 0 with ϵ. Therefore, it does not contribute to the integral (Theorem 6.1).

Together with the additivity with respect to the domain (Theorem 4.21 (b)), this gives

$$\int_{C_{a,b}} \partial_d f = \int_L \partial_d f + \int_{A_+} \partial_d f + \int_{A_-} \partial_d f.$$

By separating the variables in the integral (Theorem 6.10), the integral over the lateral part L of $C_{a,b}$ becomes

$$\int_L \partial_d f = \int_{C^*_{a,b}} \Big(\int_{-\Upsilon_b(x_*)}^{\Upsilon_b(x_*)} \partial_d f(x_*, x_d)\, \mathrm{d}x_d \Big)\, \mathrm{d}x_*,$$

that is, with the calculation of the integral of a derivative from Theorem 6.4 (b),

$$\int_L \partial_d f = \int_{C^*_{a,b}} f(x_*, \Upsilon_b(x_*)) - f(x_*, -\Upsilon_b(x_*))\, \mathrm{d}x_*.$$

Similarly,

$$\int_{A_+} \partial_d f = \int_{\mathring{B}^*_a} f(x_*, \Upsilon_b(x_*)) - f(x_*, \Upsilon_a(x_*))\, \mathrm{d}x_*,$$
$$\int_{A_-} \partial_d f = \int_{\mathring{B}^*_a} f(x_*, -\Upsilon_a(x_*)) - f(x_*, -\Upsilon_b(x_*))\, \mathrm{d}x_*.$$

Adding these three equalities together and using the additivity with respect to balls and crowns from Theorem 10.2 (b), here in $\mathbb{R}^{d-1}$ gives

$$\int_{C_{a,b}} \partial_d f = \int_{\mathring{B}^*_b} f(x_*, \Upsilon_b(x_*)) - f(x_*, -\Upsilon_b(x_*))\, \mathrm{d}x_* -$$
$$- \int_{\mathring{B}^*_a} f(x_*, \Upsilon_a(x_*)) - f(x_*, -\Upsilon_a(x_*))\, \mathrm{d}x_*.$$

With the expression of the surface integral of Theorem 10.7, this reads

$$\int_{C_{a,b}} \partial_d f = \int_{S_b} f\nu_d \, \mathrm{d}s - \int_{S_a} f\nu_d \, \mathrm{d}s.$$

This gives the stated result for $i = d$, because $\nu = n$ on S_b and $\nu = -n$ on S_a.

For $i \neq d$, simply replace x_d by x_i and x_* by $(x_j)_{j \neq i}$ in this proof. □

Appendix

Reminders

This chapter gives precise statements of any definitions and results from the first volume, *Banach, Fréchet, Hilbert and Neumann Spaces* of the *Analysis for PDEs* set that are used in this book in order that it may remain self-contained. The proofs can be found in the cited volume, referenced by [Vol. 1, Theorem A.n]. Some statements are restricted to the cases that will be useful to us if this results in a simplification; others are extended[1].

A.1. Notation and numbering

Notation of set theory. We will use the usual notation (presented in § 1.1 of Vol. 1). Let us simply recall the difference in notation between the set $\{a, b, \ldots, z\}$ and the ordered set $(a, b, \ldots, z)$. Thus, $(a, b) \neq (b, a)$ if $a \neq b$, whereas we always have $\{a, b\} = \{b, a\}$.

Countability. We assume that the sets[2] $\mathbb{N} = \{1, 2, \ldots\}$ and $\mathbb{N}^* = \{0, 1, 2, \ldots\}$ of natural numbers, $\mathbb{Z} = \{\ldots, -2, -1, 0, 1, 2, \ldots\}$ of integers and $\mathbb{Q}$ of rational numbers are familiar, together with their operations of addition, subtraction, multiplication, absolute value and their orderings. For $m \in \mathbb{N}$ and $n \in \mathbb{N}$, we denote $[\![m, n]\!] = \{i \in \mathbb{N} : m \leq i \leq n\}$ and $[\![m, \infty]\!] = \{i \in \mathbb{N} : i \geq m\} \cup \{\infty\}$.

A **sequence** of a set U is defined by specifying an element $u_n \in U$ for every $n \in \mathbb{N}$. This is an ordered set that we denote by $(u_n)_{n \in \mathbb{N}}$. A **subsequence** of $(u_n)_{n \in \mathbb{N}}$ is any sequence of the form $(u_{\sigma(n)})_{n \in \mathbb{N}}$ such that $\sigma(n) \in \mathbb{N}$ and $\sigma(n+1) > \sigma(n)$ for every $n \in \mathbb{N}$. We say that a real sequence is **increasing** if $u_{n+1} \geq u_n$.

Let us define countable sets.

DEFINITION A.1.– *A set U is **countable** if there exists a bijection from U onto a subset of $\mathbb{N}$.* □

1 **Historical notes.** The reader will find historical notes in Volume 1 for many of the definitions and properties recalled here.

2 **Caution with French notation.** In France, Italy and some other countries, authors denote by $\mathbb{N}$ the set that we denote $\mathbb{N}^*$, and they denote by $\mathbb{N}^*$ our $\mathbb{N}$!

Let us state some properties of countability [SCHWARTZ, 68, pp. 104–108].

THEOREM A.2.– (a) *Every subset of a countable set is countable.*

(b) *Every image of a countable set under a mapping is countable.*

(c) *Every finite product of countable sets is countable.*

(d) *The set* $\mathbb{Q}$ *is countable.* □

Real numbers. We will assume that the set $\mathbb{R}$ of real numbers is familiar, together with its operations of addition, subtraction, multiplication, absolute value and its ordering.

A **real interval** is any set of one of the following forms[3]: $(a, b) = \{x \in \mathbb{R} : a < x < b\}$ (open interval), $[a, b] = \{x \in \mathbb{R} : a \leq x \leq b\}$ (closed interval), $(a, b] = \{x \in \mathbb{R} : a < x \leq b\}$ and $[a, b) = \{x \in \mathbb{R} : a \leq x < b\}$ (half-open intervals), where a and b belong to $\mathbb{R} \cup \{-\infty, +\infty\}$.

We say that $m \in \mathbb{R}$ is an **upper bound** of a subset V of $\mathbb{R}$ if every $v \in V$ satisfies $v \leq m$. We say that m is the **supremum** (or least upper bound) of V if m is an upper bound of V and every other upper bound v of V satisfies $m \leq v$. If it exists, the supremum is unique and is denoted by $\sup V$. The notions of **lower bound** and **infimum** (or greatest lower bound) are obtained by replacing $\leq$ by $\geq$ in the above. If it exists, the infimum is denoted by $\inf V$.

Let us define the notion of a convergent real sequence.

DEFINITION A.3.– *We say that a sequence* $(x_n)_{n \in \mathbb{N}}$ *of* $\mathbb{R}$ ***converges** to a **limit*** $x \in \mathbb{R}$ *if, for every* $\epsilon > 0$, *there exists* $m \in \mathbb{N}$ *such that:* $n \geq m$ *implies* $|x_n - x| \leq \epsilon$. *If so, we denote* $x_n \to x$.

We say that $(x_n)_{n \in \mathbb{N}}$ ***tends to infinity*** *if, for every* $c \in \mathbb{R}$, *there exists* $m \in \mathbb{N}$ *such that:* $n \geq m$ *implies* $x_n \geq c$. *If so, we denote* $x_n \to \infty$. □

Every increasing real sequence that is bounded above converges [Vol. 1, Theorem 1.3]:

THEOREM A.4.– *Every increasing real sequence that is bounded above and every decreasing real sequence that is bounded below converges.* □

A.2. Semi-normed spaces

Vector spaces and semi-norms

A **vector space** is a set E endowed with an **addition** operation from $E \times E$ into E, denoted $+$, a **multiplication** operation from $\mathbb{R} \times E$ into E, and an element $0_E \in E$ such that, for every u, v, and w in E and every s and t in $\mathbb{R}$:

$$(u + v) + w = u + (v + w), \quad u + v = v + u, \quad u + 0_E = u,$$
$$t(u + v) = tu + tv, \quad (s + t)u = su + tu, \quad s(tu) = (st)u, \quad 1u = u,$$
$$\text{there exists an element } -u \in E \text{ such that } \quad -u + u = 0_E.$$

A **semi-norm** is a mapping p from a vector space E into $\mathbb{R}$ such that, for every $u \in E$, $v \in E$, and $t \in \mathbb{R}$:

$$p(u) \geq 0, \quad p(tu) = |t|\, p(u), \quad p(u + v) \leq p(u) + p(v).$$

A **norm** is a semi-norm p such that $p(u) > 0$ whenever $u \neq 0_E$.

Every semi-norm is a *contraction* [Vol. 1, Theorem 2.5]:

---

3 **French notation.** French authors denote the open interval by]a,b[. They denote the closed interval by [a,b], like we do, and furthermore the half-open intervals by]a,b] and [a,b[.

THEOREM A.5.– *Let p be a semi-norm on a vector space E, $u \in E$, and $v \in E$. Then*

$$|p(v) - p(u)| \leq p(v - u). \; \square$$

The upper envelope of semi-norms is a semi-norm whenever it is finite [Vol. 1, Theorem 12.2]:

THEOREM A.6.– *Let $\mathcal{P}$ be a family of semi-norms on a vector space E such that, for every $u \in E$, $\sup_{p \in \mathcal{P}} p(u) < \infty$. Then the mapping $u \mapsto \sup_{p \in \mathcal{P}} p(u)$ is a semi-norm on E.* □

Open, closed and bounded sets, sequences, connectedness

Definition of open, closed and bounded sets. Let us define the bounded, open and closed sets of a separated semi-normed space.

DEFINITION A.7.– *Let U be a subset of a separated semi-normed space E.*

(a) *We say that U is* **bounded** *if, for every $\nu \in \mathcal{N}_E$, we have $\sup_{u \in U} \|u\|_{E;\nu} < \infty$.*

(b) *We say that U is* **open** *if, for every $u \in U$, there exists a finite set N of $\mathcal{N}_E$ and $\eta > 0$ such that $v \in E$ and $\sup_{\nu \in N} \|v - u\|_{E;\nu} \leq \eta$ imply $v \in U$.*

(c) *We say that U is* **closed** *if $E \setminus U$ is open.*

(d) *We say that U is* **sequentially closed** *if every sequence of U that converges in E has a limit in U.* □

Let us define the interior and the closure [Vol. 1, Definition 2.15 and its justification].

DEFINITION A.8.– *Let U be a subset of a separated semi-normed space. The* **interior** *of U is the largest open set included in U, denoted by $\overset{\circ}{U}$. The* **closure** *of U is the smallest closed set containing U, denoted by $\overline{U}$. The* **boundary** *of U is the set $\overline{U} \setminus \overset{\circ}{U}$, denoted by ∂U.* □

Properties of open, closed and bounded sets. Closedness implies sequential closedness [Vol. 1, Theorem 2.10]:

THEOREM A.9.– *Every closed set of a separated semi-normed space is sequentially closed.* □

Let us state some properties of unions and intersections [Vol. 1, Theorems 2.11 and 2.12].

THEOREM A.10.– *In a separated semi-normed space, every (finite or infinite) union and every finite intersection of open sets is open. Every finite union and every intersection of closed sets is closed. Every finite union and every intersection of bounded sets is bounded.* □

Let us characterize the closure of a set of a separated semi-normed space [Vol. 1, Theorem 2.17].

THEOREM A.11.– *For every subset U of a separated semi-normed space E,*

$$\overline{U} = \Big\{ u \in E : \forall N \subset \mathcal{N}_E \text{ finite, } \forall \eta > 0, \exists v \in U \text{ such that } \sup_{\nu \in N} \|v - u\|_{E;\nu} \leq \eta \Big\}. \; \square$$

Properties of sequences. Let us state some properties of sequences [Vol. 1, Theorems 2.7 and 2.8].

THEOREM A.12.– *Every convergent sequence of a separated semi-normed space is bounded, Cauchy, and has a unique limit. In addition, every Cauchy sequence is bounded.*

Every subsequence of a convergent sequence converges to the same limit. □

Let us state some topological inclusion properties [Vol. 1, Theorem 3.8].

THEOREM A.13.– *Let E and F be two separated semi-normed spaces such that $E \subsetneq F$, $(u_n)_{n\in\mathbb{N}}$ a sequence of E, and $u \in E$. If $(u_n)_{n\in\mathbb{N}}$ satisfies one of the following properties in E, then it satisfies the same property in F: convergent, convergent to u, Cauchy, bounded.* □

Denseness. Let U and V be two subsets of a separated semi-normed space. We say that V is **dense** in U if $V \subset U \subset \overline{V}$. We say that V is **sequentially dense** in U if $V \subset U$ and every $u \in U$ is the limit of a sequence of V.

Let us compare denseness and sequential denseness [Vol. 1, Theorem 2.20].

THEOREM A.14.– *In a separated semi-normed space, any set that is sequentially dense in another set is dense in the latter set.* □

We say that a set U of a separated semi-normed space is **sequentially separable** if it contains a countable subset V that is sequentially dense in U.

Connectedness. Let us define connected sets [Vol. 1, Definition 2.30 and its justification].

DEFINITION A.15.– *A set U of a separated semi-normed space is said to be* **connected** *if it cannot be covered by two open sets whose intersections with U are disjoint and non-empty.*

The **connected component** *of U generated by a point u of U is the largest connected set included in U that contains u.* □

Let us state some connectedness properties [Vol. 1, Theorems 2.31 and 2.34].

THEOREM A.16.– *Every real interval is connected.*

Every set U of a separated semi-normed space is the union of its connected components. The latter are connected and pairwise disjoint (or equal), and they are open if U itself is open. □

Compactness

Definitions of compact sets. Let us first define some of the notions of compactness that are related to covering properties. A **covering** of a set U is a family of sets whose union contains U. It is said to be **open** if it consists of open sets, and **finite** if it has finitely many elements. A **subcovering** is a subfamily that is also a covering of U.

DEFINITION A.17.– *Let U be a subset of a separated semi-normed space E.*

(a) *We say that U is* **compact** *if every open covering of U has a finite subcovering.*

(b) *We say that U is* **relatively compact** *if $\overline{U}$ is compact.*

(c) *We say that U is* **precompact** *if, for every $\nu \in \mathcal{N}_E$ and $\epsilon > 0$, there exist finitely many elements $u \in E$ such that U is covered by the semi-balls $B_{E;\nu}(u, \epsilon) = \{v \in E : \|v - u\|_{E;\nu} \leq \epsilon\}$.* □

Now, let us define some notions of compactness related to the convergence of subsequences.

DEFINITION A.18.– *Let U be a subset of a separated semi-normed space E.*

(a) *We say that U is* **sequentially compact** *if every sequence of U has a convergent subsequence whose limit belongs to U.*

(b) *We say that U is* **relatively sequentially compact** *if every sequence of U has a convergent subsequence (in E).* □

Properties of compact sets. Let us compare the various notions of compactness [Vol. 1, Theorem 2.25].

THEOREM A.19.– *For every subset of a separated semi-normed space:*

(a) *Compact $\Rightarrow$ relatively compact $\Rightarrow$ precompact $\Rightarrow$ bounded.*

(b) *Sequentially compact $\Rightarrow$ relatively sequentially compact $\Rightarrow$ precompact.* □

Recall that, in a general separated semi-normed space, sequential compactness is neither stronger nor weaker than compactness [Vol. 1, Properties (2.6) and (2.7), p. 43].

Let us state a property of the subsets of a precompact set [Vol. 1, Theorem 2.26].

THEOREM A.20.– *Every subset of a precompact set of a separated semi-normed space is precompact.* □

Let us state a **separation theorem** for a closed set and a compact set that are disjoint [Vol. 1, Theorem 3.19].

THEOREM A.21.– *Let U be a closed set and K a compact subset of a normed space E such that $U \cap K = \emptyset$. Then there exists a ball $B = \{u \in E : \|u\|_E \leq r\}$ of E, where $r > 0$, such that $(U + B) \cap K = \emptyset$.* □

Let us deduce a **strong inclusion theorem** for a compact subset of an open set.

THEOREM A.22.– *Let U be an open set and K a compact subset of a normed space E such that $K \subset U$. Then there exists a ball $B = \{u \in E : \|u\|_E \leq r\}$ of E, where $r > 0$, such that $K + B \subset U$.* □

Proof. Since the set $E \setminus U$ is closed, the separation theorem (Theorem A.21) gives $r > 0$ such that $((E \setminus U) + B) \cap K = \emptyset$, which is equivalent to $(E \setminus U) \cap (K + B) = \emptyset$ and therefore to $K + B \subset U$. □

The space $\mathbb{R}^d$, metrizable spaces, Fréchet spaces and Neumann spaces

The space $\mathbb{R}^d$. The space $\mathbb{R}^d = \{x = (x_1, \ldots, x_d) : \forall i,\ x_i \in \mathbb{R}\}$ is endowed with the **Euclidean norm** defined by $|x| = (x_1^2 + \cdots + x_d^2)^{1/2}$ and the scalar product defined by $x \cdot y = x_1 y_1 + \cdots + x_d y_d$. These operations satisfy the **Cauchy–Schwarz inequality** [Vol. 1, Theorem 5.15]: for every $x \in \mathbb{R}^d$ and $y \in \mathbb{R}^d$,

$$|x \cdot y| \leq |x||y|. \tag{A.1}$$

Let us state some properties of $\mathbb{R}^d$ (and hence of $\mathbb{R}$). These results are established in the following places:
— For (a) [Vol. 1, Theorems 4.15 and 5.12].
— For the **Borel–Lebesgue theorem** (b) [Vol. 1, Theorem 5.13].
— For the **Bolzano–Weierstrass theorem** (c) [Vol. 1, Theorems 4.17 and 5.14].

THEOREM **A.23.**– (a) *$\mathbb{R}^d$ is a sequentially separable Banach space.*

(b) *In $\mathbb{R}^d$, being compact is equivalent to being closed and bounded, and being precompact is equivalent to being bounded.*

(c) *Every bounded sequence of $\mathbb{R}^d$ has a convergent subsequence.* □

Let us observe that the sum of two compact subsets of $\mathbb{R}^d$ is compact.

THEOREM **A.24.**– *The sum $U + V$ of two compact subsets of $\mathbb{R}^d$ is compact.* □

Proof. Since U and V are compact, they are bounded (Theorem A.23 (b)), so $U + V$ is also bounded. Moreover, since U is closed, $U + V$ is closed [Vol. 1, Theorem 3.20] and therefore compact (Theorem A.23 (b)). □

Metrizable, Fréchet and Neumann spaces. In a metrizable space, sequential properties coincide with topological properties. Among other results [Vol. 1, Theorems 4.6 and 4.7]:

THEOREM **A.25.**– *In a metrizable semi-normed space:*

(a) *Closedness is equivalent to sequential closedness, and relative compactness is equivalent to relative sequential compactness.*

(b) *The closure $\overline{U}$ of a set U coincides with the set of limits of its convergent sequences.*

(c) *Every subset of a sequentially separable set is sequentially separable.* □

Every product of Fréchet spaces is a Fréchet space [Vol. 1, Theorem 6.7]:

THEOREM **A.26.**– *Every product $E_1 \times \cdots \times E_d$ of Fréchet spaces is a Fréchet space.* □

Every closed subspace of a Neumann space is a Neumann space [Vol. 1, Theorem 4.12]:

THEOREM **A.27.**– *Every closed, or more generally sequentially closed, topological vector subspace of a Neumann space is a Neumann space.* □

A.3. Continuous mappings and duality

Continuous mappings

Mapping. A **mapping** from a set X into a set Y is defined by specifying an element $T(u) \in Y$ for each $u \in X$. We say that $T(u)$ is the **image** of the point u under T. The **image** of a subset U of X under T is the set $T(U) = \{T(u) : u \in U\}$.

We say that T is **injective** if $T(u) = T(v)$ implies $v = u$, **surjective** if $T(X) = Y$ and **bijective** if it is injective and surjective. We also say that T is an **injection**, **surjection** or **bijection**.

The **preimage** of a subset V of Y under T is the set $T^{-1}(V) = \{u \in X : T(u) \in V\}$. If T is bijective, there exists a unique mapping T^{-1} from Y into X, called the **inverse mapping** of T, such that $T^{-1}(T(u)) = u$ for every $u \in X$. Thus, T is the inverse mapping of T^{-1}, since $T(T^{-1}(v)) = v$.

If T is a mapping from X into Y and S is a mapping from Y into G, we denote by $S \circ T$ or $S(T)$ the **composite mapping** (or composition) defined from X into G by $(S \circ T)(u) = S(T(u))$.

For mappings taking values in a vector space, addition and multiplication by a real number t are defined by

$$(S+T)(u) \stackrel{\text{def}}{=} S(u) + T(u), \quad (tS)(u) \stackrel{\text{def}}{=} tS(u).$$

A mapping T taking values in a separated semi-normed space is said to be **bounded** if the image $T(X)$ of its domain X is bounded.

Functions. We reserve the term **function** for mappings defined on a subset of $\mathbb{R}$ or $\mathbb{R}^d$.

A real function is said to be **positive**, denoted $f \geq 0$, if $f(t) \geq 0$ for every t. We denote $f \geq g$ if $f - g \geq 0$ and $f \leq g$ if $g - f \geq 0$. We say that f is **strictly positive**, denoted $f > 0$, if $f(t) > 0$ for every t.

A real function on an interval is said to be **increasing** if $s \leq t$ implies $f(s) \leq f(t)$, and **decreasing** if $s \leq t$ implies $f(t) \leq f(s)$. It is said to be **strictly increasing** if $s < t$ implies $f(s) < f(t)$, and **strictly decreasing** if $s < t$ implies $f(t) > f(s)$.

Continuity and sequential continuity. Let us define the notion of continuity and its variants.

DEFINITION **A.28.**– *Let E and F be two separated semi-normed spaces, $X \subset E$, and T a mapping from X into F.*

(a) *We say that T is **continuous at the point** u of X if, for every $\mu \in \mathcal{N}_F$ and $\epsilon > 0$, there exists a finite set N of $\mathcal{N}_E$ and $\eta > 0$ such that:*

$$v \in X, \ \sup_{\nu \in N} \|v - u\|_{E;\nu} \leq \eta \quad \Rightarrow \quad \|T(v) - T(u)\|_{F;\mu} \leq \epsilon.$$

*We say that T is **continuous** if it is continuous at every point of X.*

(b) *We say that T is **uniformly continuous** if, for every $\mu \in \mathcal{N}_F$ and $\epsilon > 0$, there exists a finite set N of $\mathcal{N}_E$ and $\eta > 0$ such that:*

$$u \in X, \ v \in X, \ \sup_{\nu \in N} \|v - u\|_{E;\nu} \leq \eta \quad \Rightarrow \quad \|T(v) - T(u)\|_{F;\mu} \leq \epsilon.$$

(c) *We say that T is **sequentially continuous at the point** u if, for every sequence $(u_n)_{n \in \mathbb{N}}$ of X such that $u_n \to u$ in E, we have $T(u_n) \to T(u)$ in F. We say that T is **sequentially continuous** if it is sequentially continuous at every point of X.* □

A bijection is said to be **bicontinuous** if it is continuous and its inverse mapping is also continuous.

Continuity properties. Let us compare continuity and sequential continuity [Vol. 1, Theorems 7.2 and 9.1].

THEOREM A.29.– *Every continuous mapping from a subset of a separated semi-normed space E into a separated semi-normed space F is sequentially continuous.*

If E is metrizable, sequential continuity is equivalent to continuity. □

Addition and multiplication of a separated semi-normed space are continuous [Vol. 1, Theorem 7.24]:

THEOREM A.30.– *For any separated semi-normed space E, the mapping $(u, v) \mapsto (u + v)$ is linear and continuous from $E \times E$ into E, and the mapping $(t, u) \mapsto tu$ is continuous and bilinear from $\mathbb{R} \times E$ into E.* □

Let us characterize continuous mappings [Vol. 1, Theorem 8.2].

THEOREM A.31.– *A mapping T from an open set of a separated semi-normed space E into a separated semi-normed space F is continuous if and only if, for every open subset V of F, the set $T^{-1}(V)$ is open in E.* □

Let us change the space of values [Vol. 1, Theorem 7.6].

THEOREM A.32.– *Let E, F_1 and F_2 be separated semi-normed spaces such that $F_1 \subsetneq F_2$, and let $X \subset E$. Then every continuous mapping from X into F_1 is continuous from X into F_2.* □

Continuous mappings on a compact or connected set. Let us state some properties of images [Vol. 1, Theorems 8.7 and 8.6].

THEOREM A.33.– *Let T be a continuous mapping from a subset X of a separated semi-normed space into a separated semi-normed space, and let $U \subset X$.*

If U is compact, then $T(U)$ is compact. If U is connected, then $T(U)$ is connected.

If T is uniformly continuous and U is precompact, then $T(U)$ is precompact. □

Let us state **Heine's theorem** [Vol. 1, Theorem 8.10].

THEOREM A.34.– *Every continuous mapping from a compact subset of a separated semi-normed space into a separated semi-normed space is uniformly continuous and bounded.* □

Continuity of composite mappings. Let us begin with a general result [Vol. 1, Theorem 7.9].

THEOREM A.35.– *Let T be a mapping from a set of a separated semi-normed space into a set Y of a separated semi-normed space, and let S be a mapping from Y into a separated semi-normed space.*

If S and T are continuous, or uniformly continuous, or sequentially continuous, then so is $S \circ T$. □

Let us state a property of composition with linear mappings [Vol. 1, Theorem 9.4].

THEOREM A.36.– *If T is a uniformly continuous mapping from a metrizable semi-normed space into a separated semi-normed space F, and if L is a sequentially continuous linear mapping from F into a separated semi-normed space, then $L \circ T$ is uniformly continuous.* □

Continuous extension. Let us extend an equality or inequality by continuity [Vol. 1, Theorem 10.1].

THEOREM A.37.– *Let T_1 and T_2 be two continuous mappings from a set X of a separated semi-normed space taking values in a separated semi-normed space F. Then, for any subset U of X:*

(a) *If $T_1 = T_2$ on U, then $T_1 = T_2$ on $\overline{U} \cap X$.*

(b) *If $F = \mathbb{R}$ and $T_1 \leq T_2$ on U, then $T_1 \leq T_2$ on $\overline{U} \cap X$.* □

Let us state the **continuous extension theorem** [Vol. 1, Theorem 10.3].

THEOREM A.38.– *Let X and V be two sets of a separated semi-normed space E such that X is sequentially dense in V. Then every uniformly continuous mapping from X into a Neumann space F has a unique uniformly continuous extension from V into F.* □

Linear or multilinear mappings

Continuous linear or multilinear mappings. Let us define linearity and multilinearity.

DEFINITION A.39.– *A mapping L from a vector space E into another is said to be **linear** if, for every $u \in E$, $v \in E$ and $t \in \mathbb{R}$,*

$$L(u+v) = L(u) + L(v), \quad L(tu) = t\,L(u).$$

*A mapping T defined on a product of vector spaces is said to be **multilinear** if the partial mappings $u_i \mapsto T(u_1, \ldots, u_d)$ are linear. When $d = 2$, we say that T is **bilinear**.* □

Let us characterize continuous multilinear mappings [Vol. 1, Theorem 7.20].

THEOREM A.40.– *A multilinear mapping T from a product $E_1 \times \cdots \times E_d$ of separated semi-normed spaces into a separated semi-normed space F is continuous if and only if, for every $\mu \in \mathcal{N}_F$, there exist finite sets N_1 of $\mathcal{N}_{E_1}$, ..., N_d of $\mathcal{N}_{E_d}$ and $c \geq 0$ such that: for every $(u_1, \ldots, u_d) \in E_1 \times \cdots \times E_d$,*

$$\|T(u_1, \ldots, u_d)\|_{F;\mu} \leq c \sup_{\nu_1 \in N_1} \|u_1\|_{E_1;\nu_1} \cdots \sup_{\nu_d \in N_d} \|u_d\|_{E_d;\nu_d}. \ \square$$

These mappings are uniformly continuous and bounded on bounded sets [Vol. 1, Theorems 7.22 and 7.23]:

THEOREM A.41.– *Every continuous multilinear mapping from a product of separated semi-normed spaces into a separated semi-normed space is uniformly continuous and bounded on every bounded set.* □

Banach–Steinhaus theorem. This theorem, which states that every set of simply bounded continuous linear mappings is equicontinuous, can be expressed as follows in terms of semi-norms [Vol. 1, Theorem 10.11]:

THEOREM A.42.– *Let $\mathcal{T}$ be a set of continuous linear mappings from a Fréchet space E into a separated semi-normed space F such that, for every $u \in E$ and $\mu \in \mathcal{N}_F$, $\sup_{L \in \mathcal{T}} \|L(u)\|_{F;\mu} < \infty$.*

Then, for every $\mu \in \mathcal{N}_F$, there exists a finite set N of $\mathcal{N}_E$ and $c \in \mathbb{R}$ such that, for every $L \in \mathcal{T}$ and $u \in E$,

$$\|L(u)\|_{F;\mu} \leq c \sup_{\nu \in N} \|u\|_{E;\nu}. \quad \square$$

Let us state an analogous result for bilinear mappings [Vol. 1, Theorem 10.12].

THEOREM A.43.– *Let $\mathcal{T}$ be a set of continuous bilinear mappings from a product of Fréchet spaces $E_1 \times E_2$ into a separated semi-normed space F such that, for every $u_1 \in E_1$, $u_2 \in E_2$ and $\mu \in \mathcal{N}_F$, we have $\sup_{T \in \mathcal{T}} \|T(u_1, u_2)\|_{F;\mu} < \infty$.*

Then, for every $\mu \in \mathcal{N}_F$, there exists $c \in \mathbb{R}$ and finite sets N_1 of $\mathcal{N}_{E_1}$ and N_2 of $\mathcal{N}_{E_2}$ such that, for every $T \in \mathcal{T}$, $u_1 \in E_1$ and $u_2 \in E_2$,

$$\|T(u_1, u_2)\|_{F;\mu} \leq c \sup_{\nu_1 \in N_1} \|u_1\|_{E_1;\nu_1} \sup_{\nu_2 \in N_2} \|u_2\|_{E_2;\nu_2}. \quad \square$$

Dual and weak topology

Definitions. Let us define the dual.

DEFINITION A.44.– *The **dual** of a separated semi-normed space E is the vector space E' of continuous **linear forms** on E, that is, of continuous linear mappings from E into $\mathbb{R}$ endowed with the semi-norms, indexed by the bounded sets B of E,*

$$\|e'\|_{E';B} \stackrel{\text{def}}{=} \sup_{e \in B} |\langle e', e\rangle|. \quad \square$$

We denote $\langle e', e\rangle = e'(e)$ when $e' \in E'$ and $e \in E$. Let us define the weak topology.

DEFINITION A.45.– *Let E be a separated semi-normed space. We denote by E-weak the vector space E endowed with the semi-norms, indexed by $e' \in E'$,*

$$\|e\|_{E\text{-weak};e'} \stackrel{\text{def}}{=} |\langle e', e\rangle|. \quad \square$$

Properties of E**-weak.** Let us compare the weak topology and the original topology [Vol. 1, Theorem 15.2].

THEOREM A.46.– *For every separated semi-normed space E, the space E-weak is a separated semi-normed space and $E \subsetneq E$-weak.* $\square$

Let us state the **Banach–Mackey** theorem [Vol. 1, Theorem 16.1].

THEOREM A.47.– *A set is bounded in a separated semi-normed space E if and only if it is bounded in E-weak.* □

Let us state **Mazur's theorem** [Vol. 1, Theorem 16.4].

THEOREM A.48.– *A convex subset of a separated semi-normed space E is closed in E if and only if it is closed in E-weak.* □

Extractability. Let us define extractability.

DEFINITION A.49.– *A separated semi-normed space E is said to be **extractable** if every bounded sequence of E has a subsequence that converges in E-weak.* □

Every extractable space is "weakly" sequentially complete [Vol. 1, Theorem 18.15].

THEOREM A.50.– *If a separated semi-normed space E is extractable, then E-weak is a Neumann space.* □

A.4. Differentiable mappings and differentiable functions

Differentiation of a composite mapping. Let us state the expression of the differential together with conditions that guarantee its continuity [Vol. 1, Theorems 19.17 and 19.18].

THEOREM A.51.– *Let T be a mapping from an open set X of a separated semi-normed space E into an open set Y of a separated semi-normed space F, and let S be a mapping from Y into a separated semi-normed space.*

(a) *If T is differentiable at the point u of X and S is differentiable at the point $T(u)$, then $S \circ T$ is differentiable at the point u and*

$$\mathrm{d}(S \circ T)(u) = \mathrm{d}S(T(u)) \circ \mathrm{d}T(u).$$

(b) *If E is metrizable or F is a normed space, and if in addition S and T are continuous differentiable, then $S \circ T$ is continuously differentiable.* □

Differentiability properties of a function of a single real variable. Let us characterize differentiability [Vol. 1, Theorem 21.3].

THEOREM A.52.– *A function f from an open set I of $\mathbb{R}$ taking values in a separated semi-normed space E is differentiable at the point s of I with derivative $f'(s)$ if and only if $(f(s+t) - f(s))/t$ has the limit $f'(s)$ in E when $t \neq 0$, $s + t \in I$, $t \to 0$.* □

Let us state some growth properties of a differentiable real function [Vol. 1, Theorem 21.8].

THEOREM A.53.– *Let f be a differentiable function from (a,b) into $\mathbb{R}$, where $-\infty \leq a < b \leq \infty$. If $f' \geq 0$, then f is increasing; if $f' \leq 0$, then f is decreasing; if $f' = 0$, then f is constant.* □

Let us proceed to consider the finite increment theorem [Vol. 1, Theorem 22.7].

THEOREM A.54.– *Let f be a continuous function from $[a,b]$ into $\mathbb{R}$, where $-\infty < a < b < \infty$, that is differentiable on (a,b). Then*

$$(b-a) \inf_{a<s<b} f'(s) \leq f(b) - f(a) \leq (b-a) \sup_{a<s<b} f'(s). \ \square$$

Let us state the expression of the derivative of the inverse function [Vol. 1, Theorem 22.9].

THEOREM A.55.– *Let f be a continuous real function that is strictly monotonic on a bounded interval (a,b) of $\mathbb{R}$, differentiable at the point $s \in (a,b)$ and which satisfies $f'(s) \neq 0$. Then f is invertible, f^{-1} is continuous, strictly monotonic and differentiable at the point $f(s)$, and $(f^{-1})'(f(s)) = 1/f'(s)$.* □

Powers, exponentials and logarithms. Let us state some properties of real powers [Vol. 1, Theorems 22.14 and 22.15].

THEOREM A.56.– *For every $s \in \mathbb{R}$, the function $x \mapsto x^s$ is infinitely differentiable from $(0,\infty)$ into $\mathbb{R}$, and, for every $x > 0$ and $y > 0$,*

$$\frac{\mathrm{d}x^s}{\mathrm{d}x} = s x^{s-1}, \quad |x^s - y^s| \leq |s||x-y| \sup\{x^{s-1}, y^{s-1}\}. \ \square$$

Let us bound the sum of a geometric series from above [Vol. 1, Property (1.10)].

THEOREM A.57.– *Let $x \in \mathbb{R}$, $0 < x < 1$, and consider two integers such that $n \leq m$. Then:*

$$x^n + x^{n+1} + \cdots + x^m = (x^n - x^m)/(1-x) < x^n/(1-x). \ \square$$

Let us state some properties of exponentials [Vol. 1, Theorem 22.20 and Property (22.9)].

THEOREM A.58.– *For every $x > 0$, the function $s \mapsto x^s$ is infinitely differentiable from $\mathbb{R}$ into itself,*

$$\frac{\mathrm{d}x^s}{\mathrm{d}s} = x^s \log x, \quad \frac{\mathrm{d}\mathrm{e}^s}{\mathrm{d}s} = \mathrm{e}^s, \quad 0 < \mathrm{e}^{-x} \leq s^s x^{-s}. \ \square$$

Let us state some properties of logarithms.

THEOREM A.59.– *For every $x > 0$ and $t > 0$,*

$$\log \frac{1}{x} = -\log x, \quad |\log x| \leq \frac{x^{-t}}{t} \text{ if } x \leq 1, \quad |\log x| \leq \frac{x^t}{t} \text{ if } x \geq 1. \ \square$$

Proof. Theorem 22.17 of Volume 1 gives the inequality stated for $x \leq 1$ and, for every x, it gives $(1 - x^{-t})/t \leq \log x \leq (x^t - 1)/t$. For $x \geq 1$, this gives $0 \leq \log x \leq (x^t - 1)/t \leq x^t/t$.

We have $\log(1/x) = -\log x$ since $\log(xy) = \log x + \log y$ [Vol. 1, Theorem 22.18] and $\log 1 = 0$. □

Bibliography

[1] ARCHIMEDES, *Opere omnia*, three volumes, J.-L. Heiberg ed., Teubner, 1913–1915 (= *Les Œuvres complètes*, Desclée de Brouwer, Paris, 1921).

[2] ARZELÀ, C., Funzioni di linee, *Rend. Atti R. Accad. Lincei*, (4) 5 (1889), 342–348.

[3] ARZELÀ, C., Sulle funzioni di linee, *Mem. R. Accad. Sci. Ist. Bologna*, (5) 5 (1895), 55–74.

[4] ASCOLI, G., Le curve limiti di una varietà data di curve, *Atti R. Accad. Lincei*, Memorie della Cl. Sci. Fis. Mat. Nat., (3) 18 (1883–1884), 521–586.

[5] BARROW, I., *Lectiones Geometricae*, Londini, 1670 (= *Geometrical Lectures*, 1916).

[6] BERNOULLI, J. (Jacob), Analysis problematis antehac propositi. De Inventione Lineæ descendus a corpore gravi percurrendæ uniformiter, *Opera*, Cramer and Philibert, Geneva, (1744), 421–426.

[7] BERNOULLI, J. (Johann), Lectiones matematicæ de methodo integralium allisque (1691–1692), *Opera omnia III*, 385–559. Original edition, Bousquet, Lausanne and Geneva, 1742 (= Olms, Hildesheim, 1968).

[8] BOCHNER, S., *Vorlesungen über Fouriersche Integrale*, Leipzig, Akad. Verlagsgesellschaft, 1932.

[9] BOCHNER, S., Integration von Funktionen deren Werte die Elemente eines Vektorraumes sind, *Fundamenta Mat.*, 20 (1933), 262–276.

[10] BOLZANO, B., Ein Analytischer Beweis der Lehrsatzes, daß zwischen je zwei Werthen, die ein entgegengesetzes Resultat gewähren, wenigstens eine reelle Wurzel der Gleichung liege, *Abhandlungen K. Böhm. Gesell. Wissen*, (3) 5 (1817), 1–6.

[11] BOURBAKI, N., *Intégration*, Hermann, 1965.

[12] BOURBAKI, N., *Espaces vectoriels topologiques*, Hermann, new edition, 1967.

[13] BOURBAKI, N., *Fonction d'une variable réelle*, Hermann, new edition, 1976.

[14] BOURBAKI, N., *Éléments d'histoire des mathématiques*, Hermann, new edition, 1974.

[15] BOURBAKI, N., *Variétés différentielles et analytiques : fascicule de résultats*, Hermann, 1971.

[16] CAJORI, F., *A History of Mathematical Notations*, two volumes, Open Court (Chicago), 1974.

[17] CANTOR, G., *Gesammelte Abhandlungen*, Springer, 1932.

[18] CARTAN, E., *Œuvres complètes*, 6 volumes, Gauthiers–Villars, 1953–1955.

[19] CARTAN, H., *Cours de calcul différentiel*, Hermann, revised edition, 1977.

[20] CAUCHY, A., *Cours d'analyse de l'École Royale Polytechnique*, Debure, Paris, 1821.

[21] CAUCHY, A., *Résumé des leçons données à l'École Royale Polytechnique sur le calcul infinitésimal*, Debure, Paris, 1823.

[22] CAUCHY, A., *Mémoire sur les intégrales définies*, Bures frères, Paris, 1825.

[23] CLARKSON, J. A., Uniformly convex spaces, *Trans. Amer. Math. Soc.*, 40 (1936), 396–414.

[24] CLIFFORD, W. K., *Elements of Dynamic*, 1878 (= *Historical Math Monographs*, Cornell University).

[25] CONDORCET, N., *Mémoire sur les équations aux différences partielles*, Mémoires de l'Académie Royale des Sciences, année 1770, Paris, 1773.

[26] CORIOLIS, G.-G., *Du calcul de l'effet des machines, ou considérations sur l'emploi des moteurs et sur leur évaluation, pour servir d'introduction à l'étude spéciale des machines*, Carilian–Gœury, Libraire des corps royaux des ponts et chaussées et des mines, Paris, 1829.

[27] D'ALEMBERT, J., *Réflexions sur la cause générale des vents*, David, Paris, 1747.

[28] DE RHAM, G., *Variétés différentiables*, Hermann, 1955.

[29] DREYFUSS, P., *Formes différentielles exactes et fermées dans* $\mathbb{R}^d$, Université Côte d'Azur.

[30] DU BOIS-REYMOND, P., Fortsetzung der Erläuterungen zu den Anfangsgründen der Variationsrechnung, *Mathematische Annalen*, 15, no 2 (1879), 564–576.

[31] DUGUNDJI, J., *Topology*, Allyn and Bacon, Inc., Boston, 1966.

[32] EDWARDS, R., *Functional Analysis*, Winston, 1965.

[33] EUCLIDE, *Elementorum Libri* XV, Apud Hieronymum de Marnef & Gulilmum Canellat, Paris, 1573 (= *Les œuvres d'Euclide*, trans. F. Peyrard (1819), A. Blanchard, Paris, 1966).

[34] EULER, L., *Institutiones calculi differentialis*, Acad. Imper. Sci. Petropoli, 1755.

[35] EULER, L., *Institutionum calculi integralis, Volumen Primum*, Acad. Imper. Sci. Petropoli, 1768.

[36] EULER, L., De formulis integrabilibus duplicatis, *Novi comentarii academiæ scientiarum Petropolitanæ*, 14 (1770), 72–103.

[37] FOURIER, J., *Théorie analytique de la chaleur*, Didot, Paris, 1822.

[38] FRÉCHET, M., Sur quelques points du calcul fonctionnel (thesis), *Rend. Palermo*, 22 (1906), 1–74.

[39] FUBINI, G., Sugli integrali multipli, *Atti Accad. Naz. Lincei*, 16 (1907), 608–614.

[40] GIRAULT, V. and RAVIART, P.-A., *Finite Element Methods for Navier–Stokes Equations*, Springer, 1986.

[41] GRÉGOIRE DE SAINT-VINCENT, P. (P. GERGORII A SANCTO VICENTIO), *Opus geometricum quadraturæ circuli et sectionum coni*, two volumes, Antverpiæ, 1647.

[42] GREGORY, J., *James Gregory Tercentenary Memorial Volume, Containing his Correspondance with John Collins and his Hitherto Unpublished Mathematical Manuscripts*, H. W. Turnbull ed., London, 1939.

[43] HAAR, A., Über die Variation der Doppelintegrale, *Journal für die reine und angewandte Mathematik*, 149, nos 1/2 (1926), 1–18.

[44] HAAR, A., Zur Variationsrechnung, *Abhandlungen aus dem Mathematischen Seminar der Universität Hamburg*, 8, no 1 (1931/1932), 1–27.

[45] HATCHER, A., *Algebraic Topology*, Cambridge University Press, 2002 (= free version licensed by the publisher: http://www.math.cornell.edu/~hatcher).

[46] HEINE, E., Über trigonometrische Reihen, *J. Reine Angew. Math. (J. de Crelle)*, 71 (1870), 353–365.

[47] HEINE, E., Aus brieflichen Mittelheilungen (namentlich über Variationsrechnung), *Mathematische Annalen*, 2 (1870), 187–191.

[48] HEINE, E., Die Elemente der Functionenlehre, *J. Reine Angew. Math. (J. de Crelle)*, 74 (1872), 172–188.

[49] KOLMOGOROV, A. N., Zur Normierbarkeit eines allgemeinen topologischen linearen Raumes, *Studia Math.*, 5 (1934), 29–33 (= in English, *Selected Works*, vol. 1, Dordrecht, Kluwer (1991), 183–186).

[50] LADYZHENSKAYA, O. A., *The Mathematical Theory of Viscous Incompressible Flow*, Gordon and Breach, 1963.

[51] LAGRANGE, J., *Sur une nouvelle espèce de calcul relatif à la différentiation et à l'intégration des quantités variables*, Nouveaux mémoires de l'Académie royale des sciences et belles-lettres de Berlin, année 1772, Berlin, 185–221, 1774.

[52] LANG, S., *Real Analysis*, Addison-Wesley, 1969.

[53] LEBESGUE, H., *Leçons sur l'intégration et la recherche des fonctions primitives*, Gauthiers-Villars, 1904.

[54] LEIBNIZ, W. G., Nova metodus pro maximis et minimis, itemque tangentibus, quae nec fractas nec irrationales quantitates moratur, et singulare pro illis calculi genus, *Acta Eruditorum*, 1884, 467–473

[55] LERAY, J., Sur le mouvement d'un liquide visqueux emplissant l'espace, *Acta Math.*, 63 (1934), 193–248.

[56] LIONS, J.-L., *Quelques méthodes de résolution des problèmes aux limites non linéaires*, Dunod & Gauthier–Villars, 1969.

[57] MERCATOR, N., *Logarithmotechnia... cui nunc accedit vera quadratura hyperbolæ...*, Londini, 1668.

[58] NEPER, J., *Mirifici logarithmorum canonis descriptio*, Lyon, 1619.

[59] NEUMANN, J. VON, On complete topological linear spaces, *Trans. Amer. Math. Soc.*, 37 (1935), 1–20.

[60] NEWTON, I., *The Mathematical Papers, vol. III, 1670–1673*, Cambridge University Press, 1969.

[61] OSTROGRADSKY, M., Dissertation (in Russian), *Memoirs Imperial Acad. Sci. St. Petersburg, Sc. Math., Phys., Nat.*, 1 (1831), 19–53.

[62] PIER, J.-P., *Histoire de l'intégration*, Masson, 1996.

[63] POINCARÉ, H., *Les méthodes nouvelles de la mécanique céleste*, vol. 3, Gauthiers–Villars, 1899.

[64] RIEMANN, B., *Gesammelte mathematische Werke*, Teubner, second ed., 1892.

[65] RIEMANN, B., Über die Darstellbarkeit einer Function durch eine trigonometrische Reihe, *Abh. K. Gesell. Wiss. Göttingen, Math. Classe*, 3 (1866–1867), 87–132

[66] RUDIN, W., *Principles of Mathematical Analysis*, McGraw–Hill, 1953.

[67] SCHWARTZ, L., *Analyse, topologie générale et analyse fonctionnelle*, Hermann, 1970.

[68] SCHWARTZ, L., *Analyse I. Théorie des ensembles et topologie*, Hermann, 1991.

[69] SCHWARZ, H. A., Zur integration der partiellen Differentialgleichung, *Journal reine angew. Math.*, 74 (1872), 218–253.

[70] SIMON, J., Démonstration constructive d'un théorème de G. de Rham, *C. R. Acad. Sci. Paris, sér. I*, 316 (1993), 1167–1172.

[71] SIMON, J., Primitives de distributions et applications, Research Report 93-11, Laboratoire de Mathématiques Appliquées, Université Blaise Pascal, 1993.

[72] SIMON, J., Representation of distributions and explicit antiderivatives up to the boundary, *Progress in Partial Differential Equations: The Metz Surveys 2*, M. Chipot ed., Longman, 1993, 201–205.

[73] SIMON, J., *Banach, Fréchet, Hilbert and Neumann Spaces*, Analysis for PDEs set, vol. 1, ISTE Ltd, London and John Wiley & Sons, New York, 2017.

[74] SIMON, J., *Espaces semi-normés*, in progress.

[75] SIMON, J., *Plus sur les distributions*, in progress.

[76] SMIRNOV, S. K., Decomposition of solenoidal vector charges into elementary solenoids and the structure of normal one-dimensional currents, *St. Petersburg Math. J.*, 5 (1993), no. 4, 841–867.

[77] STEGMANN, F. L., *Lehrbuch der Variationsrechnung und ihrer Anwendung bei Untersuchungen über das Maximum und Minimum*, J.G. Luckhardt, Kassel, 1854.

[78] TARTAR, L., *Topics in Nonlinear Analysis*, Mathematical Publications, Université d'Orsay, 1978.

[79] THOMAE, J., *Abriss einer Theorie der complexen Funktionen und Thetafunktionen einer Veränderlichen*, Halle (Nebert), 1870.

[80] URYSOHN, P. S., Über die Mächtigkeit zusammenhängenden Mengen, *Math. Ann.*, 94 (1925), 262–295.

[81] VOLTERRA, V., *Leçons sur les fonctions de lignes*, Gauthier-Villars, 1913.

Index

U

V, W

Other titles from

iSTE

in

Mathematics and Statistics

2020

KOROLIOUK Dmitri
Dynamics of Statistical Experiments

MANOU-ABI Solym Mawaki, DABO-NIANG Sophie, SALONE Jean-Jacques
Mathematical Modeling of Random and Deterministic Phenomena

2019

BANNA Oksana, MISHURA Yuliya, RALCHENKO Kostiantyn, SHKLYAR Sergiy
Fractional Brownian Motion: Approximations and Projections

GANA Kamel, BROC Guillaume
Structural Equation Modeling with lavaan

KUKUSH Alexander
Gaussian Measures in Hilbert Space: Construction and Properties

LUZ Maksym, MOKLYACHUK Mikhail
Estimation of Stochastic Processes with Stationary Increments and Cointegrated Sequences

MICHELITSCH Thomas, PÉREZ RIASCOS Alejandro, COLLET Bernard, NOWAKOWSKI Andrzej, NICOLLEAU Franck
Fractional Dynamics on Networks and Lattices

VOTSI Irene, LIMNIOS Nikolaos, PAPADIMITRIOU Eleftheria, TSAKLIDIS George
Earthquake Statistical Analysis through Multi-state Modeling
(Statistical Methods for Earthquakes Set – Volume 2)

2018

AZAÏS Romain, BOUGUET Florian
Statistical Inference for Piecewise-deterministic Markov Processes

IBRAHIMI Mohammed
Mergers & Acquisitions: Theory, Strategy, Finance

PARROCHIA Daniel
Mathematics and Philosophy

2017

CARONI Chysseis
First Hitting Time Regression Models: Lifetime Data Analysis Based on Underlying Stochastic Processes
(Mathematical Models and Methods in Reliability Set – Volume 4)

CELANT Giorgio, BRONIATOWSKI Michel
Interpolation and Extrapolation Optimal Designs 2: Finite Dimensional General Models

CONSOLE Rodolfo, MURRU Maura, FALCONE Giuseppe
Earthquake Occurrence: Short- and Long-term Models and their Validation
(Statistical Methods for Earthquakes Set – Volume 1)

D'AMICO Guglielmo, DI BIASE Giuseppe, JANSSEN Jacques, MANCA Raimondo
Semi-Markov Migration Models for Credit Risk
(Stochastic Models for Insurance Set – Volume 1)